OXFORD MONOGRAPHS ON GEOLOGY AND GEOPHYSICS NO. 9

Series editors

H. Charnock
J. F. Dewey
S. Conway Morris
A. Navrotsky
E. R. Oxburgh
R. A. Price
B. J. Skinner

OXFORD MONOGRAPHS ON GEOLOGY AND GEOPHYSICS

1. De Verle P. Harris: *Mineral resources appraisal: mineral endowment, resources, and potential supply: concepts, methods, and cases*
2. J. J. Veevers (ed.): *Phanerozoic earth history of Australia*
3. Yang Zunyi, Cheng Yuqi, and Wang Hongzhen: *The geology of China*
4. Lin-gun Liu and William A. Bassett: *Elements, oxides, and silicates: High-pressure phases with implications for the Earth's interior*
5. Antoni Hoffman and Matthew H Nitecki (eds.): *Problematic fossil taxa*
6. S. Mahmood Naqvi and John J. W. Rogers: *Precambrian geology of India*
7. Chih-Pei Chang and T. N. Krishnamurti (eds.): *Monsoon meteorology*
8. Zvi Ben-Avraham (ed.): *The evolution of the Pacific Ocean margins*
9. Ian McDougall and T. Mark Harrison: *Geochronology and Thermochronology by the $^{40}Ar/^{39}Ar$ method*

Geochronology and Thermochronology by the $^{40}Ar/^{39}Ar$ Method

IAN McDOUGALL
The Australian National University
Canberra

AND

T. MARK HARRISON
State University of New York
Albany

OXFORD UNIVERSITY PRESS · New York
CLARENDON PRESS · Oxford · 1988

Oxford University Press

Oxford New York Toronto
Delhi Bombay Calcutta Madras Karachi
Petaling Jaya Singapore Hong Kong Tokyo
Nairobi Dar es Salaam Cape Town
Melbourne Auckland

and associated companies in

Berlin Ibadan

Copyright © 1988 by Oxford University Press, Inc.

Published by Oxford University Press, Inc.,
200 Madison Avenue, New York, New York 10016

Oxford is a registered trademark of Oxford University Press

All rights reserved. No part of this publication may be reproduced,
stored in a retrieval system, or transmitted, in any form or by any means,
electronic, mechanical, photocopying, recording, or otherwise,
without the prior permission of Oxford University Press.

Library of Congress Cataloging-in-Publication Data
McDougall, Ian.
  Geochronology and thermochronology by the $^{40}Ar/^{39}Ar$ Method.
  (Oxford monographs on geology and geophysics: no. 9.)
  Bibliography: p.
  Includes index.
  1. Radioactive dating. 2. Argon–Isotopes.
3. Earth temperature. I. Harrison, T. Mark. II. Title.
III. Series.
QE508.M38 1988   551.7'01   87-28232
ISBN 0-19-504302-2

9 8 7 6 5 4 3 2 1

Printed in the United States of America
on acid-free paper

To Pamela and Susan

PREFACE

Results from geochronological studies based upon measurements of naturally occurring radioactive elements and their daughter products have been central to many notable advances in earth sciences in modern times. A spectacular example was the delineation and calibration of the geomagnetic polarity time scale, which formed one of the essential foundations for development of the model of plate tectonics for the Earth. Indeed, much of our knowledge of rates of geological processes stems from studies based upon geochronological measurements. Time scales ranging from those appropriate to the age of the solar system—including the Earth, Moon, and meteorites—to those more in keeping with direct human experience are accessible by application of appropriate isotopic systems.

Since the discovery of radioactivity nearly a century ago, many geochronological methods have been developed and successfully applied to a great diversity of geological situations. Continued development of new and existing methods and increased understanding of the meaning of results obtained are required to sustain progress in the geological sciences. The ^{40}Ar/^{39}Ar isotopic dating method is an excellent example of a technique that, from serendipitous beginnings, has been developed and utilized with marked success to investigate a wide range of geological problems. In particular, it has opened a new dimension in the application of the K–Ar dating scheme to the determination of thermal histories of both terrestrial and extraterrestrial rock samples. The combination of isotopic closure temperature concepts with knowledge obtained from ^{40}Ar/^{39}Ar dating measurements has provided a uniquely powerful approach to the study of kinetic processes in geological terranes.

This work is a response to our belief that there is a need for a monograph on ^{40}Ar/^{39}Ar dating to provide concise knowledge concerning the application of this method to geological studies. We have striven for a reasonably comprehensive but by no means exhaustive coverage of the principles and practices of ^{40}Ar/^{39}Ar dating, with emphasis on interpretation of results. We hope that the monograph will be useful to practicing earth scientists, as well as to students, who wish to understand the advantages and limitations of the technique, or who wish to gain some insight into how to interpret ^{40}Ar/^{39}Ar age data.

In attempting to provide an overview of the current state of knowledge, we commonly cite examples from the available literature. It is perhaps inevitable that we draw rather heavily upon our own work, in large measure because we feel comfortable with examples we know so well.

During the writing of this book we have received much help from colleagues, especially those who have kindly read and commented upon one or more chapters in draft form. In this regard we wish to thank S. L. Baldwin, J. W. Delano, M. H. Dodson, R. A. Duncan, P. L. McFadden, F. S. Spear, S. R. Taylor, G. Turner, and E. B. Watson. We are particularly grateful to M. Honda, J. C. Roddick, and P. K. Zeitler, who nobly undertook the task of reading the whole manuscript. The reviews, detailed comments, and constructive criticism received have been invaluable in helping us to improve the content and presentation. The comments also showed us that this book could have been written in several different ways. Grenville Turner was most helpful in providing information on aspects of the history of the ^{40}Ar/^{39}Ar dating method, and Leo Kristjansson kindly provided some details about the early Icelandic work. We thank Matt Heizler and Robyn Maier for assistance with drafting of figures. Typing of the manuscript was cheerfully and effectively carried out by Julie Stringer, Melissa Stevenson, and Diana Paton. We thank authors and publishers who have granted us permission to reproduce figures. One of us (T.M.H.) acknowledges the Australian National University for the award of a visiting fellowship, during tenure of which the initial planning for the book and some writing were undertaken.

Canberra   I. McD
Albany   T.M.H.
September 1987

CONTENTS

Chapter 1 Historical introduction   3
1.1   General comments   3
1.2   Historical outline   4

Chapter 2 Basis of the ^{40}Ar/^{39}Ar dating method   9
2.1   Potassium–argon dating   9
2.2 ^{40}Ar/^{39}Ar dating   12
2.3   Potassium and its isotopes   14
2.4   The ^{40}K decay scheme and constants   15
2.5   Derivation of age equations   17
2.6   Atmospheric argon   19
2.7   Materials suitable for dating   21
    2.7.1   Introduction   21
    2.7.2   Feldspars   22
       2.7.2.1   Alkali feldspar   22
       2.7.2.2   Plagioclase   23
    2.7.3   Feldspathoids   24
    2.7.4   Micas   25
       2.7.4.1   Biotite and phlogopite   25
       2.7.4.2   Muscovite   26
       2.7.4.3   Lepidolite   26
       2.7.4.4   Glauconite   26
    2.7.5   Amphiboles   27
    2.7.6   Pyroxenes   28
    2.7.7   Whole rocks   28
       2.7.7.1   Volcanic rocks   29
       2.7.7.2   Metamorphic rocks   32
    2.7.8   Glass   33
    2.7.9   Clay minerals   34
    2.7.10   Evaporites   35
2.8   Range of applicability   35

Chapter 3 Technical aspects   40
3.1   Introduction   40
3.2   Sample preparation   40
3.3   Monitor minerals   41
3.4   Nuclear reactions   43
3.5   Nuclear reactors as a neutron source   44
3.6   The ^{39}K(n, p)^{39}Ar reaction   47
3.7   Interfering nuclear reactions and correction factors   51
    3.7.1   General   51
    3.7.2   Reactions on calcium   51
    3.7.3   Reactions on potassium   55
    3.7.4   Reactions on chlorine   57
3.8   Optimization of irradiation parameters   57
    3.8.1   Introduction   57
    3.8.2   Sample size   58
    3.8.3   Production of sufficient ^{39}Ar   59
    3.8.4   Minimization of reactor-induced ^{40}Ar interference   60
    3.8.5   Minimization of interference from ^{40}Ca(n, nα)^{36}Ar reaction   60
    3.8.6   Interference from the ^{42}Ca(n,α)^{39}Ar reaction   62
    3.8.7   Other interferences   62
    3.8.8   Conclusions   63
3.9   Neutron flux gradients   63
3.10   Self-shielding   66
3.11   Temperature effects   66
3.12   Lattice damage   66
3.13   Decay factors   67
3.14   Total sample activity   67
3.15   Safety aspects   68
3.16   Argon extraction systems   69
    3.16.1   Introduction   69
    3.16.2   Pumps and pressure measurement   69
    3.16.3   Furnaces   70
    3.16.4   Temperature monitoring   72
    3.16.5   Gas cleanup   72
    3.16.6   Extraction system blanks   73
3.17   Mass spectrometry   73
    3.17.1   Introduction   73
    3.17.2   Basic concepts   74
    3.17.3   Ion sources   78
    3.17.4   Ion detection and collection   78
    3.17.5   Static mode operation   79
    3.17.6   Machine calibration   80
    3.17.7   Orifice correction   81
    3.17.8   Data acquisition   82
    3.17.9   Calculation of ^{40}Ar*/^{39}Ar$_K$ 83
    3.17.10   Error estimates   85

Chapter 4 Interpretation of results: age spectrum and isochron approaches   86
4.1   Introduction   86
4.2   Single-site diffusion (Turner model)   86
4.3   Age spectra conforming to a single-site diffusion model   93
4.4   Slow cooling models   96
4.5   Numerical solutions   97
4.6   Mixed phases   99

4.7	Resolution within the age spectrum	106		5.7.6	Summary of Dodson's closure temperature model	138
4.8	Excess argon	106	5.8	Solutions of the heat flow equation		139
	4.8.1 Introduction	106		5.8.1 Introduction		139
	4.8.2 Excess ^{40}Ar uptake by a homogeneous phase	107		5.8.2 Simple uplift		139
	4.8.3 Excess ^{40}Ar uptake by a mixture	108		5.8.3 Finite tabular pluton without uplift		140
	4.8.4 Excess ^{40}Ar and flat release patterns	110		5.8.4 Dike with uplift and heat generation		141
	4.8.5 Excess ^{40}Ar and variable age spectra	110		5.8.5 Model parameters		142
4.9	Recoil distribution of ^{39}Ar	110		5.8.5.1 Introduction		142
4.10	Grain size and distribution	112		5.8.5.2 Thermal properties and dimensions		143
4.11	Phase changes in vacuum	113	5.9	Diffusion studies and results		143
4.12	Behavior of minerals and whole rocks	114		5.9.1 Background		143
	4.12.1 High-temperature alkali feldspar	114		5.9.2 Experimental criteria		145
	4.12.2 Low-temperature alkali feldspar	115		5.9.3 Laboratory studies of argon diffusion in natural silicates		147
	4.12.3 Plagioclase	116		5.9.3.1 Introduction		147
	4.12.4 Feldspathoids	116		5.9.3.2 Biotite-phlogopite		148
	4.12.5 Biotite	116		5.9.3.3 Hornblende		151
	4.12.6 Muscovite and phengite	117		5.9.3.4 Muscovite		153
	4.12.7 Amphiboles	117		5.9.3.5 Feldspars		153
	4.12.8 Pyroxene	117	5.10	Coupling of argon diffusion and heat flow		154
	4.12.9 Whole rocks	117		5.10.1 Introduction		154
	4.12.10 Evolution of illite to muscovite	118		5.10.2 ^{40}Ar* loss from microcline at Inyo Domes, California		154
4.13	Isotope correlation diagrams	120		5.10.3 The model		155

Chapter 5 Diffusion theory, experiments, and thermochronology   **127**

5.1	Introduction	127	Appendix A.5.1	Separation of variables solution for plane sheet	156
5.2	The process of diffusion	127	Appendix A.5.2	Laplace transform solution for semi-infinite medium	158
5.3	Phenomenological basis of diffusion theory	128	Appendix A.5.3	Translation to spherical coordinates	159
5.4	Methods of solution for constant D (or κ)	130	Appendix A.5.4	Closure temperature for first-order loss	160
	5.4.1 Plane sheet geometry	130	Appendix A.5.5	Sample diffusion calculation	162
	5.4.2 Semi-infinite medium	131			
	5.4.3 Spherical and cylindrical geometries	131	**Chapter 6 Applications and case histories**		**163**
5.5	Numerical approaches for variable D	131	6.1	Overview	163
5.6	Calculation of episodic ^{40}Ar* loss	132	6.2	Lunar geochronology	163
5.7	Slow cooling	133		6.2.1 General comment	163
	5.7.1 Dodson's model	133		6.2.2 Mare Tranquillitatis geochronology	163
	5.7.2 The mathematics of slow cooling	134		6.2.2.1 Introduction	163
	5.7.3 Closure temperature for first-order loss and volume diffusion	134		6.2.2.2 *Apollo 11* basalts	164
	5.7.4 Dodson's method of solution of the accumulation–diffusion–cooling equation in terms of heat conduction with variable boundary conditions	136		6.2.2.3 Age spectra, low-K basalts	164
				6.2.2.4 Age spectra, high-K basalts	169
				6.2.2.5 Argon loss models	171
			6.2.3	*Apollo 12* basalts	171
	5.7.5 Closure profiles in minerals	136	6.2.4	Significance of mare basalt dating	173

	6.2.5	Geochronology of the lunar highlands	174
6.3	Age spectra reflecting episodic argon loss: examples and applications	178
	6.3.1	Introduction	178
	6.3.2	Contact aureole studies	178
	6.3.3	Sedimentary basin thermal histories	179
	6.3.4	Application of isochron analysis to partially outgassed xenoliths	181

6.4	Uplift and cooling studies		182
	6.4.1	Introduction	182
	6.4.2	Simple cooling of an igneous intrusion	182
	6.4.3	Metamorphic cooling histories	184
6.5	Geochronology and paleomagnetism		186
References			191
Index			209

GEOCHRONOLOGY AND THERMOCHRONOLOGY BY THE ^{40}Ar/^{39}Ar METHOD

1. HISTORICAL INTRODUCTION

1.1 GENERAL COMMENTS

Scientists have sought ways and means of determining the numerical age of the Earth and its constituent rocks since the study of geology along modern scientific lines began more than two centuries ago. In the latter half of the nineteenth century, there was a lively debate regarding the age of the Earth which centered about William Thomson (later Lord Kelvin), a prominent British physicist. Kelvin adopted Fourier's (1820) model of conductive heat flow from an initially molten earth, from which he concluded that the age of the Earth could not be too far in excess of 100 million years (Ma) (Thomson, 1863). He later reduced this estimate to between 20 and 40 Ma (Kelvin, 1899). Although many prominent geologists acceded to this outlook, several argued that the complexity of the geological record demanded a much longer time scale than Kelvin was advocating. However, other contemporary physicists (Perry, 1895a,b,c; Heaviside, 1899), using models incorporating the possibility of mantle convection, were able to accommodate time scales as great as 5 Ga (1 Ga = 10^9 years). Detailed discussion of these debates is given in Burchfield (1975) and Hallam (1983).

The discovery of radioactivity by Becquerel (1896a,b,c), and the subsequent recognition that radioactive decay involves the release of substantial energy, provided an apparent reconciliation of these divergent views, which were the result of overestimated values of radioactive heat production (e.g., Strutt, 1906; Holmes, 1928). From our present perspective, it is apparent that true resolution of this controversy occurred only recently as a result of recognition of the importance of mantle convection and the greatly increased understanding of Earth behavior, especially through the plate tectonics model (Richter, 1986; Harrison, 1987). However, it was the discovery of radioactivity which laid the essential foundation for the development of techniques for direct measurement of the numerical or physical age of rocks, based upon the occurrence in nature of a number of long-lived radioactive isotopes. This led to the development of the completely new field of geochronology by isotopic dating, which subsequently has provided quantitative estimates for the age of many kinds of rocks within the Earth. Thus, the means became available by which the age of geological events could be inferred and rates of geological processes could be determined. Applications of these isotopic dating methods to rocks have provided much important information contributing to the very rapid development of knowledge concerning the origin and evolution of the Earth, as well as other planetary bodies. The revolution that has taken place, especially over the last two decades, in our understanding as to how the Earth works, owes much to results obtained from geochronological studies.

Although potassium was shown to be radioactive from experiments conducted during the first decade of this century, it was not until the late 1940s that the decay of radioactive parent ^{40}K to stable daughter ^{40}Ar was successfully utilized as a means of measuring the age of rocks. Rapid development of techniques followed, leading to the widespread application of the conventional K–Ar isotopic method to the dating of many rock types of diverse age. Indeed the K–Ar method quickly became established as an extremely versatile dating technique, and remains preeminent in the isotopic dating of rocks of Cenozoic age, as well as being used extensively for measuring ages on older rocks. In the mid 1960s, Merrihue (1965) and Merrihue and Turner (1966) recognized that, through production of ^{39}Ar from neutron interactions on ^{39}K, an entirely new approach to K–Ar dating was possible, which could provide a much more powerful means of measuring ages and evaluating their significance. This variant of the K–Ar isotopic dating technique, known as the ^{40}Ar/^{39}Ar dating method, is the subject of the present monograph. Discussion of the conventional K–Ar dating method is restricted to basic information necessary for presentation of the principles and practices of ^{40}Ar/^{39}Ar dating, because earlier works have covered this field comprehensively. The book by Dalrymple and Lanphere (1969) is particularly useful in this regard, and some other publications dealing with the K–Ar dating technique include those by Schaeffer and Zähringer (1966), York and Farquhar (1972), and Faure (1977, 1986). The geochemistry of argon and the other noble gases recently was reviewed by Ozima and Podosek (1983).

1.2 HISTORICAL OUTLINE

The K–Ar dating method involves isotopes of two very different elements that were discovered at widely separated times. Thus, potassium (together with sodium) was isolated by Humphrey Davy in the early nineteenth century as a result of his researches into electrolysis (Davy, 1808). Argon was found in the last decade of the same century by Rayleigh and Ramsay (1895), who separated, documented, and named this noble gas element.

Within a decade of Becquerel's discovery of radioactivity, the possibility of using this phenomenon as a basis for measuring the age of geological samples was realized. Thus, Rutherford (1906) reported ages for uranium-bearing minerals based upon the determination of their uranium and helium contents, the latter known to be a product of the radioactive decay of the parent uranium. Similarly, Boltwood (1907) developed the U–Pb chemical dating method, which depends upon the accumulation of daughter lead from decay of uranium. These pioneering studies not only opened up the field of geochronology, but demonstrated that some minerals and rocks in the Earth's crust were hundreds of millions of years old.

During this heroic period of discovery, Thomson (1905) showed that an alloy of the alkali metals was radioactive, emitting negatively charged particles. Shortly thereafter Campbell and Wood (1906) demonstrated that the radioactivity was an intrinsic and atomic property of potassium, although Campbell (1908) continued to explore, without success, the possibility that the activity was associated with an impurity of some other element.

Aston (1921), using a mass spectrograph, showed that potassium consists of at least two isotopes, ^{39}K and ^{41}K, but as nuclear theory developed it became increasingly clear that neither of these isotopes was likely to be radioactive. Major advances were made in the mid 1930s when, virtually simultaneously, Klemperer (1935) and Newman and Walke (1935) persuasively argued that the radioactive isotope was probably ^{40}K, yielding ^{40}Ca as the daughter product by β^- decay. In addition, Newman and Walke (1935) noted that decay of ^{40}K to ^{40}Ar also was a distinct possibility. In response to these predictions, Nier (1935) successfully instituted a search for this isotope of potassium by means of mass spectrometry, showing that the ^{39}K/^{40}K ratio in nature was 8600 ($\pm 10\%$). This was confirmed almost immediately by Brewer (1935) who used a different type of ion source in the mass spectrometer employed in the measurements. Hevesy (1935) produced some experimental evidence indicating that ^{40}K was the radioactive isotope of potassium. Following the discovery of ^{40}K, Sitte (1935) noted that because ^{40}Ca and ^{40}Ar are both stable, ^{40}K could be expected to decay with emission of positrons as well as electrons, but the lack of detection of the former led him to conclude that it was unlikely that ^{40}K was radioactive. In view of these doubts it became necessary to measure the radioactivity associated with each isotope of potassium. Thus, by means of mass spectrometry, Smythe and Hemmendinger (1937) separated the three known isotopes of potassium, and their results demonstrated conclusively that, indeed, ^{40}K is responsible for the known activity of potassium.

In the same year Von Weizsäcker (1937) argued that ^{40}K may have a dual mode of decay to ^{40}Ar and ^{40}Ca, apparently unaware that Newman and Walke (1935) had raised this possibility previously. Von Weizsäcker's ingenious reasoning leading to this conclusion was based upon the observation that ^{40}Ar is about three orders of magnitude greater in abundance in the Earth's atmosphere compared with neon, krypton, and xenon, relative to the cosmic abundances of these noble gases. Thus Von Weizsäcker argued that the apparent gross excess of ^{40}Ar in the atmosphere demanded that there be another source for ^{40}Ar, which he inferred was derived from the decay of ^{40}K in rocks of the Earth. Because positron emission had not been detected from ^{40}K, he concluded that the ^{40}Ar was produced by means of electron capture by the ^{40}K nucleus. Furthermore, Von Weizsäcker suggested that this hypothesis could be tested by looking for an excess of ^{40}Ar in old potassium-bearing rocks. Bramley (1937) was able to argue the case for the dual decay of ^{40}K even more strongly by combining Von Weizsäcker's arguments with the earlier observation that gamma rays were associated with the decay of potassium, first recognized by Kolhörster (1930). Bramley (1937) showed that the emission of γ rays should be expected to occur if ^{40}K decayed by an electron capture process. Additional evidence favoring the view that electron capture occurs derived from cloud chamber experiments reported by Thompson and Rowlands (1943). They showed that potassium almost certainly emits X-radiation during decay of ^{40}K. They reasoned that capture of an electron by the nucleus of a ^{40}K atom is likely to occur from the orbital adjacent to the nucleus (the K-shell), and that its place would be taken immediately by an electron from an outer shell, resulting in the emission of X-radiation.

The long quest for confirmation that ^{40}Ar indeed was a decay product of ^{40}K finally was realized when Aldrich and Nier (1948) conclusively demonstrated that argon extracted from a number of potassium-rich minerals was significantly enriched in ^{40}Ar relative to ^{36}Ar when compared with the ^{40}Ar/^{36}Ar ratio in atmospheric argon. They noted that with improvements in technique, together with determination of accurate decay constants for ^{40}K, the ^{40}K to ^{40}Ar branch might become useful for the measurement of the age of rocks.

Following the work of Aldrich and Nier, there was rapid development of the necessary techniques, such that the dating method based upon decay of ^{40}K to ^{40}Ar quickly became established as a means of determining the age of rocks. It was soon recognized, however, that there was considerable variation in the degree of retentivity of radiogenic argon in minerals.

An important precursor to the development of the ^{40}Ar/^{39}Ar dating technique was the study by Wänke and König (1959), who showed that K–Ar ages could be measured after irradiating samples in a nuclear reactor. Interactions with neutrons caused transformation of a proportion of the ^{39}K to ^{39}Ar and a proportion of the ^{40}Ar to ^{41}Ar. Subsequent to extraction of the argon from the sample, the ratio of ^{40}Ar/K was determined by a counting technique, as both ^{39}Ar and ^{41}Ar are unstable and decay with half lives of 269 years (a) and 1.83 hours (h), respectively, by emission of β^- particles. In the absence of atmospheric ^{40}Ar, this ratio is proportional to the ^{40}Ar*/^{40}K in the sample and hence proportional to the K–Ar age, where ^{40}Ar* represents the radiogenic argon. This approach was not very satisfactory, however, as it was not possible to correct for ^{40}Ar originating from sources other than the *in situ* decay of ^{40}K. In addition, corrections for interferences from other nuclear reactions could not be effected.

A significant step in developing the ^{40}Ar/^{39}Ar dating method was taken by Craig Merrihue, a graduate student in John H. Reynolds' laboratory in the Department of Physics, University of California, Berkeley. This work was reported in an abstract (Merrihue, 1965), which was published at about the time he met his untimely death in a climbing accident. Merrihue (1965) explained that ^{39}Ar generated in a nuclear reactor from ^{39}K in a sample could be measured mass spectrometrically, instead of by a counting technique, after extraction of the argon from the irradiated sample. This ^{39}Ar derived from ^{39}K is designated ^{39}Ar$_K$. In addition other isotopes of argon could be measured in the mass spectrometer, including ^{40}Ar and ^{36}Ar, the latter facilitating correction for nonradiogenic ^{40}Ar present in the gas. From these isotope abundance measurements, the ^{40}Ar*/^{39}Ar$_K$ ratio could be derived, and as this ratio is proportional to ^{40}Ar*/^{40}K (the

^{40}K/^{39}K being essentially constant in nature), it is also proportional to the K–Ar age. The age was calculated by comparison with the ^{40}Ar*/^{39}Ar$_K$ ratio found for a standard sample of accurately known K–Ar age, irradiated at the same time as the sample to be dated. Use of a standard sample as a neutron flux monitor meant it was unnecessary to know the actual neutron dose received by the samples. It was also unnecessary to measure absolute abundances of either argon or potassium in the sample whose age was to be determined.

The origins of the ^{40}Ar/^{39}Ar dating method can be traced to the serendipitous discovery of the presence of ^{39}Ar from isotopic analyses of argon extracted from meteorites that previously had been irradiated with neutrons in a nuclear reactor. The research being undertaken was in connection with iodine–xenon dating studies of the kind initiated by Jeffery and Reynolds (1961; see also Reynolds, 1963). In Reynolds' laboratory at Berkeley it was normal practice, at least as early as 1963, to isotopically analyze the gases extracted from irradiated meteorites for He, Ne, Ar, and Kr in addition to Xe, which was the element of major interest at the time. Craig Merrihue recognized in late 1963 or early 1964 that ^{39}Ar observed in the argon isotope spectra was generated from neutron interactions on ^{39}K (Grenville Turner, personal communication). This led directly to the development of the ^{40}Ar/^{39}Ar dating method. A portion of a chart recording showing the relative abundances of the argon isotopes in a gas fraction from the meteorite Pantar measured in December 1963 at Berkeley is shown in Fig. 1.1; results from this experiment were given in Merrihue and Turner (1966).

The truly seminal work in relation to ^{40}Ar/^{39}Ar dating was reported in the classic paper by Merrihue and Turner (1966). This paper was written by Turner subsequent to Merrihue's death, and, following a Berkeley tradition, the authors were listed alphabetically. Grenville Turner held a postdoctoral appointment in Reynolds' laboratory in Berkeley from August 1962 until June 1964. Both Turner and Merrihue were closely associated with studies of noble gases in meteorites during this period. From about the middle of 1963 Turner was concerned with measurement of the isotopic composition of noble gases extracted from meteorites that had been neutron irradiated in relation to iodine–xenon dating. Jeffery and Reynolds (1961) had introduced the technique of releasing xenon in steps at progressively higher temperatures from an irradiated sample to examine the correlation of reactor-produced ^{128}Xe from ^{127}I with excess ^{129}Xe found in the meteorites and thought to have originated from the decay of extinct ^{129}I. As previously remarked, it was established practice to measure isotope ratios of the other noble gases, including argon, as well as xenon. Thus, once it was recognized that ^{39}Ar was generated from ^{39}K during irradiation, and could form the basis of a dating method, there was a considerable body of data accumulated by Turner and Merrihue already available for analysis.

The first ^{40}Ar/^{39}Ar age results were given in the landmark paper by Merrihue and Turner (1966) together with many of the concepts and approaches now widely applied in ^{40}Ar/^{39}Ar geochronological studies. It was emphasized that in this new technique the potassium and argon effectively are being determined simultaneously by measurement of the argon isotope ratios in a mass spectrometer. Because isotope ratios can be measured more precisely than concentrations of potassium and argon, the method not only can yield precise ages but also can be used for measuring very small samples. Ages calculated from the argon released by direct fusion of several irradiated meteorite samples generally were found to be in good agreement with K–Ar ages measured by conventional techniques. Of even greater significance was that Turner recognized and demonstrated the advantages of releasing the gas in stages by heating the sample at successively higher temperatures, starting well below the beginning of melting, based upon the data accumulated as an adjunct to the iodine–xenon studies. These argon data were plotted on an isotope correlation diagram. For two of the meteorites, Bjurböle and Pantar, the results yielded reasonable straight lines, the slopes of which provided an estimate of age, and also allowed assessment of some of the underlying assumptions. From the less straightforward

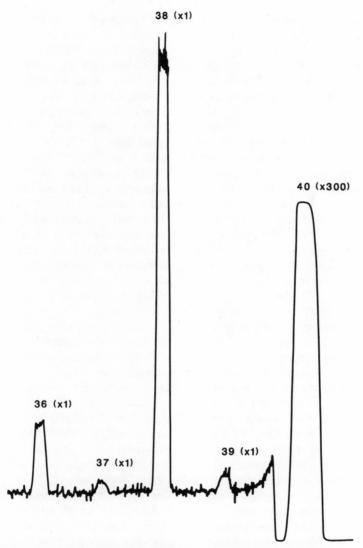

Fig. 1.1. Tracing of portion of the argon mass spectrum measured on gas extracted at 900°C from the meteorite Pantar in December 1963 by Grenville Turner in John Reynolds' laboratory, University of California, Berkeley. This sample had been irradiated in a nuclear reactor in connection with I–Xe studies, but isotope ratios for all the noble gases were routinely measured. The argon isotopes were recorded on the same sensitive scale ($\times 1$) except for ^{40}Ar, which was measured on a more attenuated scale ($\times 300$) owing to its larger size. The ^{40}Ar is dominantly radiogenic, the ^{39}Ar and ^{37}Ar were produced during neutron irradiation from potassium (^{39}K) and calcium (^{40}Ca), respectively, and the ^{38}Ar and ^{36}Ar consist of a mixture of trapped, neutron-induced, and atmospheric argon components. The first ^{40}Ar/^{39}Ar dating results were published in Merrihue and Turner (1966), and included data from Pantar and a number of other meteorites. Mass spectrum by courtesy of Grenville Turner.

results obtained on the meteorite Bruderheim, Turner suggested that a ^{40}Ar/^{39}Ar step heating experiment potentially could provide information relating to the spatial distribution of radiogenic argon in a sample, as the release of argon is likely to be controlled by diffusion. This raised the prospect of being able to identify samples that had been thermally disturbed subsequent to original crystallization, as well as determining the age of the thermal event. Thus, the possibility of elucidating thermal histories by employing the step heating approach was clearly identified. And all these contributions were made in a single five-page paper!

In an earlier laboratory report written in

Icelandic, Sigurgeirsson (1962), outlined many of the principles of the $^{40}Ar/^{39}Ar$ method of dating rocks. Not only did Sigurgeirsson recognize the advantage of generating ^{39}Ar from ^{39}K by neutron irradiation in a nuclear reactor, with subsequent measurement of the argon isotopes by mass spectrometry, but also that a step heating approach to argon extraction from rock samples might be of considerable value. He noted some of the problems likely to arise from interfering reactions during neutron irradiation and suggested means whereby corrections might be made. Unfortunately, this remarkably prescient internal report became known only after the $^{40}Ar/^{39}Ar$ dating technique was firmly established, and thus it played no significant role in the development of the method as far as can be determined.

During the period 1962–1964 Sigurgeirsson received some financial support from the Science Fund in Iceland specifically to facilitate the development of the proposed new dating technique. In 1963 Sigurgeirsson purchased an omegatron, which he planned to use for measurement of the isotopic composition of argon, and steps were taken to construct a vacuum line suitable for the extraction and purification of argon from samples. In addition, Sigurgeirsson arranged for the neutron irradiation of a batch of potassium and calcium salts, and two basalt samples in the Risö nuclear reactor in Denmark in April 1964. The argon extraction system and the omegatron mass spectrometer apparently were never fully commissioned, so that Sigurgeirsson did not succeed in testing his ideas.

Following publication of the Merrihue and Turner paper in 1966, development of techniques and applications of $^{40}Ar/^{39}Ar$ dating gathered momentum. Further measurements by Turner et al. (1966) on the chondritic meteorite, Bruderheim, were interpreted as indicating marked diffusional loss of radiogenic argon by a heating event at a much younger time than that of original crystallization. Models of argon loss, based upon diffusion theory, were developed in greater detail by Turner (1968, 1969, 1970a), and form the foundation for most subsequent interpretations of $^{40}Ar/^{39}Ar$ age spectra. Turner's (1970d) application of the $^{40}Ar/^{39}Ar$ dating technique to lunar basalts provided an even more convincing demonstration of the utility of the stepwise heating technique. Evaluation of interference effects from other argon isotopes produced during the irradiation, and determination of appropriate correction factors were undertaken by Mitchell (1968a), Brereton (1970), and Turner (1971a). The first applications of the $^{40}Ar/^{39}Ar$ dating method to terrestrial rocks were concerned mainly with the total fusion approach whereby the gas is released in one step by melting the sample. In general these total fusion $^{40}Ar/^{39}Ar$ ages agreed with ages determined by the conventional K–Ar method (Mitchell, 1968a; Dunham et al., 1968; York and Berger, 1970; Dalrymple and Lanphere, 1971). But stepwise heating experiments on terrestrial samples were soon undertaken (Fitch et al., 1969; Miller et al., 1970; York et al., 1971; Lanphere and Dalrymple, 1971; Brereton, 1972). Some of the early results, or at least the interpretation of them, were controversial. These pioneering studies provided a firm foundation for further development in both the technical aspects of measuring precise and accurate $^{40}Ar/^{39}Ar$ ages and in the application of the method to geological samples.

2. BASIS OF THE ^{40}Ar/^{39}Ar DATING METHOD

2.1 POTASSIUM–ARGON DATING

The ^{40}Ar/^{39}Ar method of dating rocks has its foundations in the potassium–argon (K–Ar) isotopic dating method, a widely used technique for measuring numerical ages on minerals and rocks. Since the K–Ar method was developed over 30 years ago, it has been applied to a diverse range of geological samples to help elucidate many important geological problems of local, regional, or global significance. Particularly notable successes, dependent largely upon dating by the K–Ar method, include the development of the geomagnetic polarity time scale and the numerical calibration of the Phanerozoic geological or relative time scale.

The method is based upon the occurrence in nature of the radioactive isotope of potassium, ^{40}K, which has a half life of 1250 million years (Ma). This isotope of potassium has a dual decay to ^{40}Ca and to ^{40}Ar, and the branch yielding radiogenic argon (^{40}Ar*) as daughter product provides the basis for the K–Ar dating technique through its accumulation over geological time. In the simplest case of an igneous rock, for example, an unaltered lava, the K–Ar method normally yields an age that is equal to the time that has elapsed since its eruption and cooling. At the high temperature of a magma, the argon contained within the melt will tend to equilibrate with the ambient gas phase, which is likely to be atmospheric in composition at or near the Earth's surface. Thus, argon that partitions into the melt during its generation in the source region for the magma, and possibly significantly enriched in radiogenic argon, is expected to exchange with argon of atmospheric composition as it approaches the surface. If complete equilibrium is attained, then effectively all trace of preexisting radiogenic argon that may have been present will be lost. However, subsequent to cooling of the lava, the ^{40}Ar* generated from the decay of ^{40}K begins to accumulate quantitatively within the crystal structures of the minerals comprising the rock (Fig. 2.1). At ambient temperature, the radiogenic argon remains trapped within the crystals indefinitely because of its relatively large atomic size of about 1.9 Å. Subsequent measurement of the amount of parent ^{40}K and daughter ^{40}Ar* contained within the rock or mineral, combined with the known rate of decay of ^{40}K to ^{40}Ar*, enables an age to be calculated, reflecting the time since eruption and cooling of the lava.

The K–Ar dating method was found to give reliable ages on many rapidly cooled igneous rocks, but in some cases it was noted that ages on potassium-bearing minerals from the same rock, whether igneous or metamorphic, were discordant. This initially puzzling phenomenon is now relatively well understood in terms of differences in diffusion behavior for radiogenic argon in different mineral structures during slow cooling or during thermal events subsequent to original crystallization. As the radiogenic argon is trapped within crystal lattices as neutral atoms, increased temperature causes diffusive transfer, with the rate of diffusion increasing exponentially with temperature (see Chapter 5). Thus, rocks that have experienced elevated temperatures after crystallization may partially or completely lose accumulated radiogenic argon from their constituent minerals, depending upon the diffusion behavior, the temperature, and the time involved. A K–Ar age therefore may register the time since crystallization and cooling below a critical temperature, the time since cooling after a metamorphic event, or an intermediate age that does not date a particular event, but simply reflects partial diffusion loss of radiogenic argon during a metamorphism. These aspects of the K–Ar dating method can be

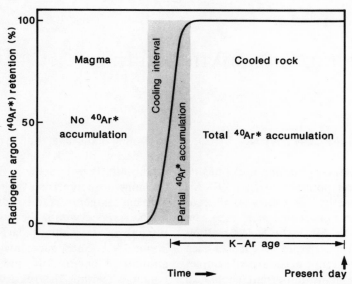

Fig. 2.1. Diagram illustrating the commencement of accumulation of radiogenic argon in an igneous rock as it cools from magmatic temperatures. From Potassium-Argon Dating by G. B. Dalrymple and M. A. Lanphere, copyright 1969 W. H. Freeman and Company, with permission.

partly explored by judicious choice of samples for measurement, but can be much more fully exploited using the $^{40}Ar/^{39}Ar$ dating technique to decipher the detailed thermal history of a given region.

The K–Ar method is one of the most versatile and widely applied of the various geochronometers available for dating rocks, and with the advent of the $^{40}Ar/^{39}Ar$ technique the applications are being progressively broadened. In part this is because potassium is the eighth most abundant element in the Earth's continental crust, comprising about 1 wt% (Taylor and McLennan, 1985). Minerals in which potassium is an essential element are fairly common in nature, and include many of the micas and the potassic alkali feldspars. Potassium also is present in a range of other minerals as a major or minor element, so that the K–Ar method, in principle, is applicable to many rocks and individual minerals. Another reason for its popularity as a dating method is that, with current techniques, there is a very high sensitivity for detection of radiogenic argon. In favorable circumstances, the technique can be applied to igneous rocks as young as a few thousand years, with no older limit in terms of the physical measurements. All these advantages apply, and indeed are reinforced, when the $^{40}Ar/^{39}Ar$ dating method is employed.

In the K–Ar dating method, the usual practice is to measure the potassium and argon on separate portions of the sample. Potassium invariably is measured as total potassium and the amount of ^{40}K calculated from the known $^{40}K/K$ ratio in nature. The most commonly used techniques of measurement are flame photometry, atomic absorption spectrometry, and isotope dilution, all of which can yield precise and accurate measurements provided due care and attention are taken in preparation of standards and unknowns and in calibration. Precision of better than 1% can be readily achieved and accuracy of the same order is possible, if not always attained. Argon normally is determined by isotope dilution using ^{38}Ar as a tracer with extraction of the gas from the sample by fusion in an ultrahigh vacuum system. Following purification of the argon, isotopic analysis is carried out by means of mass spectrometry, from which the content of $^{40}Ar^*$ can be calculated. Again, dependent upon proper calibrations of the tracer and mass spectrometer, the precision and accuracy of measurement allow determination of $^{40}Ar^*$ to about 1% in those cases where the proportion of $^{40}Ar^*$ is greater than $\sim 10\%$, with a

TABLE 2.1. *Argon nomenclature*

Atmospheric argon: Argon with the isotopic composition of that found in the present-day atmosphere

Radiogenic argon: Argon formed from *in situ* decay of ^{40}K in a rock or mineral

Trapped argon: This refers to the argon that is trapped or incorporated within a rock or mineral at the time of its formation or during a subsequent event. For terrestrial samples, the trapped argon component commonly, but not necessarily, has atmospheric composition. In extraterrestrial samples, the trapped argon is very different in composition from atmospheric argon, often having ^{40}Ar/^{36}Ar ~1

Cosmogenic argon: Argon produced from cosmic ray interactions with target nucleii such as calcium, titanium, and iron, mainly involving spallation reactions, but also through neutron capture because secondary neutrons are produced by the cosmic ray bombardment. Normally only of concern when dealing with extraterrestrial samples

Neutron-induced argon: Argon produced in a sample during irradiation in a nuclear reactor, owing to neutron interactions on chlorine, potassium, and calcium

Extraneous (including excess and inherited) argon: In those cases in which trapped argon in terrestrial samples has ^{40}Ar/^{36}Ar >295.5, the value of this ratio in atmospheric argon, the additional ^{40}Ar commonly is referred to as extraneous argon (cf. Damon, 1968; Dalrymple and Lanphere, 1969). Excess argon is that component of ^{40}Ar incorporated into samples by processes other than by *in situ* radioactive decay of ^{40}K. Inherited argon probably is best defined as that ^{40}Ar, essentially radiogenic, introduced into a rock or mineral sample by physical contamination from older material

progressive degradation in precision as the proportion of nonradiogenic argon increases toward 100%, owing to error magnification effects (Baksi et al., 1967; Cox and Dalrymple, 1967; Dalrymple and Lanphere, 1969).

Because there are several possible sources and types of argon of different isotopic composition in geological samples, we have summarized the nomenclature followed in this monograph in Table 2.1.

As with all isotopic dating methods, there are a number of assumptions that must be fulfilled for a K–Ar age to relate to events in the geological history of the region being studied. These same assumptions also apply to the ^{40}Ar/^{39}Ar method, although this latter technique provides greatly increased opportunities for their testing. The principal assumptions are given below with brief comment as to their validity, and will be further discussed implicitly or explicitly in the following chapters.

1. The parent nuclide, ^{40}K, decays at a rate independent of its physical state and is not affected by differences in pressure or temperature. This is a major assumption, common to all dating methods based on radioactive decay; the available evidence suggests that it is well founded (Friedlander et al., 1981). Although changes in the electron capture partial decay constant for ^{40}K possibly may occur at high pressures, theoretical calculations by Bukowinski (1979) indicate that for pressures experienced within a body of the size of the Earth the effects are negligibly small.

2. The ^{40}K/K ratio in nature is constant. As ^{40}K is rarely determined directly when ages are measured, this is an important underlying assumption. Isotopic measurements of potassium in terrestrial and extraterrestrial samples indicate that this assumption is valid, at least to the extent that no differences greater than about 3‰ have been reported in the ^{39}K/^{41}K ratio. The evidence for the essentially constant isotopic ratios for the potassium isotopes will be presented in more detail subsequently.

3. The radiogenic argon measured in a sample was produced by *in situ* decay of ^{40}K in the interval since the rock crystallized or was recrystallized. Violations of this assumption are not uncommon. Well-known examples of incorporation of extraneous ^{40}Ar include chilled glassy deep-sea basalts that have not completely outgassed preexisting ^{40}Ar*, and the physical contamination of a magma by inclusion of older xenolithic material. Further examples will be discussed later, as the ^{40}Ar/^{39}Ar method allows the presence of extraneous argon to be recognized in some cases.

4. Corrections can be made for nonradiogenic ^{40}Ar present in the rock being dated. For terrestrial rocks the assumption generally is made that all such argon is atmospheric in composition with ^{40}Ar/^{36}Ar $= 295.5$, and although this commonly is so, there are exceptions. Various ways of assessing this assumption are available including the use of isotope correlation diagrams. Extraterrestrial samples such as meteorites and lunar rocks have nonradiogenic argon of quite different composition to that of atmospheric argon, but corrections often can be made satisfactorily, particularly as

the nonradiogenic contributions usually are minor.

5. The sample must have remained a closed system since the event being dated. Thus, there should have been no loss or gain of potassium or $^{40}Ar^*$, other than by radioactive decay of ^{40}K. Departures from this assumption are quite common, particularly in areas of complex geological history, but such departures can provide useful information that is of value in elucidating thermal histories.

These basic assumptions must be tested and assessed in each study that is undertaken. This is usually best done by measuring a suite of rocks or minerals from the area under study. The consistency or lack of consistency of the results, together with knowledge of the geology of the area, allows assessment of some of these assumptions, and provides the basis for conclusions as to the reliability and meaning of the measured ages. As will become evident later, an important advantage of the $^{40}Ar/^{39}Ar$ dating method is that the assumptions underlying calculation and interpretation of an age are more readily assessed than is the case for conventional K–Ar age measurements.

2.2 $^{40}Ar/^{39}Ar$ DATING

In the $^{40}Ar/^{39}Ar$ method the sample to be dated is first irradiated in a nuclear reactor to transform a proportion of the ^{39}K atoms to ^{39}Ar through the interaction of fast neutrons. Following irradiation, the sample is placed in an ultrahigh vacuum system, and the argon extracted from it by fusion is purified and analyzed isotopically in a mass spectrometer. The relative abundances of ^{40}Ar, ^{39}Ar, ^{37}Ar, and ^{36}Ar are measured, and, in some cases, ^{38}Ar also. The $^{40}Ar^*/^{39}Ar_K$ is determined, where $^{40}Ar^*$ is the radiogenic argon, and $^{39}Ar_K$ is the ^{39}Ar produced from ^{39}K during the irradiation. This ratio is derived after correcting for some interferences, mainly by using the measured ^{37}Ar, and by utilizing the ^{36}Ar to correct for the presence of nonradiogenic ^{40}Ar. The $^{40}Ar^*/^{39}Ar_K$ ratio is proportional to the $^{40}Ar^*/^{40}K$ ratio in the sample, and therefore is proportional to age. This is so because the $^{39}Ar_K$ is dependent upon the amount of ^{39}K present in the sample, and the $^{39}K/^{40}K$ ratio is essentially constant in nature. Rather than determining the absolute dose of fast neutrons the sample has received during the irradiation, a standard sample of accurately known K–Ar age is irradiated together with the unknown, and the age of the unknown is derived by comparison with the $^{40}Ar^*/^{39}Ar_K$ of the flux monitor standard. The $^{40}Ar/^{39}Ar$ age so determined normally agrees to within experimental error with the conventional K–Ar measured on the same sample.

Thus, in the $^{40}Ar/^{39}Ar$ method the great advantage is that the ratio of daughter ($^{40}Ar^*$) to parent (^{40}K) is measured in a single isotopic analysis, obviating the need for a separate potassium analysis, overcoming problems of sample inhomogeneity, and, in principle, allowing smaller samples to be measured. Another advantage of this approach is that isotope ratios can be measured more precisely than separate determinations of potassium and argon, and therefore in principle a more precise age determination is possible than by the conventional K–Ar method. Although a $^{40}Ar/^{39}Ar$ total fusion age measurement is potentially more precise and, with proper control, more accurate than a K–Ar age, it provides little or no additional information compared with the latter, and has the same inherent problems of interpretation as a K–Ar age.

The major advantage of the $^{40}Ar/^{39}Ar$ method over the conventional K–Ar method, however, is that after irradiation a sample need not be directly fused to release the argon, but can be heated in steps, starting at temperatures well below that of fusion. The argon extracted at each step can be analyzed isotopically and thus a series of apparent ages determined on a single sample. This approach, known as the step heating or incremental heating technique, introduced by Merrihue and Turner (1966), provides a wealth of additional information that can provide insights into the distribution of $^{40}Ar^*$ in the sample, relative to the distribution of ^{39}K and hence ^{40}K. The method relies upon the release of the argon by thermal diffusion processes in the vacuum system as the sample is heated at successively higher temperatures. Thus, for a sample that has quantitatively retained its $^{40}Ar^*$ since crystallization, both ^{40}Ar* and ^{39}Ar are likely to occur in the same proportions in similar sites within crystals as they have both been derived from potassium. During a step heating experiment the two isotopes normally will be released in proportion because of their similar diffusion coefficients, yielding an essentially constant ^{40}Ar*/^{39}Ar$_K$ ratio in each gas fraction extracted, so that the derived ages also will be constant. A plot of the apparent ^{40}Ar*/^{39}Ar$_K$ age for each step against cumulative proportion of argon released, usually the ^{39}Ar, known as an age spectrum, will yield a flat pattern often termed a plateau (Fig. 2.2a). A flat age spectrum of this kind is readily interpreted as indicating that the sample has remained a closed system and, thus, thermally undisturbed since crystallization.

A sample that has lost a proportion of its ^{40}Ar* some time after its initial crystallization, as, for example, during a thermal metamorphism, will have sites within its lattice that have different ratios of daughter radiogenic argon (^{40}Ar*) to parent ^{40}K. During a step heating experiment such differences will be revealed by variations in the ^{40}Ar*/^{39}Ar$_K$ ratio measured on the gas fractions successively released by diffusion from the sample, yielding an age spectrum that is not flat. Thus, a sample that has suffered a thermal event subsequent to

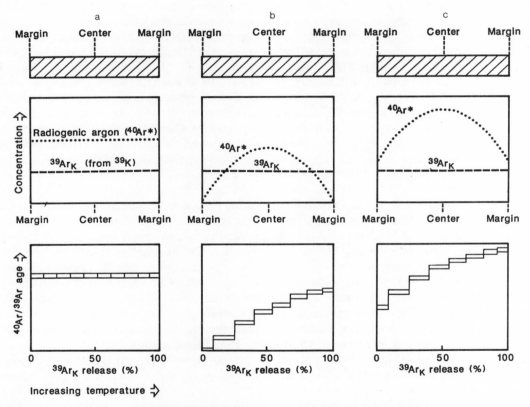

Fig. 2.2. Schematic diagrams showing model ^{40}Ar/^{39}Ar age spectra. Top diagram portrays an idealized crystal in cross section, the middle diagram in each panel shows the concentration of radiogenic argon and neutron-induced ^{39}Ar$_K$ across the crystal, and the lower diagram illustrates the ^{40}Ar/^{39}Ar age spectrum expected from measurement of argon extracted in successive steps at progressively higher temperature from the idealized crystal. (a) The case of a crystal undisturbed subsequent to initial crystallization and rapid cooling. (b) The case in which partial loss of radiogenic argon has occurred from the crystal in geologically recent times, so that there is a marked gradient of radiogenic argon across the crystal from essentially zero at the grain boundary. (c) The same case as in b, except that significant accumulation of radiogenic argon has occurred since the reheating event owing to the passage of time. A maximum age for the time of the reheating is given by the ^{40}Ar/^{39}Ar age for the gas released in the first step of the experiment, and a minimum age for the primary crystallization of the crystal is provided by the apparent age measured on the gas released at the highest temperature. Thickness of bars in model age spectra indicates nominal uncertainty in the individual ages.

crystallization will have lost some ^{49}Ar* from its more easily outgassed sites, generally those adjacent to grain or subgrain boundaries, so that the ratio of ^{40}Ar* to ^{40}K in such sites will be low relative to the more retentive sites. The least retentive sites, however, are likely to have retained all of their potassium, as empirical evidence suggests that this element normally has very much lower diffusion transport rates than ^{40}Ar*. Thus, following irradiation, these sites will have their full complement of ^{39}Ar generated from the ^{39}K. A step heating experiment on this sample will yield low ^{40}Ar*/^{39}Ar$_K$ ratios for the gas extracted at the lower temperatures, as the argon will be diffusing from the more readily outgassed sites initially. At higher temperatures the gas will be released from progressively more retentive sites increasingly remote from grain surfaces, and the ^{40}Ar*/^{39}Ar$_K$ ratio will therefore increase. Age spectra of the kind illustrated in Fig. 2.2b and c will be obtained, from which an estimate of the age of the thermal event may be derived from the apparent age given by the initially released argon, and a minimum age for the primary crystallization of the sample can be inferred from the measurements on the gas released in the higher temperature steps. Clearly, the conventional K–Ar and the ^{40}Ar/^{39}Ar total fusion ages for this sample will be intermediate between that of crystallization and subsequent thermal disturbance and will have little geological meaning. This hypothetical example is commonly matched by age spectra measured on actual samples, and illustrates the great advantage of the ^{40}Ar/^{39}Ar dating method over the conventional K–Ar technique in determining the geological history of a sample.

Subsequent sections of this chapter are devoted to presentation of some of the relevant basic information necessary for both K–Ar and ^{40}Ar/^{39}Ar age determinations, as well as discussion of the geological materials suitable for dating by these methods.

2.3 POTASSIUM AND ITS ISOTOPES

Potassium, the element of atomic number 19, in common with other alkali elements of group 1a of the periodic table, has a singularly occupied s orbital outside a closed shell of noble gas configuration. As a very reactive element, potassium does not occur naturally in the uncombined state. It has atomic radius of 2.03 Å and ionic radius of 1.33 Å. As the eighth most abundant element in the Earth's continental crust it is widely distributed, occurring as an essential or minor element in many minerals.

Potassium is produced during nucleosynthesis by the s-process (Burbridge et al., 1957), and as shown by Nier (1935) has three naturally occurring isotopes, ^{39}K, ^{40}K, and ^{41}K. The relative abundances of these isotopes can be explained in terms of their production by the s-process.

Nier's (1950) classic study on the isotopic composition of a number of elements, including potassium and argon, indicated that ^{40}K comprised 0.0119 ($\pm$0.0001) atom% of potassium, with ^{39}K the major isotope (93.09 $\pm$ 0.04 atom%), and ^{41}K making up the balance (6.91 $\pm$ 0.04 atom%). Subsequently, a more detailed investigation of the relative abundances of the potassium isotopes was reported by Garner et al. (1975b); the results are summarized in Table 2.2, and yield an atomic weight of 39.0983 $\pm$ 0.0001 for potassium.

In both the K–Ar and ^{40}Ar/^{39}Ar dating techniques, a basic underlying assumption is that the relative abundances of the isotopes of potassium are constant in rock and mineral samples. This assumption is necessary because ^{40}K is not measured directly in either method. In the K–Ar method, the total potassium is measured, and in the ^{40}Ar/^{39}Ar technique the ^{39}K (through ^{39}Ar) is determined; the ^{40}K is then derived by assuming an isotopic composition for potassium.

Several investigations have been undertaken to specifically assess the isotopic composition of potassium in natural materials, and although considerable differences have been

TABLE 2.2. Isotopic abundances of potassium[a]

Isotope	Atomic abundance (%)
^{39}K	93.2581 $\pm$ 0.0029
^{40}K	0.01167 $\pm$ 0.00004
^{41}K	6.7302 $\pm$ 0.0029

[a]After Garner et al. (1975b).

reported, it would appear that in fact the range of authenticated variations in the ^{40}K/^{41}K ratio is restricted to a maximum of a few percent. Thus, as summarized by Garner et al. (1975b), ^{39}K/^{41}K ratios have been reported that range from 13.5 ± 0.1 to 14.3 ± 0.1; much of this variation is attributed to isotopic fractionation effects in the mass spectrometers used for the measurements, and to difficulties associated with their absolute calibration.

Kendall (1960) measured the potassium isotopic composition in a variety of natural terrestrial materials, including some minerals, and concluded that within the errors (<1%) he could not detect any variation in composition. The ^{41}K/^{40}K ratio was measured by Burnett et al. (1966) on terrestrial samples of plagioclase, basalt, an olivine–hornblende mixture, and a potassium salt, without finding any detectable variation. However, in 3 of 11 meteorites measured, enrichments in ^{40}K as great as several percent were reported. This was interpreted as being caused by the addition of some ^{40}K generated from ^{40}Ca by cosmic radiation. Verbeek and Schreiner (1967) reported variations of up to 3% in the ^{39}K/^{41}K ratio across a granite–amphibolite contact, ascribed to isotopic fractionation effects related to thermal diffusion of potassium along a concentration gradient. Morozova and Alferovskiy (1974) found similar variations in the ^{39}K/^{41}K ratio across a gneiss–amphibolite contact zone; they also suggested the variations were caused by diffusion processes. The most recent work involving particularly careful control in the mass spectrometric analysis of more than 70 terrestrial samples, including samples from the same locality from which Verbeek and Schreiner (1967) reported anomalous isotopic compositions, revealed no potassium isotopic variation greater than 0.15% (Garner et al., 1975a,b; S. S. Goldich, personal communication, 1987). However, depletions of up to 0.8% in the ^{39}K/^{41}K ratio were found in a number of lunar samples by the same authors.

It can be concluded that the isotopic composition of potassium in nature is essentially constant, with variations probably rarely exceeding 1% in terms of the ^{40}K/K ratio. For terrestrial samples the variation in isotopic composition generally appears to be <0.15%. Thus, the constancy of potassium isotope ratios in samples used for dating, based upon the decay of ^{40}K, seems to be a very well-founded assumption.

2.4 THE ^{40}K DECAY SCHEME AND CONSTANTS

A useful summary of the principles underlying the radioactive decay of ^{40}K is given by Faure (1977, 1986). The three naturally occurring isotopes of potassium, ^{39}K, ^{40}K, and ^{41}K, have 19 protons ($Z = 19$) and 20, 21, and 22 neutrons (N), respectively, in their nucleii (Fig. 2.3). Both ^{39}K and ^{41}K, with odd numbers of protons and even numbers of neutrons, are stable, but in keeping with the empirical observation that nuclides with an odd number of both neutrons and protons are rarely stable, ^{40}K is radioactive.

In this context, Mattauch (1934) formulated a rule stating that two nuclides of the same mass number, known as isobars, cannot both be stable unless they are separated by more than one atomic number. Under this rule, the three isobars $^{40}_{18}$Ar, $^{40}_{19}$K, and $^{40}_{20}$Ca cannot all be stable and thus ^{40}K is expected to be radioactive. The other two nuclides are stable as the difference in atomic number is two. Note that ^{40}Ca has so-called magic numbers of both protons and neutrons, and thus is especially stable (Friedlander et al., 1981).

A diagrammatic representation of the decay scheme of ^{40}K is given in Fig. 2.4, after Beckinsale and Gale (1969) and Faure (1977). It is seen that ^{40}K has a dual decay, with 89.5% of the decays yielding ^{40}Ca by electron (β^-) emission, and loss of 1.33 MeV of energy, and the remainder of the decays (10.5%) producing ^{40}Ar, both daughter products being stable isotopes. The ^{40}K → ^{40}Ca branch rarely is used for age determination of rocks, because calcium is very common in nature, and ^{40}Ca is its most abundant isotope (96.9%), so that enrichments owing to ^{40}K decay are difficult to detect. However, in favorable circumstances, where common calcium is relatively low in abundance (e.g., in lepidolite micas, potassium-rich evaporites, or high-potassium granites), the method can be used effectively (Marshall and DePaolo, 1982; Obradovich et al., 1982).

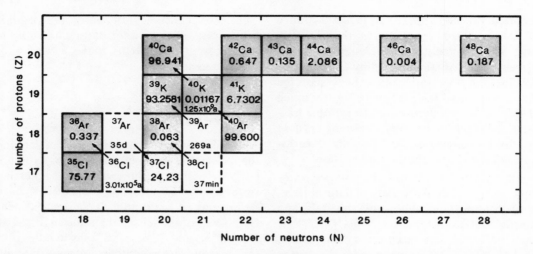

Fig. 2.3. Chart of the naturally occurring isotopes of chlorine, argon, potassium, and calçium (boxes with solid outlines), together with a number of the radioactive isotopes that are produced artificially (boxes with dashed outlines) and are relevant to the $^{40}Ar/^{39}Ar$ dating technique. Arrows indicate the decay paths for the radioactive nuclides. For each isotope the mass number is shown as a superscript to the chemical symbol; the natural abundance (in atomic percent) is listed below the symbol, and the half life for each unstable isotope is given at the bottom of each relevant box. Format and data adapted from Walker et al. (1983); see also Ozima and Podosek (1983).

The branch of the decay scheme of immediate interest involves the production of $^{40}Ar^*$ from ^{40}K. The decay is dominantly by orbital electron capture with release of 1.51 MeV of energy (Fig. 2.4). This process occurs by capture of an extranuclear electron by a proton, converting it to a neutron, with emission of a neutrino. Thus, electron capture produces an isobaric daughter nuclide; that is a nuclide of the same mass number as the parent. In the great majority of cases, electron capture produces $^{40}_{18}Ar$ in the excited state, followed by emission of a γ-ray to reach the ground state of ^{40}Ar. A minor proportion of the ^{40}K decays ($\sim 0.001\%$) proceed by emission of a positron (β^+), having an end point energy of 0.49 MeV, with annihilation of the positron and release of an additional 1.02 MeV of energy. A third mode of decay of ^{40}K to ^{40}Ar takes place by electron capture directly to the ground state and comprises about 0.16% of the total ^{40}K decays.

Beckinsale and Gale (1969) summarized and assessed the available determinations of the specific activities for natural potassium, and their preferred values are given in Table 2.3. The partial decay constants (λ_i) are derived

TABLE 2.3. *Specific activities for natural potassium*[a]

Activity	Disintegrations per second per gram K
$\dfrac{dn_{\beta^-}}{dt}$	28.27 ± 0.05
$\dfrac{dn_{\gamma}}{dt}$	3.26 ± 0.02
$\dfrac{dn_{\beta^+}}{dt}$	$3.25(\pm 0.37) \times 10^{-4}$
$\dfrac{dn_{ec}}{dt}$	$\sim 5.0(\pm 1.0) \times 10^{-2}$

[a]After Beckinsale and Gale (1969).

TABLE 2.4. *Decay constants for* ^{40}K[a]

Quantity	Value
λ_{β^-}	$4.962(\pm 0.009) \times 10^{-10} a^{-1}$
λ_e	$0.572(\pm 0.004) \times 10^{-10} a^{-1}$
λ'_e	$0.0088(\pm 0.0017) \times 10^{-10} a^{-1}$
$\lambda = \lambda_{\beta^-} + \lambda_e + \lambda'_e$	$5.543(\pm 0.010) \times 10^{-10} a^{-1}$
Branching ratio $= \dfrac{\lambda_e + \lambda'_e}{\lambda_{\beta^-}}$	0.117 ± 0.001
$\dfrac{\lambda}{\lambda_e + \lambda'_e}$	9.540
$t_{1/2} = \dfrac{\ln 2}{\lambda}$	$1.250(\pm 0.002) \times 10^9 a$

[a]From Beckinsale and Gale (1969), Garner et al. (1975b), and Steiger and Jäger (1977).

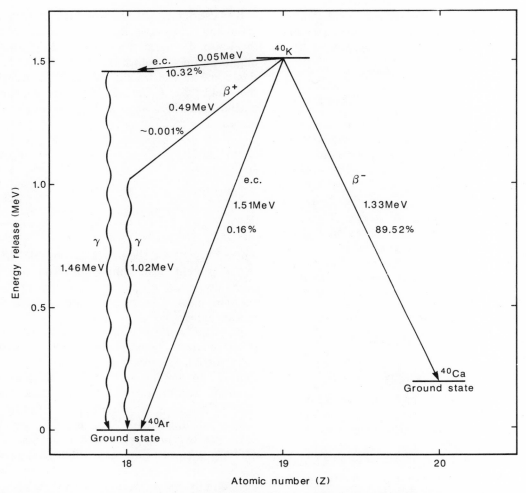

Fig. 2.4. Decay scheme for ^{40}K, illustrating the dual decay to ^{40}Ca (89.5%) and ^{40}Ar (10.5%). Note that the ^{40}K to ^{40}Ar branch is dominated by electron capture (e.c.). Adapted from Beckinsale and Gale (1969), Dalrymple and Lanphere (1969), Garner et al. (1975b), and Faure (1977).

using the following relationship:

$$(2.1) \quad \lambda_i = \frac{dn_i}{dt} \frac{AY}{fN_0} \quad \text{per year}$$

where dn_i/dt is the specific activity in disintegrations per second per gram of natural potassium, A is the atomic weight of potassium (39.0983) on the ^{12}C scale (Garner et al., 1975b), f is the atomic abundance of ^{40}K in natural potassium (0.0001167; Garner et al., 1975b), N_0 is Avogradro's number (6.0225 × 10^{23}), and Y is the number of seconds in a mean solar year (3.1558 × 10^7). Using this equation the constants given in Table 2.4 are derived. Note that these values differ from those calculated by Beckinsale and Gale (1969), simply because the revised ^{40}K abundance given by Garner et al. (1975b) is employed. The values are those recommended also by the IUGS Subcommission on Geochronology (Steiger and Jäger, 1977), and are now virtually universally applied. Prior to 1977, several different sets of decay constants for ^{40}K were in use.

2.5 DERIVATION OF AGE EQUATIONS

Soon after the discovery of radioactivity, Rutherford and Soddy (1903) demonstrated empirically that the rate of decay of a radioactive substance follows an exponential law, with the activity at any instant proportional to the number of radioactive atoms present. Thus,

(2.2) $\qquad -dN/dt \propto N$

and

(2.3) $\qquad dN/dt = -\lambda N$

where N is the number of radioactive atoms present at time t, and λ is the constant of proportionality known as the decay constant, which is the probability of any particular atom decaying per unit time. Thus the decay constant can be thought of as the fraction of parent radioactive atoms decaying per unit time. The half life ($t_{1/2}$) is the time required for a given number of radioactive atoms to decay to half that number and is related to the decay constant as follows:

(2.4) $\qquad t_{1/2} = \ln 2/\lambda = 0.693/\lambda$

Rearrangement and integration of Eq. (2.3) yields

(2.5) $\qquad N = N_0 \exp(-\lambda t)$

where N represents the number of radioactive atoms present at time t, and N_0 is the number of radioactive atoms present at time $t = t_0$, some time in the past.

For a simple decay scheme in which radioactive parent (N) decays to daughter product (D),

(2.6) $\qquad N_0 = N + D$

where N and D are the number of parent and daughter atoms present at a given time.

Substituting in Eq. (2.5) we obtain

(2.7) $\qquad N = (N + D) \exp(-\lambda t)$

Rearranging and taking natural logarithms yields the basic equation used in geochronology:

(2.8) $\qquad t = \dfrac{1}{\lambda} \ln\left(1 + \dfrac{D}{N}\right)$

By inspection of Eq. (2.8), one can appreciate how sensitive a function the daughter to parent ratio is to age. Note that when half the original parent has decayed (i.e., $D/N = 1$), the expression collapses to Eq. (2.4), the half life definition.

Because of the dual decay of ^{40}K to ^{40}Ar and ^{40}Ca, and because we are concerned here with the K–Ar dating method, the basic equation requires some modification to be applied in this context. The fraction of the ^{40}K decays that yield ^{40}Ar, f_{Ar}, is the ratio of the relevant partial decay constants ($\lambda_e + \lambda'_e$) to the decay constant for ^{40}K (λ):

(2.9) $\qquad f_{Ar} = \dfrac{\lambda_e + \lambda'_e}{\lambda_e + \lambda'_e + \lambda_\beta} = \dfrac{\lambda_e + \lambda'_e}{\lambda}$

Substituting in Eq. (2.8) we find

(2.10) $\qquad t = \dfrac{1}{\lambda} \ln\left(1 + \dfrac{\lambda}{\lambda_e + \lambda'_e} \dfrac{^{40}Ar^*}{^{40}K}\right)$

where $^{40}Ar^*/^{40}K$ is the ratio of radiogenic ^{40}Ar to ^{40}K present in the sample now. Thus as the decay constants are known, measurements of the amounts of $^{40}Ar^*$ and ^{40}K enable an age to be calculated.

For the $^{40}Ar/^{39}Ar$ dating method, the ^{39}Ar generated from ^{39}K during the irradiation with fast neutrons is proportional to the ^{40}K present in the sample, as the $^{40}K/^{39}K$ ratio is essentially constant in nature; thus the $^{40}Ar^*/^{39}Ar_K$ ratio is proportional to age. But simple substitution of $^{39}Ar_K$ for ^{40}K in Eq. (2.10) clearly is not appropriate, as the amount of $^{39}Ar_K$ generated obviously will depend upon the duration of the neutron irradiation, the neutron flux, the proportion of neutrons above the threshold energy capable of producing the nuclear reaction in question, and other factors, so that a somewhat different approach is adopted.

Equation (2.10) can be rearranged and expressed in terms of $^{40}Ar^*$:

(2.11) $\qquad ^{40}Ar^* = {}^{40}K \dfrac{\lambda_e + \lambda'_e}{\lambda} [(\exp \lambda t) - 1]$

Following the derivation of Mitchell (1968a), the amount of ^{39}Ar that is produced from ^{39}K in a sample during irradiation with neutrons is given by

(2.12) $\qquad ^{39}Ar_K = {}^{39}K \, \Delta \displaystyle\int \phi(E)\sigma(E)\,dE$

where $^{39}Ar_K$ is the number of atoms of ^{39}Ar produced from ^{39}K in the sample, ^{39}K is the original number of atoms of ^{39}K present, Δ is the duration of the irradiation, $\phi(E)$ is the neutron flux at energy E, and $\sigma(E)$ is the neutron capture cross section at energy E for the ^{39}K(n,p)^{39}Ar reaction. Combining Eqs. (2.11) and (2.12) it follows that for an irradiated sample of age t:

$$(2.13) \quad \frac{^{40}\text{Ar}^*}{^{39}\text{Ar}_K} = \frac{^{40}\text{K}}{^{39}\text{K}} \frac{\lambda_e + \lambda'_e}{\lambda} \frac{1}{\Delta} \frac{[(\exp \lambda t) - 1]}{\int \phi(E)\sigma(E)\,dE}$$

As shown by Grasty and Mitchell (1966) and Mitchell (1968a), it is convenient to define a dimensionless irradiation parameter, J, as follows:

$$(2.14) \quad J = \frac{^{39}\text{K}}{^{40}\text{K}} \frac{\lambda}{\lambda_e + \lambda'_e} \Delta \int \phi(E)\sigma(E)\,dE$$

Substituting Eq. (2.14) in Eq. (2.13) gives

$$(2.15) \quad \frac{^{40}\text{Ar}^*}{^{39}\text{Ar}_K} = \frac{(\exp \lambda t) - 1}{J}$$

which, upon rearrangement, allows calculation of the age, t, of the sample:

$$(2.16) \quad t = \frac{1}{\lambda} \ln\left(1 + J \frac{^{40}\text{Ar}^*}{^{39}\text{Ar}_K}\right)$$

Note that this equation is of similar form to Eq. (2.10). Provided that the irradiation parameter, J, can be determined, it is apparent from Eq. (2.16) that only measurement of the $^{40}\text{Ar}^*/^{39}\text{Ar}_K$ ratio in gas extracted from the irradiated sample is required for an age to be calculated. Thus modification of Eq. (2.10) for calculation of a $^{40}\text{Ar}/^{39}\text{Ar}$ age [Eq. (2.16)] simply involves a proportionality factor where

$$(2.17) \quad {}^{40}\text{K} = {}^{39}\text{Ar}_K \left(\frac{\lambda}{\lambda_e + \lambda'_e} \frac{1}{J}\right)$$

From Eqs. (2.12) and (2.14) the parameter, J, relating to the production of ^{39}Ar from ^{39}K during irradiation, is dependent upon the duration of the irradiation, the neutron flux, and the neutron capture cross section. Because of the difficulties encountered in accurately determining the relevant integrated fast neutron dose a sample has received, Merrihue and Turner (1966) suggested that a mineral of accurately known K–Ar age be irradiated together with the unknown to monitor the dose. Rearrangement of Eq. (2.15) yields

$$(2.18) \quad J = \frac{(\exp \lambda t) - 1}{{}^{40}\text{Ar}^*/{}^{39}\text{Ar}_K}$$

As the age, t, of the standard sample is known from conventional K–Ar age measurement, the parameter J can be determined from Eq. (2.18) by simply measuring the $^{40}\text{Ar}^*/^{39}\text{Ar}_K$ in the gas extracted from the standard sample after irradiation. This value of J is then used in Eq. (2.16), together with the $^{40}\text{Ar}^*/^{39}\text{Ar}_K$ ratio measured on the unknown sample irradiated at the same time, so that its age can be determined. It will be clear from the foregoing that the function of the standard sample is that of a neutron dosimeter, commonly termed the flux monitor, allowing the determination of the irradiation parameter, J.

An implication of Eq. (2.16) is that because only isotope ratios need to be known to calculate an age, a sample with a uniform distribution of radiogenic argon need not be completely outgassed in order to obtain a valid age, contrasting with the conventional K–Ar dating method.

2.6 ATMOSPHERIC ARGON

As a mass spectrometer (see Section 3.17) does not normally measure absolute abundances or absolute ratios, some standardization procedure needs to be adopted to calibrate the machine. For gas source mass spectrometers used for the isotopic analysis of argon, the standard generally employed is atmospheric argon, which is conveniently available, comprising 0.937% by volume of dry air (Kellas, 1895). The isotopic composition of atmospheric argon was measured by Nier (1950), and his results are given in Table 2.5. The derived value for atmospheric $^{40}\text{Ar}/^{36}\text{Ar}$ of 295.5 is included by Steiger and Jäger (1977) in their list of recommended values for use in geochronology.

It is essential to measure atmospheric argon on a regular basis to determine the mass discrimination of the mass spectrometer, particularly as the discrimination tends to change with time, albeit rather slowly, generally on a time scale of months. Thus the $^{40}\text{Ar}/^{36}\text{Ar}$ ratio determined for atmospheric argon in a mass spectrometer might be 299.7. This machine discriminates, therefore, in favor of the ^{40}Ar relative to ^{36}Ar in terms of Nier's (1950) measurements; a correction factor of 0.9860 needs to

TABLE 2.5. *Isotope composition of atmospheric argon*[a]

Isotope	Abundance (atom%)
^{40}Ar	99.600
^{38}Ar	0.0632 ± 0.0001
^{36}Ar	0.3364 ± 0.0006

[a] After Nier (1950), who reported the atmospheric argon isotope ratio data as ^{36}Ar/^{40}Ar = 0.003378 ± 0.000006 and ^{38}Ar/^{40}Ar = 0.000635 ± 0.000001. From these data ^{40}Ar/^{36}Ar = 296.0 ± 0.5. The value of ^{40}Ar/^{36}Ar = 295.5 for atmospheric argon, recommended by Steiger and Jäger (1977) for use in geochronology, is derived from the rounded values of atomic abundance given by Nier (1950) for ^{40}Ar, ^{38}Ar, and ^{36}Ar of 99.600, 0.063, and 0.337%, respectively.

be applied to obtain the standard value. Appropriate correction factors for other argon isotopic ratios can be calculated readily, assuming the correction is linear with mass. These corrections are made to measured isotopic ratios on a routine basis prior to further data processing.

In principle the atmospheric ^{38}Ar also could be compared with ^{36}Ar or ^{40}Ar for the determination of the mass discrimination of the machine. However, for a mass spectrometer that has been used for argon measurements by isotope dilution, involving ^{38}Ar as a tracer, the small proportion of ^{38}Ar in atmospheric argon becomes difficult to measure precisely and accurately because of memory problems, resulting from the progressive release of ^{38}Ar atoms previously implanted into various parts of the mass spectrometer, a consequence of operating the machine in the static mode, isolated from its pump.

Knowledge of the atmospheric argon isotopic composition is essential for successful K-Ar and ^{40}Ar/^{39}Ar age measurements, as corrections must be made for any contaminating atmospheric argon contained within the sample or contributed from the vacuum system in which the gas is extracted from the sample. This correction is made by assuming that all of the ^{36}Ar is of atmospheric argon origin after corrections are applied for mass discrimination of the mass spectrometer and after due allowance is made for other known sources of ^{36}Ar, such as minor ^{36}Ar in the ^{38}Ar tracer in the case of measurement of argon by isotope dilution, or minor amounts of neutron-induced ^{36}Ar in the case of dating by the ^{40}Ar/^{39}Ar method. For terrestrial samples the corrected ^{36}Ar is assumed to be derived from atmospheric argon. As

(2.19) $\qquad (^{40}\text{Ar}/^{36}\text{Ar})_A = 295.5$

then

(2.20) $\qquad (^{40}\text{Ar})_A = 295.5 \times (^{36}\text{Ar})_A$

where the subscript A indicates the isotope is from the atmosphere. As the total argon of mass 40, ^{40}Ar$_T$ is the sum of the atmospheric and radiogenic,

(2.21) $\qquad ^{40}\text{Ar}_T = (^{40}\text{Ar})_A + {}^{40}\text{Ar}^*$

then the radiogenic argon, ^{40}Ar*, can be derived by difference so that

(2.22) $\qquad ^{40}\text{Ar}^* = (^{40}\text{Ar})_T - 295.5(^{36}\text{Ar})_A$

For ^{40}Ar/^{39}Ar measurements, the proportions of the total ^{40}Ar that are atmospheric and radiogenic are determined rather than the actual amounts. For extraterrestrial samples the situation is somewhat more complex because trapped nonradiogenic argon normally will have a composition quite different from that of atmospheric argon. In such cases, $(^{40}\text{Ar}/^{36}\text{Ar})_{\text{trapped}}$ generally has a value of about unity. However, corrections usually can be made appropriately by use of correlation diagrams or other means, including by assumption of a particular ^{40}Ar/^{36}Ar ratio. As extraterrestrial samples commonly have highly radiogenic argon, uncertainties in the application of corrections for trapped and atmospheric argon often have little effect on the calculated age.

It is good practice to normalize all results to the atmospheric value, so that measurements between different laboratories can be compared directly. In addition, it is immediately obvious when deduced initial ^{40}Ar/^{36}Ar values found in a sample or suite of samples differ from the normally expected atmospheric value for terrestrial rocks. Note that should the reference value for the ^{40}Ar/^{36}Ar ratio in atmospheric argon determined by Nier (1950) be slightly in error, then this will not significantly affect the calculated age whether measured by the conventional K–Ar or by the ^{40}Ar/^{39}Ar method.

2.7 MATERIALS SUITABLE FOR DATING

2.7.1 Introduction

In principle any potassium-bearing mineral or rock can be used for K–Ar and ^{40}Ar/^{39}Ar dating. In practice the range of samples that can be usefully dated is rather more restricted. The techniques are applied mainly to igneous and metamorphic rocks, although measurements on sedimentary rocks can be useful in some cases. For a K–Ar determination to be feasible, a sample obviously must have measurable potassium and radiogenic argon, and for an age to be of any value it must be on a sample that is able to retain radiogenic argon quantitatively at temperatures experienced in at least some geological environments. Minerals most suitable for K–Ar age measurement are those in which potassium is located in lattice sites. The potassium may be an essential cation, but some minerals useful for dating contain relatively minor amounts of potassium substituting for other cations. Generally there is little point in measuring minerals that do not accept potassium into lattice sites. For example, potassium does not substitute in the quartz structure, and although potassium may be measurable in this mineral, it normally will be associated with impurities or inclusions within the quartz. Nevertheless there may be some circumstances under which measurements might be undertaken on potassium-poor minerals such as quartz, beryl, and olivine, for example, to examine whether excess argon is present or not.

In Table 2.6 the most commonly used materials for K–Ar dating are listed, showing that many common rock-forming minerals can be utilized for this purpose. A brief discussion of the suitability of the most widely employed minerals will be given here, augmenting the

TABLE 2.6. *Commonly utilized materials for K–Ar and ^{40}Ar/^{39}Ar dating and rock types in which they occur*[a]

	Rock type			
	Volcanic	Hypabyssal and plutonic	Metamorphic	Sedimentary
Feldspars				
High-temperature alkali feldspar	b			
Low-temperature alkali feldspar		b	b	c
Plagioclase	b	b	b	
Feldspathoids				
Leucite	b	b		
Nepheline	b	b		
Mica				
Biotite	b	b	b	c
Phlogopite			b	
Muscovite		b	b	c
Phengite			b	
Lepidolite		b		
Glauconite				d
Amphibole	b	b	b	
Pyroxene	e	e		
Whole rock	b		e	
Volcanic glass	d			
Clays				c,d
Evaporites				d

[a] Expanded from Dalrymple and Lanphere (1969).
[b] Wide application.
[c] Useful for provenance studies.
[d] Not generally useful for ^{40}Ar/^{39}Ar dating.
[e] Useful in some circumstances.

earlier summary presented by Dalrymple and Lanphere (1969). Most of these minerals also can be used for ^{40}Ar/^{39}Ar age determinations, but there are important exceptions. Thus, some clays and mica-like minerals (e.g., glauconite) are not suitable owing to their inability to quantitatively retain ^{39}Ar generated during irradiation, and in addition they may begin to lose radiogenic argon in conditions little above ambient temperature.

As will be evident from Table 2.6, the K–Ar method is applicable to a wide range of minerals and rock types forming in different environments. Nevertheless the utility of individual minerals from different rock types is quite variable, and the interpretation of the meaning of a measured age must rest on careful assessment of the geological environment as well as other factors. Thus, a measurement on a mineral from a metamorphic rock is likely to record the age of cooling below the closure temperature for argon retention in that particular mineral specimen, rather than the actual age of the peak of the metamorphism. Similarly, a measurement on a mineral from a sedimentary rock may reflect the provenance of the detrital material comprising the rock, rather than the age of deposition. In all cases, the ages obtained must be interpreted in the context of the geology of the region being investigated.

2.7.2 Feldspars

Minerals of the feldspar group collectively comprise the most abundant constituent of crystalline rocks in the Earth's crust, and they also are an important component in detrital sediments. Feldspars occur in a diverse range of compositions and structural states, and the reader is referred to Deer et al. (1963), Smith (1974), Ribbe (1975), and Brown (1984) for full details. In broad terms, two solid solution series are recognized, comprising the plagioclase series ($NaAlSi_3O_8$–$CaAl_2Si_2O_8$) and the alkali feldspar series ($NaAlSi_3O_8$–$KAlSi_3O_8$) within the ternary system that has Na, K, and Ca end members. The structural state of a feldspar is very dependent upon the temperature of crystallization and upon the subsequent thermal history. A feldspar that has crystallized at high temperature and has been cooled rapidly, as for example in the case of feldspar in a lava, generally will have a high-temperature structural state. Low-temperature feldspars have structures produced by slow cooling from high temperature, as is commonly found in plutonic and metamorphic rocks. Intermediate structural states also are known. Although the structures are all monoclinic or triclinic, there is a bewildering array of structural states, twinning, and exsolution features in the feldspars that cannot be considered in detail here. As might be expected, therefore, the feldspars show a range of behavior in terms of their suitability for age determination by the K–Ar and ^{40}Ar/^{39}Ar dating methods, summarized briefly below.

2.7.2.1 Alkali Feldspar

Because of their abundance in nature, especially in silicic rocks, and because of their relatively high potassium contents (up to 14%), alkali feldspars commonly have been used for K–Ar age measurements. Alkali feldspars were an obvious choice for this purpose from the earliest stages of development of the K–Ar dating method, but the results often were difficult to interpret, for reasons that are now becoming increasingly understood.

Alkali feldspars in volcanic rocks, commonly occurring as phenocrysts in silicic lavas, are preserved metastably in their high-temperature structural state owing to rapid cooling. These feldspars of the sanidine–anorthoclase–high albite solid solution series may be quite homogeneous, especially for compositions richer in potassium than Or_{67}. Volcanic alkali feldspars in the composition range Or_{60}–Or_{25}, however, may exhibit unmixing of sodium-rich and potassium-rich phases on a submicroscopic scale to yield cryptoperthites or cryptoantiperthites.

As a generalization, minerals of the high-temperature alkali feldspar series are ideal for K–Ar and ^{40}Ar/^{39}Ar dating, in many cases yielding ages that are readily interpreted in terms of the timing of volcanism, especially for rocks that have not been reheated since extrusion, because diffusion rates for radiogenic argon remain negligible at low temperature. As alkali feldspar remains stable during heating in a vacuum system to high temperature, it is an excellent mineral for ^{40}Ar/^{39}Ar age spectrum measurements; virtually ideal flat age spectra often are obtained on samples that have remained thermally undisturbed since crystallization (McDougall, 1981, 1985).

In plutonic rocks and in many metamorphic rocks, alkali feldspar usually has equilibrated in the subsolidus region of the stability field (Smith, 1974), resulting in unmixing or exsolution to potassium-rich and sodium-rich phases. Depending upon the original composition of the alkali feldspar and the cooling history, a range of exsolution textures is observed, often with development of perthites or microperthites. In the early stages of application of the K-Ar dating method, alkali feldspars from plutonic and metamorphic rocks commonly were measured. It was found, however, that the ages in many cases were less than those obtained on coexisting biotite, so that most workers subsequently abandoned use of this mineral for dating purposes. Indeed Dalrymple and Lanphere (1969) and Faure (1977) concluded that these low-temperature feldspars probably do not retain radiogenic argon quantitatively, even at room temperature. It is now becoming evident that this behavior can be understood in terms of diffusion of radiogenic argon from alkali feldspars during protracted cooling of plutonic and metamorphic rocks. Unmixed alkali feldspars have a low closure temperature for radiogenic argon, about 150°C (Foland, 1974; Harrison and McDougall, 1982), probably owing to the very short distance an argon atom has to diffuse before it can effectively leave the structure. Thus, during slow cooling, the mineral continues to lose radiogenic argon above the closure temperature, and may remain in the temperature interval of partial argon retention for a considerable time as the temperature approaches the ambient value. This characteristic of low-temperature alkali feldspars is beginning to be put to good use in elucidating thermal histories of geological terranes, especially by application of the $^{40}Ar/^{39}Ar$ age spectrum technique. Similarly, $^{40}Ar/^{39}Ar$ age measurements on alkali feldspars separated from detrital sedimentary rocks also may be used to obtain age information relating to the provenance of the source material, as well as providing control on the thermal history of the sedimentary sequence (Harrison and Bé, 1983). In addition, unambiguous recognition of excess argon in unmixed alkali feldspars, from distinctive U-shaped $^{40}Ar/^{39}Ar$ age spectra, has been documented by Zeitler and Fitz Gerald (1986). A more detailed treatment of these questions will be presented in Section 4.6.

2.7.2.2 Plagioclase

Members of the plagioclase series of feldspars commonly have been utilized for dating by the K-Ar method. As plagioclases comprise a solid solution series between a sodium end member (albite) and a calcium end member (anorthite), they have relatively low potassium contents, generally in the $<0.1-\sim1\%$, range, in part depending upon the composition of their host rocks. There is almost complete solid solution between end members in the case of the high-temperature series, representatives of which are found in some volcanic rocks. Plagioclase belonging to the low-temperature albite–anorthite series occurs in many igneous and metamorphic rocks, and also in sedimentary rocks as detrital minerals as well as authigenically. Studies using X-ray techniques show that members of the low-temperature series structurally are very complex, with submicroscopic domains of different composition (Smith, 1974). Plagioclases with a structural state intermediate between those of the high- and low-temperature forms also are common.

Plagioclase has been widely used for K-Ar dating of igneous rocks, and, to a lesser degree, metamorphic rocks, but relatively little is known about its retention properties for radiogenic argon, and no detailed laboratory studies have been undertaken of the argon diffusion characteristics of this mineral. From empirical evidence, however, the closure temperature for retention of radiogenic argon in plagioclase probably is somewhat lower than that for biotite ($\sim 300°C$), possibly in the 200–250°C range. Owing to its relatively low potassium content, incorporation of excess argon can have dramatic effects on the age measured, yielding anomalously old values, a problem that has been recognized in plagioclase from igneous as well as metamorphic rocks (Livingstone et al., 1967; Damon et al., 1967; Dalrymple et al., 1975). Analysis by the

^{40}Ar/^{39}Ar age spectrum technique is particularly useful in recognizing the presence of excess argon in plagioclase, as distinctive saddle-shaped (U-shaped) age spectra commonly are found (Dalrymple and Lanphere, 1974; Dalrymple et al., 1975; Lanphere and Dalrymple, 1976; Harrison and McDougall, 1981). Although K–Ar and ^{40}Ar/^{39}Ar dating of plagioclase from terrestrial rocks not uncommonly has given results that are difficult to interpret, plagioclase has proved to be a very reliable mineral for ^{40}Ar/^{39}Ar age measurement of lunar rocks; this will be discussed in more detail in Chapter 6.

2.7.3 Feldspathoids

Leucite and nepheline are the only two minerals of this diverse group that have been used with any degree of success for K–Ar dating.

Leucite ($KAlSi_2O_6$) is a relatively rare mineral, found mainly in potassium-rich, silica-undersaturated lavas, although it has been recognized in some hypabyssal rocks. It is unknown in plutonic rocks, but there is evidence in some rocks from deep-seated environments, as well as in some hypabyssal and volcanic rocks, that it initially crystallized from the magmas, but is now represented by a mixture of alkali feldspar and nepheline, termed pseudoleucite, or by alteration products. Because of its very high potassium content (up to 17.9%), with only limited substitution of sodium for potassium, leucite is a good candidate for K–Ar dating, but its rarity and the ease with which it alters mean that it has not been used extensively for dating purposes. Studies in which leucite has been employed to date rocks, mainly volcanic, include those by Evernden and Curtis (1965), Wellman et al. (1970), Wellman (1973), Radicati di Brozolo et al. (1981), and Tingey et al. (1983). The limited evidence available suggests that leucite is a reliable geochronometer for dating rocks not subsequently reheated. Experiments on leucite *in vacuo* by Evernden et al. (1960) showed that most of the argon is lost in 1 or 2 days at $\sim 550°C$, probably owing to its transition from pseudoisometric to isometric structure.

Because of the high potassium content of leucite, relatively large amounts of radiogenic argon are produced per unit time, and this, taken together with the modest amount of atmospheric argon found associated with the mineral, means that it is particularly useful for dating very young rocks. Thus, Tingey et al. (1983) measured leucite separated from two samples of leucitite lava on Gaussberg, a volcano on the coast of East Antarctica, and obtained conventional K–Ar ages of 52 ± 3 and 59 ± 2 ka, respectively; the proportion of radiogenic argon to total argon ranged up to 42%. The most comprehensive dating study on leucite so far made was by Radicati di Brozolo et al. (1981), who measured ^{40}Ar/^{39}Ar age spectra on separates of this mineral from pyroclastics in the Albion Hills volcanic complex of the Roman province in Italy. Excellent results were obtained, showing essentially flat age spectra, with a mean age of 358 ± 8 ka for five of the samples. Biotite from three of the samples gave ^{40}Ar/^{39}Ar results concordant with the coexisting leucite, but one sample yielded an older apparent age of 380 ± 22 ka and a fifth sample gave a younger age of 278 ± 15 ka. Measurements by the Rb–Sr method on leucite, biotite, and pyroxene on three samples gave ages that are less precise but in good agreement with the ^{40}Ar/^{39}Ar leucite ages. Villa (1986), however, recently reported the presence of excess argon in some leucites from the Roman volcanic province.

Nepheline is a relatively common mineral in alkaline rocks, occurring as a primary phase in many igneous rocks; it is also present in some metamorphic rocks. Although its composition often is close to $Na_3KAl_4Si_4O_{16}$, more sodic or potassic compositions are found, especially in volcanic rocks. Nepheline has not been widely used for K–Ar dating, but it appears to yield reliable ages and is quite retentive of radiogenic argon from studies made by Macintyre et al. (1966) on rocks from the Precambrian Grenville Province in Canada. York and Berger (1970) reported a single ^{40}Ar/^{39}Ar total fusion age on nepheline concordant with its K–Ar age. Similarly, Shanin et al. (1967) and Gerling et al. (1969) found that nepheline was suitable for K–Ar dating.

2.7.4 Micas

Minerals of the mica group exhibit a wide range of compositions and are characterized by a layered atomic structure that accounts for their platy habit and perfect basal cleavage. Micas occur in a very wide range of igneous and metamorphic rocks as well as in sediments as recycled detrital grains, and glauconite grows authigenically in the sedimentary environment. As a number of the micas contain relatively high potassium contents, they have been used for K–Ar dating from the earliest stages of development of the method. Those species most commonly utilized for dating purposes are biotite, phlogopite, muscovite, phengite, lepidolite, and glauconite, and brief comments on their suitability are given below.

2.7.4.1 Biotite and Phlogopite

The biotite group of trioctahedral micas encompasses an extensive range of compositions from the Mg-rich end member, phlogopite, $K_2Mg_6Si_6Al_2O_{20}(OH)_4$, to annite, $K_2Fe_6Si_6Al_2O_{20}(OH)_4$, but often with some additional substitution of Al in octahedral and tetrahedral sites toward end-member compositions of eastonite, $K_2Mg_5AlSi_5Al_3O_{20}(OH)_4$, and siderophyllite, $K_2Fe_5AlSi_5Al_3O_{20}(OH)_4$ (see Deer et al., 1962). Other substitutions are possible, especially F^- replacing OH^-. Generally phlogopite is the term used in those cases in which the atomic ratio Mg/Fe > 2; the more Fe-rich members are usually called biotites. The end-member compositions have potassium contents in the 7.4–9.1% range, but most fresh, unaltered biotites have potassium contents between about 7 and 8%. Phlogopites near the end-member composition are restricted mainly to Mg-rich rocks, especially dolomitic marbles and ultramafic rocks. Biotite is very common in igneous and metamorphic rocks; its composition in terms of the Fe/Mg ratio is determined by the bulk composition of the host in both igneous and metamorphic rocks.

Biotite group minerals are very useful for geochronological purposes. In the case of rapidly cooled igneous rocks a biotite K–Ar age will approximate closely the age of emplacement and cooling, provided that no subsequent reheating has occurred. For more slowly cooled igneous and metamorphic rocks, a biotite age usually will reflect the time since cooling below a temperature of 300–350°C, the approximate closure temperature for argon at moderate cooling rates. Thus, a biotite age normally will not provide a measure of the time since emplacement of a plutonic igneous body, nor the time since the peak of a metamorphic episode, but instead will yield information related to the cooling history. With due recognition of this factor, biotite K–Ar ages can be turned to great advantage for the determination of the thermal history of a terrane, especially when used in conjunction with measurements on other minerals that have different closure temperatures for retention of radiogenic daughter products.

In some cases, however, biotite has incorporated or trapped argon, usually termed extraneous or excess argon, from the environment. This excess argon rarely can be distinguished from radiogenic argon generated from *in situ* decay of ^{40}K within the mineral, so that anomalously old K–Ar ages may be obtained. Thus, Funkhouser et al. (1966) showed that biotite from the Mauna Kuwale Rhyodacite in the Waianae Range, Oahu, Hawaii has significant amounts of excess argon contained within small fluid inclusions in biotite separated from this relatively young volcanic body. Excess argon has been reported in phlogopite separated from kimberlitic rocks by Lovering and Richards (1964) and by Zartman et al. (1967). This phenomenon has also been recognized in biotites from metamorphic terranes (Richards and Pidgeon, 1963; Pankhurst et al., 1973; Roddick et al., 1980; Dallmeyer and Rivers, 1983; Foland, 1983). Generally the anomalously old ages found are interpreted as indicating incorporation of excess argon into the mica owing to a relatively high partial pressure of argon in the environment. As will be discussed in more detail in Section 4.12.5, $^{40}Ar/^{39}Ar$ age spectrum measurements on biotites add little additional information. Thus, Pankhurst et al. (1973) obtained a nearly flat age spectrum for a biotite from a metamorphic rock in Greenland, yielding an apparent age of about 5 Ga, clearly anomalously old. Although structure is observed in $^{40}Ar/^{39}Ar$ age spectra measured on some biotites, it is rarely possible to interpret these in terms of distribution of radiogenic argon in the crystals. The inability to produce meaningful age spectra from biotites probably is because biotite becomes unstable during heating *in vacuo*, undergoing dehydroxylation and delamination (Hanson et al., 1975), which apparently results in any diffusion gradients of radiogenic argon being destroyed during the step heating experiments.

Some of the difficulties associated with using biotite for K–Ar and $^{40}Ar/^{39}Ar$ age spectrum measurements have been emphasized. Nevertheless, it is probably true that the majority of biotites yield reliable cooling ages; when used in conjunction with age data on other phases and in the context of the local geology, biotite ages can be very helpful in elucidating thermal histories.

2.7.4.2 Muscovite

The white dioctahedral mica, muscovite, has an ideal composition $K_2Al_4Si_6Al_2O_{20}(OH, F)_4$, so that it can contain up to about 9.7% potassium in its structure. Muscovite is a relatively common mineral in regionally metamorphosed sediments, as it is stable over a wide range of temperature–pressure conditions. It also is present in some silicic granitic rocks and in pegmatites associated with granitic bodies. Sericite is the term used to describe fine-grained muscovite. Phengite is a variety that can form under relatively high pressure–low temperature, metamorphic conditions, characterized by $Si/Al > 3$ with substitution of Mg or Fe^{2+} for Al in octahedral sites to preserve electrical neutrality.

The white micas are used extensively for both K–Ar and $^{40}Ar/^{39}Ar$ dating, as they show good retention properties for radiogenic argon. For well-crystallized samples, the closure temperature for argon is significantly higher than that of biotite, probably about 350°C for moderate cooling rates (cf. Purdy and Jäger, 1976; Jäger, 1979). In contrast with biotite, muscovite and phengite commonly yield $^{40}Ar/^{39}Ar$ age spectra that can be readily interpreted in terms of distribution of radiogenic argon in the crystals, as the gradients of $^{40}Ar^*$ in the crystals do not seem to be destroyed in the vacuum system during a step heating experiment. Incorporation of extraneous argon in white micas does not normally seem to be a problem.

2.7.4.3 Lepidolite

Lithium may replace octahedral Al in muscovite giving rise to lithian muscovites, the charge balance being preserved by Si replacing tetrahedral Al. With substitution of more than about 3.5% of LiO_2 into the muscovite structure the mineral becomes known as lepidolite (Deer et al., 1962). Lepidolite is confined almost exclusively to granitic pegmatites. Because it is relatively rare, lepidolite is not widely utilized for K–Ar dating, although its high potassium content makes it very suitable for the purpose. Argon retention properties have not been studied, but it is expected to behave in a manner similar to muscovite.

2.7.4.4 Glauconite

Found in marine sediments where it grows authigenically, glauconite is a potassium-bearing dioctahedral mica that is a major component of greensands, often forming rounded green pellets. It is virtually the only material useful for direct K–Ar dating of sediments for which it has been widely employed, but with varying success. A group of informative papers on glauconite and its use in geochronology is to be found in Odin (1982a), in which the term glaucony (plural—glauconies) is advocated for the green pellets found in sedimentary rocks. This term was introduced to distinguish the pellets from the mineral species, glauconite, as they commonly contain significant amounts of clay minerals, especially smectites, and, thus, vary considerably in their mineralogical constitution.

Glauconies occur in a shallow marine sedimentary environment where they grow by progressive glauconitization of granular substrates such as carbonate fragments or a range of detrital mineral grains including mica, feldspar, and clay. Glauconitization proceeds by poorly understood diagenetic processes at or adjacent to the sediment–sea water interface, resulting in a progressive increase in the proportion of the micalike component relative to the clay component with a progressive increase in the potassium content to a maximum of about 6.5%. This diagenetic growth process may be halted at any stage, probably owing to burial by younger sediments.

Glauconite itself essentially is a hydrous, potassium, iron, magnesium aluminosilicate with additional adsorbed water, which can be removed at a temperature of $\sim 100°C$ without loss of radiogenic argon. The argon is released in the vacuum system between ~ 200 and $600°C$ as the mineral structure breaks down and releases its combined water. The physical size of the crystallites that comprise glaucony usually lies in the $0.1-5\ \mu m$ range.

As emphasized in Odin (1982a), glaucony is a material that can yield reliable K–Ar ages in only certain circumstances. Thus, glauconies with potassium contents $<5\%$ commonly contain radiogenic argon inherited from the original substrate, so that measured ages are often considerably older than the age of deposition and diagenetic growth. Evernden et al. (1960) showed that glauconite has diffusion characteristics for argon similar to those for other micas, but pointed out that the extremely small crystal size allows diffusional loss of argon to occur at temperatures of $\sim 100°C$ if maintained for a few million years. This factor, together with the low stability of the minerals comprising the glaucony pellets, means that low apparent ages commonly are found. In spite of these difficulties, glaucony has been used extensively for numerical calibration of the relative geological time scale, especially in the Cretaceous and Cenozoic, in the absence of other suitable geochronometers. Results on glaucony, however, should always be interpreted with caution, and K–Ar ages on high-temperature minerals from volcanic rocks are generally much more reliable.

Attempts to utilize glaucony for $^{40}Ar/^{39}Ar$ dating have not been successful (Yanase et al., 1975; Brereton et al., 1976; Fitch et al., 1978; Foland et al., 1984). These workers have shown than during irradiation variable loss of ^{39}Ar by recoil occurs, and, in addition, glaucony does not remain stable during incremental heating *in vacuo*.

The mineral celadonite, similar to, but distinct from glauconite (Buckley et al., 1978), usually is found associated with basalts, especially submarine basalts. Little is known of its suitability for K–Ar dating, but it is likely to behave rather like glauconite. A brief report by Duncan and Staudigel (1986) suggests that dating of celadonite may help constrain the age of cessation of hydrothermal activity in the seafloor environment.

2.7.5 Amphiboles

Minerals of this diverse group are of widespread occurrence in igneous and metamorphic rocks. Some varieties accept potassium into their structures to a maximum of about 2%, but more generally in the $0.1-1.0\%$ range. Since the pioneer studies by Hart (1961, 1964), amphiboles have been used increasingly for K–Ar age measurement, as it has been shown both empirically and experimentally that some well-crystallized hornblendes are extremely retentive of radiogenic argon, with a closure temperature for moderate cooling rates of $\sim 500°C$ (Hanson and Gast, 1967; Harrison, 1981).

Hornblende is the most widely used amphibole for dating by the K–Ar method and has a general formula $(Ca,Na,K)_{2-3}(Mg,Fe,Al)_5(Si_7AlO_{22})(OH)_2$, but kaersutite, barkevikite, actinolite, and the alkali amphiboles, glaucophane, richterite, and potassium richterite also may accept some potassium into their structures (Deer et al., 1963), and have been employed successfully for dating.

Because of its excellent argon retention properties, hornblende is a particularly useful mineral for K–Ar dating of the cooling of igneous and metamorphic rocks. There is some empirical evidence that suggests that argon retentivity of hornblende varies with composition (O'Nions et al., 1969; Berry and McDougall, 1986), although preliminary laboratory experiments fail to confirm this finding (Harrison, 1981).

In blueschist terranes glaucophane commonly yields K–Ar ages somewhat lower than those measured on hornblende and white mica (Coleman and Lanphere, 1971; Suppe and Armstrong, 1972; McDowell et al., 1984). However, Sisson and Onstott (1986) found that for a crossite (subglaucophane) separated from a blueschist in southern Alaska, most of the radiogenic argon is derived from small white mica inclusions. If this is a general phenomenon the variability of potassium content and the relatively low closure temperature for argon in blue amphibole concentrates can readily be explained.

Expecially in rocks that have been metamorphosed, the presence of fine-scale exsolution or unmixing may have dramatic effects on the ability of amphibole to retain radiogenic argon quantitatively (cf. Harrison and Fitz Gerald, 1986). Presumably this is because radiogenic argon then has only small distances to diffuse to an exsolution boundary from which it can readily escape. Thus, the evidence points toward a range of closure temperatures for argon in amphiboles, depending upon the structure of the minerals.

Amphiboles are found to be useful for $^{40}Ar/^{39}Ar$ dating studies, as it is possible to obtain detailed information from age spectrum measurements of the distribution of radiogenic argon in both thermally undisturbed and disturbed samples. Thus, Dallmeyer (1974) and Dallmeyer et al. (1975) demonstrated that hornblendes in thermally disturbed terranes often yield monotonically rising age spectra interpreted in terms of argon loss as a result of reheating. Harrison and McDougall (1980b, 1981) and Dallmeyer and Rivers (1983) have also documented argon uptake profiles in amphiboles from metamorphic rocks. Indeed, the presence of excess argon in amphiboles appears to be much more common that previously suspected.

Overall, amphibole is one of the most useful minerals for K–Ar dating, especially when combined with $^{40}Ar/^{39}Ar$ age spectrum measurement. In conjunction with ages on other minerals, results from amphibole can provide good control on the cooling history of geological terranes. For example, in granitic rocks that contain hornblende, biotite, and alkali feldspar, ages on these minerals may provide detailed information on the cooling history and allow conclusions to be reached as to whether the cooling interval was short or protracted.

2.7.6 Pyroxenes

Potassium substitutes in the pyroxene structure only in very minor amounts, generally comprising $<0.1\%$, and often very much lower than this. Such behavior is expected from crystal chemistry considerations, as pyroxene has a composition dominated by the doubly charged cations Ca, Mg, and Fe. Even the sodic pyroxenes, aegerine and jadeite, rarely contain much potassium, although values of $\sim 0.2\%$ have been reported (Deer et al., 1963) in aegerine. In spite of the low potassium contents, and the possibility that at least some of the potassium is carried in inclusions or contaminants in the mineral separate, pyroxene has been used for K–Ar dating of igneous rocks with some success (Hart, 1961; McDougall, 1961). Because of the very low potassium content, potential difficulties caused by incorporation of excess argon, leading to anomalously old ages, were recognized at an early stage (Hart and Dodd, 1962), and found to be of common occurrence in pyroxenes from both igneous and metamorphic rocks (Kistler et al., 1969; Schwartzman and Giletti, 1977; McDougall and Green, 1964; Harrison and McDougall, 1981). Thus, minerals of this group are not widely employed for K–Ar dating, although pyroxenes are very retentive of radiogenic argon (Schwartzman and Giletti, 1977). In principle, pyroxenes can be dated by the $^{40}Ar/^{39}Ar$ method, but for the reasons given above, generally have been rarely used (Lanphere and Dalrymple, 1976; Harrison and McDougall, 1981).

2.7.7 Whole rocks

Potassium–argon dating is most appropriately performed on potassium-bearing phases separated from the host rock. However, for many volcanic rocks, slates, and phyllites the grain size of the individual minerals is so small that it is not practicable to undertake mineral separations. In such circumstances measurements commonly are made on whole rock samples, and there is now a vast array of data available showing the efficacy of this approach, on suitable samples. Clearly, the reliability of K–Ar ages determined on whole rock samples of volcanics is critically dependent upon complete outgassing or exchange of any preexisting radiogenic argon contained within the magma at the time of emplacement, and upon the argon retention properties of the potassium-bearing phases in the samples; analogous considerations apply in the case of metamorphic rocks.

2.7.7.1 Volcanic Rocks

For volcanic rocks it is now abundantly clear that reliable K–Ar ages can be obtained on a wide range of compositions, from mafic to silicic, provided that certain criteria are met. As a generalization, it has been found that holocrystalline, fresh lavas, which preserve their high-temperature mineralogy without significant alteration, determined by careful petrographic examination of a thin section, yield K–Ar ages that usually are readily interpreted in terms of the time elapsed since eruption and cooling. Empirically it has been shown that most modern, subaerially erupted lavas, free of xenoliths, contain trapped argon of atmospheric composition (Dalrymple, 1969; Krummenacher, 1970). Xenolithic material contained within a lava, however, may not be completely outgassed of its preexisting radiogenic argon, or equilibrated with atmospheric argon, even at the high temperature of a magma (Dalrymple, 1964, 1969; McDougall et al., 1969; Gillespie et al., 1982). In addition it should be noted that lavas, especially basaltic lavas, possessing phenocrysts of olivine, pyroxene, or plagioclase, may, in some cases, give anomalously old apparent ages, owing to incorporation of excess argon from the environment as the phenocrysts crystallized in the magma prior to its eruption, that is intratellurically. It is good practice to remove such phenocrysts from the sample, and measure the age on the fine grained groundmass that is more likely to have equilibrated with the atmosphere. Isotopic analyses of argon extracted from a minority of modern volcanic rocks have yielded $^{40}Ar/^{36}Ar$ ratios less than 295.5, the atmospheric argon value. Ratios as low as 284 have been reported, and are best explained in terms of isotopic mass fractionation of argon of atmospheric composition during exchange with the magma prior to or during eruption (Krummenacher, 1970; Kaneoka, 1980). The observed variation of the measured $^{38}Ar/^{36}Ar$ ratio supports this interpretation. Clearly, if the trapped argon isotopic composition in a rock differs significantly from that of atmospheric argon, then any K–Ar age, calculated assuming that the nonradiogenic argon is atmospheric in composition, will be in error. This is an important limitation, especially when dating young rocks. Nevertheless, it should be emphasized that the majority of lavas appear to have trapped argon indistinguishable in isotopic composition from atmospheric argon. With due recognition of the caveats, it can be stated that fresh, well-crystallized subaerial volcanic rocks provide excellent material for K–Ar dating purposes, as the high-temperature mineral phases retain radiogenic argon quantitatively at temperatures experienced within the upper part of the crust.

In contrast, whole rock samples of volcanics that show extensive alteration of the high-temperature phases, with replacement by clay, chlorite, zeolite, and so forth, generally yield aberrantly low K–Ar ages, because the alteration products often do not retain radiogenic argon quantitatively even at temperatures not much above ambient. Petrographic examination allows such samples to be rejected as unsuitable for dating.

But the major difficulty in making decisions as to whether or not samples are likely to yield reliable ages lies with those volcanic samples that show some alteration of the high-temperature phases. Minor alteration of olivine is unlikely to cause any serious problem, because this mineral contains insignificant potassium. However, alteration of plagioclase in basalt, or alkali fedspar in a more silicic volcanic rock, normally will result in a measured age being too low owing to loss of radiogenic argon from the alteration products. Rocks containing glass, very common in volcanics, also commonly yield low ages, as glass normally is enriched in potassium and often is hydrated and submicroscopically devitrified, resulting in loss of radiogenic argon, even at low temperature. Clearly, decisions as to which rocks can be accepted for age measurement remain somewhat subjective. Careful examination of samples under the petrographic microscope is essential, and the use of an electron microprobe to determine the location of potassium can be helpful in making a judgement as to the suitability of a rock for K–Ar age measurement (Mankinen and Dalrymple, 1972). Probably the most appropriate policy to adopt is to reject for dating those samples that show more than minor alteration of potassium-bearing phases.

The question of reliability of ages on fresh or slightly altered samples can best be evaluated by measuring a suite of samples in known stratigraphic relation to one another. From the consistency or lack of consistency of the results it is often possible to assess the reliability of the ages. This approach is enhanced if measurements can also be made on suitable mineral separates, for example, alkali feldspar phenocrysts, which commonly occur in more silicic volcanic rocks, for comparison with results obtained from whole rock samples (Webb and McDougall, 1967; McDougall and Schmincke, 1977). By observing sensible precautions in choosing only those whole rock samples meeting the criteria set out above, excellent K–Ar dating results may be obtained.

There are numerous examples of the successful use of whole rock samples from subaerial volcanics for K–Ar dating purposes. A particularly important example was the development of the geomagnetic polarity time scale for the last 5 Ma. The time control was provided by K–Ar dating of volcanic rocks on which paleomagnetic polarity measurements also were made; for summaries of this work see McDougall (1979) and Mankinen and Dalrymple (1979). Detailed studies of the history of large subaerial volcanoes have been particularly successful, and have demonstrated a high degree of consistency of measured ages with the stratigraphy using a range of whole rock samples of different mineralogical and chemical composition (McDougall, 1964, 1971). Documentation of the age progression of volcanism in linear volcanic chains, particularly in the ocean basins, by K–Ar dating of whole rock lava samples, has provided information interpreted in terms of lithospheric plate motions (Duncan and McDougall, 1976; McDougall and Duncan, 1980; Dalrymple et al., 1980; Duncan and Clague, 1985). Similarly an extensive dating program on lavas in Iceland has yielded a time scale for the evolution of this large volcanic island, astride the Mid-Atlantic Ridge, and together with stratigraphic and magnetostratigraphic studies has facilitated mapping of Iceland (McDougall et al., 1984). Most of these studies have been on relatively young volcanic rocks, less than 15 Ma old. But much older whole rock samples can be dated successfully provided suitably fresh, holocrystalline samples are available.

The presence of alteration in whole rock samples, especially the development of clays, can have other important consequences for the K–Ar dating of such samples. Generally, higher contents of atmospheric argon are found in rocks containing significant amounts of alteration products compared with unaltered rocks, owing to replacement of high-temperature phases by low-temperature hydrated minerals associated with which are appreciable quantities of atmospheric argon. Apart from making radiogenic argon more difficult to detect over the larger background of atmospheric argon, there can be a more serious consequence related to laboratory-induced isotopic fractionation of atmospheric argon during pumpdown in the vacuum system, prior to argon extraction. The fractionation of atmospheric argon trapped within clays or at the end of clay-filled microcracks can occur essentially by a Rayleigh distillation process, whereby ^{36}Ar is pumped more efficiently than ^{40}Ar, so that the $^{40}Ar/^{36}Ar$ ratio in the gas remaining becomes appreciably greater than that in unfractionated atmospheric argon. Subsequently, when the argon is extracted from the sample by fusion and isotopically analyzed, corrections are made for the non-radiogenic argon assuming it has normal atmospheric argon composition. If fractionation has in fact occurred, the calculated age will be anomalously old, even to the extent of many percent. This phenomenon was first documented in detail by Baksi (1974) on samples of Picture Gorge Basalt, Oregon, of middle Miocene age, and which contain saponitic clays. Subsequently, McDougall et al. (1976) observed similar problems in some Icelandic basalts. Baking of a sample during pumpdown in the vacuum system, prior to argon extraction, may exacerbate the problem of isotopic fractionation owing to the clays losing water and contracting, thus increasing the possibility of a greater density of microcracks. By not baking the sample prior to argon extraction, the problem is reduced but not necessarily overcome, because it is caused by the pumping. It is a phenomenon that is much more common than most workers recognize.

The ^{40}Ar/^{39}Ar age spectrum method also can be applied to whole rock samples, and has been used for dating deep-sea volcanics with some success, and less commonly for dating subaerial volcanic rocks. As whole rocks consist of a number of phases that are likely to have different diffusion properties for radiogenic argon, and different grain sizes, a much less clear picture of the distribution of radiogenic argon is likely to be found by ^{40}Ar/^{39}Ar measurements compared with similar experiments on separated minerals. Nevertheless the technique has been applied to slightly altered subaerial basalts in the hope that the gas released at the higher temperatures may yield simple age spectra reflecting the crystallization age of the high-temperature minerals in the rocks. Thus Mussett et al. (1980) reported measurements on some hydrothermally altered Icelandic basalts and found nearly flat age spectra for many of the samples, interpreted as providing reliable estimates for the age of extrusion and cooling of the lavas. The results obtained are broadly consistent with the stratigraphy, although they concluded it was not possible to confirm that the results are more accurate than those measured by the conventional K–Ar method on carefully selected samples. Hall and York (1978) demonstrated the potential of employing the ^{40}Ar/^{39}Ar age spectrum technique to quite youthful basaltic rocks of the order of 50 ka old.

Few ^{40}Ar/^{39}Ar age measurements have been made on ancient whole rock samples. However, Martinez et al. (1984) have reported age spectra exhibiting well-developed plateau segments with ages of 3400–3490 Ma on metamorphosed komatiites and komatiitic basalts from the Archean Barberton Greenstone Belt in southern Africa. The plateau ages are less than 100 Ma younger than Sm–Nd ages on similar rocks and are interpreted as recording the timing of the greenschist metamorphism that has affected the rocks. The komatiites are composed almost entirely of serpentine, tremolite, chlorite, and magnetite, and the komatiitic basalts consist mainly of actinolite with some ferrohornblende, epidote, albite, chlorite, and titanomagnetite. These ancient greenstones have retained their radiogenic argon remarkably well, especially considering the mineralogy, and the results suggest that the rocks have remained at low temperatures since the greenschist metamorphism.

Application of the ^{40}Ar/^{39}Ar age spectrum approach to whole rock samples of meteorites and lunar samples has been particularly successful; some aspects of this topic are discussed in Section 6.2.

In contrast with the K–Ar dating of subaerial lavas, measurements on submarine volcanic rocks have been far less successful. Two major problems have been identified in the K–Ar dating of volcanics from the deep-sea environment. First, basaltic magma on eruption in a submarine environment commonly is quenched to a glass. Noble gases, including radiogenic argon, presumably from the source regions of the magmas, frequently are trapped within the glass, so that measured K–Ar ages are anomalously old (Dalrymple and Moore, 1968; Funkhouser et al., 1968; Noble and Naughton, 1968; Dymond, 1970). Dalrymple and Moore (1968) showed that youthful lavas, probably less than 1000 years old, erupted along the submarine extensions of the active east rift zone of Kilauea Volcano, Hawaii, yielded apparent ages ranging up to 43 Ma on material from the glassy rims of the pillow lavas. They found that the amount of excess radiogenic argon in the glass tends to increase with the depth below sea level at which the lavas were erupted, indicating a relation with hydrostatic pressure. Dalrymple and Moore (1968) and Dymond (1970) also demonstrated that in the interiors of pillows, where the magma had the opportunity to at least partly crystallize owing to a slower cooling rate, much of the preexisting argon had been lost or exchanged, as the measured ^{40}Ar/^{36}Ar ratios were indistinguishable from that of atmospheric argon. This finding suggests that samples from pillow interiors may yield reliable K–Ar ages, but the second major problem, that of alteration, often intervenes to prevent this being realized in practice. Very few igneous rocks from the deep-sea environment meet the criteria of freshness and crystallinity for K–Ar dating, as hydrothermal alteration and submarine weathering result in development of chlorite, smectite, and other clay minerals as well as calcite, at the expense of the primary high-temperature minerals. Because of loss of radiogenic argon from the alteration products and addition of potassium by submarine weathering, measured ages commonly are too young (Seidemann, 1977). Thus, the hope that reliable K–Ar ages could be determined routinely on volcanic samples from the ocean floor has largely gone unfulfilled.

Application of the $^{40}Ar/^{39}Ar$ age spectrum technique to deep-sea volcanic rocks has provided more reliable estimates of age in a number of cases, although many age spectra are complex and, thus, difficult to interpret. The rationale is that it might be expected that argon released at the higher temperatures in a $^{40}Ar/^{39}Ar$ step heating experiment is likely to be from the original high-temperature phases, and thus reflect the time since crystallization, even if argon has been lost from alteration products. In $^{40}Ar/^{39}Ar$ dating studies on altered deep-sea rocks Clague et al. (1975), Dalrymple and Clague (1976), and Seidemann (1978) found that during neutron irradiation and prior to extraction of argon from the samples, variable but significant loss of ^{39}Ar commonly occurs, probably from sites in clays and other alteration products. This loss takes place by recoil and because of the poor retentivity of argon in such material. Thus, $^{40}Ar/^{39}Ar$ total fusion ages commonly are considerably older than conventional K–Ar ages on the same samples, and in some cases are found to be consistent with K–Ar ages on plagioclase (Dalrymple and Clague, 1976). These $^{40}Ar/^{39}Ar$ total fusion ages are thought to approximate more nearly the age of eruption of the volcanics, based upon the reasonable assumption that phases losing ^{39}Ar during irradiation also are likely to have lost their radiogenic argon (Dalrymple et al., 1980). In the conventional K–Ar dating method, potassium contained within clays and other alteration products is measured along with that in the high-temperature minerals, so that if loss of radiogenic argon has occurred from the alteration products, the measured ages obviously will be too low.

Although $^{40}Ar/^{39}Ar$ age spectra on deep-sea volcanic rocks exhibit a wide range of patterns, in those cases for which essentially flat spectra are obtained, especially in the higher temperature steps, the ages commonly can be interpreted as reflecting the time since eruption (Dalrymple et al., 1980; Walker and McDougall, 1982). In summary, there has been more success in dating deep-sea basalts and other volcanic rocks using the $^{40}Ar/^{39}Ar$ method than with the conventional K–Ar method. However, problems can arise in the interpretation of age spectra measured on whole rocks because the samples consist of a number of phases of different argon retentivity, and because of alteration. Nevertheless, it is of considerable importance to pursue studies of this kind because of the clear need to establish numerical age control on igneous rocks from the deep-sea environment, particularly as most other isotopic dating methods are inapplicable to submarine volcanics.

2.7.7.2 Metamorphic Rocks

In order to provide control on the timing of metamorphism, dating of individual minerals normally is preferred. But low grade metamorphic rocks, such as slates and phyllites, are too fine grained for mineral separation, and, thus, K–Ar age measurements necessarily have to be done on whole rock samples. Information pertaining to the age of metamorphism will be obtained if all preexisting radiogenic argon contained within the original detrital minerals is outgassed during the thermal event, and provided that the rock begins to retain radiogenic argon quantitatively soon after the metamorphism ends. Loss of argon will occur by diffusion as the temperature increases and will be assisted by the crystallization of new minerals or recrystallization of existing phases, and by the deformation that accompanies the metamorphism. Petrographic examination of samples is helpful in determining how much mineralogical reconstitution has taken place. Regional metamorphism involves considerable time during which heating and subsequent cooling occur. Thus, there is unlikely to be a unique age for the metamorphic event, and, even if complete outgassing of preexisting argon occurred, the timing of subsequent closure of the minerals to argon loss will depend upon the cooling rate and the mineralogical composition of the rock. Nevertheless, Harper (1964, 1967), Dodson and Rex (1971), and Adams et al. (1975) have argued that fine grained slates and phyllites free of coarsely crystalline detrital minerals, and with muscovite (sericite) as the major potassium-bearing phase, are likely to lose preexisting argon readily during prograde metamorphism and then begin to retain radiogenic argon soon after the peak of metamorphism.

Harper (1964, 1966, 1967, 1968) and Dodson and Rex (1971) carried out detailed K–Ar dating studies on whole rock samples of slates and phyllites from various regions in the British Isles and the Appalachians. Other geological and geochronological evidence supported their claims that the Paleozoic ages found reflected cooling after metamorphism, but in some cases the results were best interpreted in terms of cooling brought about by uplift and erosion well after the peak of metamorphism. In a later study, Leitch and McDougall (1979) measured K–Ar ages on low grade metapelites from the Late Paleozoic Nambucca Slate Belt of New South Wales. They found that rocks in the prehnite–pumpellyite metagreywacke facies often showed evidence for inherited argon from detrital phases, but that higher grade rocks in the pumpellyite–actinolite and greenschist facies, which experienced temperatures of $\sim 300°C$, were essentially completely reset at 250–255 Ma ago, ascribed to metamorphism and deformation at that time, consistent with other geological data from the region.

These results illustrate that K–Ar dating of fine grained metamorphic whole rock samples can provide useful time control on cooling and uplift of terranes subsequent to a tectonothermal event, but that additional geological information usually is required to adequately assess the meaning of the results. Reynolds and Muecke (1978) applied the $^{40}Ar/^{39}Ar$ age spectrum technique to rocks of this kind with apparent success. Interpretation of results from such studies are complicated, however, because the rocks contain a number of different phases. But where muscovite is the main potassium-bearing phase, this approach may be of considerable value.

2.7.8 Glass

Cautionary comments were made previously concerning the use of samples containing glass for K–Ar dating. Two problems are recognized: (1) loss of radiogenic argon from glass as it hydrates and devitrifies, and (2) incorporation of excess argon in quenched basaltic glass in the submarine environment. As noted by Dalrymple and Lanphere (1969), fresh, unhydrated obsidian may provide reliable K–Ar ages. Kaneoka (1972) demonstrated that increasing hydration of obsidian causes a marked decrease in measured K–Ar ages. Drake et al. (1980) showed that apparently fresh glass from Late Cenozoic rhyolitic pumices from tuff beds in northern Kenya may yield K–Ar ages that are essentially concordant with coexisting high-temperature alkali feldspar. A subsequent study by Cerling et al. (1985), on similar glasses from pumices in the same region of northern Kenya, demonstrated that during hydration potassium commonly is lost, and, surprisingly, lost more readily than radiogenic argon, so that some anomalously old K–Ar ages were found. Lanford et al. (1979) explained that because of their mobility and a large concentration gradient at the surface, alkalis tend to diffuse out of glass. In the absence of a corresponding charge balance, however, the alkalis cannot leave because associated anions (and presumably argon) are immobile and, after a small loss, an electric field is set up that inhibits further diffusion. However, if H^+ (or, in fact H_3O^+) is introduced at the surface, the hydrogen can interdiffuse with the alkalis, replacing them in the glass structure, perhaps without disturbing the neutral argon. It is concluded that volcanic glasses are of marginal value for K–Ar dating, as other geochronological data usually are required for authentication of the measured ages.

An example of successful use of glass for K–Ar dating, however, concerns measurement of those interesting objects known as tektites (Glass, 1982). There is now a consensus that tektites formed by very high-temperature melting of terrestrial material as a result of a cataclysmic event, such as meteorite impact, with hypervelocity travel of the ejected material through the atmosphere and deposition in strewn fields over parts of the Earth's surface. Tektites are highly siliceous glasses, typically containing ~2% potassium, and are virtually devoid of water (<0.01%), attesting to their high-temperature origin. That the measured K–Ar ages accurately reflect the melting event seems to be well borne out by the results. Thus the bediasites of Texas yielded consistent ages of 34.0 ± 0.5 Ma (Zähringer, 1963) on eight samples, and ideal flat $^{40}Ar/^{39}Ar$ age spectra for three samples were shown by York (1984) with plateau ages that agreed with the earlier K–Ar ages. Measurements on the moldavites of Czechoslovakia yielded concordant K–Ar ages of 15.1 ± 0.2 Ma on seven samples, when adjusted to currently used decay constants (Gentner et al., 1963; Zähringer, 1963). It has long been recognized that a potential source for the moldavites was the 24-km-diameter Nördlinger Ries Crater in southern Germany, and Gentner et al. (1963) confirmed that its formation occurred at a time indistinguishable from the age of the moldavites. More recently, Staudacher et al. (1982) showed that both glass from the crater and a moldavite yielded nearly flat $^{40}Ar/^{39}Ar$ age spectra, with plateau ages of 15.0 ± 0.2 Ma, respectively, further strengthening the genetic relationship. Tektites from the large Southeast Asian strewn field, extending over at least 8000 km, gave K–Ar ages generally in the 0.6–0.9 Ma range (Reynolds, 1960; Gentner and Zähringer, 1960; Zähringer, 1963; McDougall and Lovering, 1969), on a large number of samples. Recovery of similar microtektites from deep-sea sedimentary cores indicates that deposition occurred at about the Brunhes–Matuyama chron boundary, the age of which is known to be 0.73 ± 0.02 Ma. All these results on tektites suggest that glasses of this kind are excellent for K–Ar and $^{40}Ar/^{39}Ar$ dating, in contrast to most volcanic glasses. It should also be noted that Huneke (1978) obtained good $^{40}Ar/^{39}Ar$ age spectra on lunar glass, which is also free of water, although devitrified.

2.7.9 Clay minerals

The potential for direct measurement of the depositional age of sedimentary rocks by utilizing clay minerals was recognized at an early stage in development of the K–Ar dating technique. In particular, illite was extensively investigated as a mineral for dating, but with varied success (Hurley et al., 1961, 1963; Evernden et al., 1961). Illite is compositionally similar to muscovite but contains less potassium and more silica, and may have up to ~8.5% potassium. Two main problems were identified in the studies cited above. First, measured ages commonly were much too old to be that of deposition, indicating that recycling of detrital illite from older sedimentary rocks was occurring, without complete resetting and equilibration in the depositional environment. Second, even in those cases in which the illitic clays had equilibrated or grown authigenically in the sediment, it was shown that the very fine crystal size allows loss of radiogenic argon to occur at temperatures little above ambient, causing the measured ages to be too young. These twin problems have mitigated against the employment of illites and other clay minerals, such as smectites (montmorillonites), for K–Ar dating on a routine basis.

Despite these difficulties, clay minerals have been utilized by some workers in attempts to characterize the provenance of detrital material in sediments or to determine whether authigenic growth has occurred. For example, in a recent report Clauer et al. (1984) measured K–Ar ages on smectites from deep-sea cores in the Atlantic Ocean.

$^{40}Ar/^{39}Ar$ age spectrum measurements on illitic clays generally do not give sensible results owing to loss of both ^{40}Ar and ^{39}Ar during irradiation, and because release of argon in the vacuum system during a step heating experiment probably occurs by breakdown of the clay lattice rather than by volume diffusion (Halliday, 1978). Similar problems are expected for most clay and claylike minerals, including celadonite, glauconite, and smectite. Despite these difficulties, useful information was obtained by Frank and Stettler (1979) and by Hunziker et al. (1986) using the $^{40}Ar/^{39}Ar$ step heating technique on illite separated from sediments and very low grade metasediments in the Swiss Alps and adjacent regions; these results will be discussed in more detail in Section 4.12.10.

2.7.10 Evaporites

Potassium-bearing salts such as sylvite (KCl), carnallite ($KMgCl_3 6H_2O$), polyhalite [$Ca_2MgK_2(SO_4)_4 2H_2O$], and langbeinite [$K_2Mg_2(SO_4)_3$] occur in some evaporite deposits, and technically are readily measured by the K–Ar dating method. Because of their solubility in water and the relative ease with which they recrystallize, these minerals commonly yield results that are younger than the age of deposition, due to loss of radiogenic argon. Thus Smits and Gentner (1950) measured K–Ar ages on sylvite samples from salt deposits in Germany and obtained ages that are somewhat too young to be regarded as that of formation. Later studies confirmed that sylvite is particularly prone to argon loss (Schilling, 1973; Obradovich et al., 1982). Polyhalite yielded K–Ar ages of 198–216 Ma on Late Permian (~250 Ma) evaporites from New Mexico, ascribed to diagenetic resetting (Brookins et al., 1980). Langbeinite appears to provide depositional ages in some cases (Schilling, 1973; Lippolt and Oesterle, 1977), and laboratory diffusion studies (Lippolt and Oesterle, 1977) suggest it has argon retention properties similar to that of high-temperature alkali feldspar. Although it is now clear that deposition ages rarely are obtained by K–Ar dating of minerals from evaporites, Brookins et al. (1980) and Obradovich et al. (1982) pointed out that age measurements can be very useful for determining the stability of salt deposits over geological time. The $^{40}Ar/^{39}Ar$ age spectrum technique has not been applied to potassium-bearing salts, despite the potentially useful additional information that could be obtained using this approach; in particular, experiments on langbeinite would be of interest.

2.8 RANGE OF APPLICABILITY

Because the K–Ar dating method is based upon the accumulation of radiogenic argon, measurement becomes easier technically with increasing age, and therefore there is no older limit for the method. Clearly, potassium measurement can be done on a sample irrespective of its age, and, thus, in itself, is not a limiting factor. The main limitation is at the younger end of the time scale, involving the detection of a small amount of radiogenic argon from a much larger background of atmospheric argon. As the proportion of radiogenic to total ^{40}Ar decreases, it will be evident that the error in its measurement will increase exponentially (Fig. 2.5; see Baksi et al., 1967, and Cox and Dalrymple, 1967). With current techniques of argon isotopic measurement in a gas-source mass spectrometer, error magnification begins to dominate when the proportion of $^{40}Ar^*$ is less than about 10% of the total ^{40}Ar, and that in cases where $\leq 5\%$ of the total ^{40}Ar is radiogenic, the errors become progressively increasingly large. As the $^{40}Ar^*$ decreases toward zero, small errors in calibration of the discrimination of the mass spectrometer at different gas pressures similarly can lead to large errors in the measured $^{40}Ar^*$, and, thus, in reported age. Similarly, should the trapped argon composition differ from that assumed, increasingly large systematic errors in the calculated age will occur as the proportion of $^{40}Ar^*$ decreases.

In the case of terrestrial samples there are two sources of nonradiogenic argon, usually of atmospheric composition, that ultimately determine the younger limit at which the method can be applied. First, there is always a finite amount of trapped argon, normally atmospheric in composition, associated with the sample itself, and, second, there is a contribution of nonradiogenic argon from the vacuum system in which the argon extractions are performed. The first source of contamination is inherent to the sample, whereas the second is related to the equipment used and, thus, is controllable within limits.

It has long been recognized that the amount of atmospheric argon associated with samples varies widely, but that certain types of samples have characteristic air argon contents (McDougall, 1966; Dalrymple and Lanphere, 1969). A compilation of data from the A.N.U. laboratory is given in Fig. 2.6, from which it is evident that the platy minerals, such as biotite and muscovite, generally have an order of magnitude greater atmospheric argon contamination than the prismatic minerals such as alkali feldspar and hornblende, or relatively

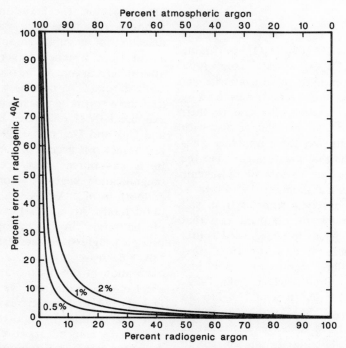

Fig. 2.5. Plot showing how the error in the measurement of radiogenic argon increases exponentially as its proportion relative to the total argon decreases toward zero. Based upon formula derived by Baksi et al. (1967). In the calculations the error in the extrapolation of the ^{40}Ar and ^{38}Ar peak heights is taken as 0.2% and the error for ^{36}Ar is as shown on the curves. Errors are standard deviations expressed as a percentage. These curves relate to argon measurement by isotope dilution, but very similar relations hold for the ^{40}Ar/^{39}Ar dating technique.

coarsely crushed (say, >0.2 mm), fresh, young volcanic rocks. For two samples with equal atmospheric argon contents, clearly that with the highest potassium content will be more readily measurable owing to the larger amount of ^{40}Ar* generated per unit time. On this basis, of the more common minerals, alkali feldspar (sanidine) probably is the most useful for dating young rocks because of its relatively low atmospheric argon and high potassium content. But fresh, whole rock, volcanic samples commonly are nearly as favorable for this purpose, especially if they have high potassium contents.

Of particular note is that measured contents of atmospheric argon in fresh whole rock samples of volcanics commonly are considerably lower than the solubility of argon in silicate melts. The experimental results show that the solubility of argon in natural silicate liquids obeys Henry's law, and that at temperatures of $\sim 1300°$C for the partial pressure of argon in air at 1 atmosphere, the solubility is in the range $2-9 \times 10^{-7}$ cm^3 STP/g (1–4 $\times 10^{-11}$ mol/g) for a number of basaltic liquids, rising to as much as 1.4×10^{-6} cm^3 STP/g (6×10^{-11} mol/g) for an andesitic liquid (Hayatsu and Waboso, 1985; Jambon et al., 1986; Hiyagon and Ozima, 1986; Lux, 1987). In keeping with the expectation that argon will partition preferentially into the melt during crystallization, Hiyagon and Ozima (1982, 1986) demonstrated experimentally that olivine has ~ 0.1–0.2 of the concentration of argon within it compared with that in the enclosing melt from which it is crystallizing. The observation that measured atmospheric argon contents in fresh, well-crystallized, whole rock samples of volcanics ranging from basalt to trachyte are as low as 5×10^{-8} cm^3 STP/g (2×10^{-12} mol/g) (see Fig. 2.6) is consistent with a relatively low distribution coefficient for argon in crystalline phases compared with the melt. These circumstances explain why fresh whole rock samples commonly have very low atmospheric argon contents, and, thus, are particularly useful for dating quite youthful volcanic rocks.

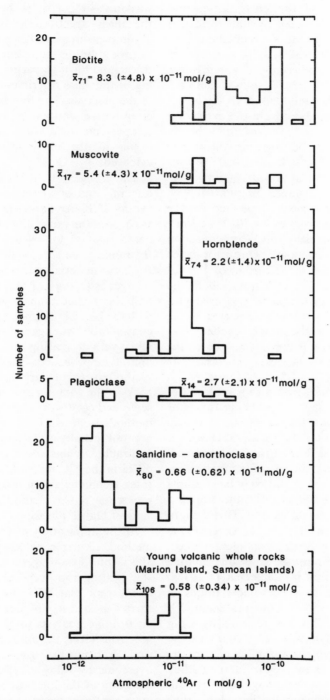

Fig. 2.6. The atmospheric argon content for a number of materials utilized for K–Ar dating shown in the form of histograms. Average value and standard deviation for each kind of sample are indicated. Data are from the laboratory at the Australian National University over a period from 1980 to 1985. Note the similarity with earlier results of McDougall (1966) and Dalrymple and Lanphere (1969).

As an example of the great power and sensitivity in measurement of numerical ages on young samples by the K–Ar method, consider the case of an alkali feldspar (sanidine) of age 10,000 years containing 10% potassium. Application of Eq. (2.11) shows that the amount of ^{40}Ar* generated is 1.73×10^{-13} mol/g. Assuming a blank from the extraction line of 5×10^{-13} mol ^{40}Ar, and atmospheric ^{40}Ar contamination associated with the sample of 1×10^{-12} mol ^{40}Ar/g, we find that the proportion of ^{40}Ar* compared with the total ^{40}Ar is 10.3% for a 1 g sample. In modern gas-source mass spectrometers operated in the static mode, a sensitivity of 3×10^{-15} mol/mV is readily achieved using ion beam collection on a Faraday cup and a 10^{11} Ω resistor in the electrometer, so that a total beam of ^{40}Ar of ~ 560 mV (5.6×10^{-12} A) is obtained in the above example, which can be measured precisely. Of critical importance, however, is the ability to measure the much smaller ^{36}Ar beam, as this isotope is used to correct for the atmospheric ^{40}Ar. In the case quoted the ^{36}Ar will be ~ 1.7 mV ($\sim 1.7 \times 10^{-14}$ A), which normally can be measured with a precision of $\leq 1\%$. On this basis, and allowing for error magnification effects, the radiogenic ^{40}Ar and hence the age should be measurable with a precision of $\leq 10\%$. The extraction line blank and the atmospheric argon contamination associated with the sample as used in the example quoted are at the lower end of values found in practice, but are not unrealistic. This demonstrates just how sensitive the K–Ar technique can be in dating young samples. Indeed, Dalrymple (1967) reported ages on sanidine from young rhyolites in California of 10,000 years or less with analytical uncertainties as low as 10%, and with ^{40}Ar* often exceeding 5% of the total ^{40}Ar. This was achieved by using large amounts of sample (~ 20 g) to reduce the effect of the atmospheric argon blank from the extraction line, and by treating the sanidine concentrates ultrasonically in weak HF to remove adhering contaminating glass, thus reducing the atmospheric argon contamination. More recently, Cassignol and Gillot (1982) have measured K–Ar ages as young as 4000 years with acceptable precision ($\sigma \approx 10\%$) on sanidine, although the proportion of ^{40}Ar* was only $\sim 1\%$. This has been realized by using the mass spectrometer manometrically for argon measurement, dispensing with the usual isotope dilution technique employing a ^{38}Ar tracer, and through particularly close monitoring of the mass discrimination and sensitivity of the machine. For this approach to succeed, quantitative extraction of argon from the sample and quantitative transfer of purified argon to the mass spectrometer must be achieved.

The range of applicability of different minerals and of whole rock samples for conventional K–Ar dating is shown in Fig. 2.7. Because of the variability of atmospheric argon contamination for particular mineral species, the range given is to be taken as a guide only. For most materials it is possible to measure K–Ar ages as young as 1 Ma, and for whole rocks and the higher potassium content minerals, with the possible exception of micas, samples of considerably younger age can be measured, albeit with decreasing precision as the younger limit is approached. Thus, the K–Ar dating method is particularly useful for measuring rocks in the younger part of the geological time scale, and covers the age range for which other methods such as U–Pb, Rb–Sr, and Sm–Nd are not usually applicable, and overlaps in favorable circumstances with the range of 0–50 ka of the ^{14}C method. Although the fission track technique also is useful for dating young rocks, the K–Ar method generally yields results of higher precision.

In the application of the ^{40}Ar/^{39}Ar dating method, similar considerations apply, especially if total fusion ages are measured. For the step heating approach the situation is analogous, except that sufficient argon must be extracted in each step of the experiment to enable an isotopic analysis to be made in the mass spectrometer at the appropriate level of precision. Problems of sensitivity, however, may be overcome by using an electron multiplier for detection of the ion beams, provided that the mass spectrometer has a sufficiently low background at the masses of interest. Thus, in principle, it should be possible to measure ^{40}Ar/^{39}Ar age spectra on samples as young as those amenable to conventional K–Ar dating. Indeed, in a step heating experiment,

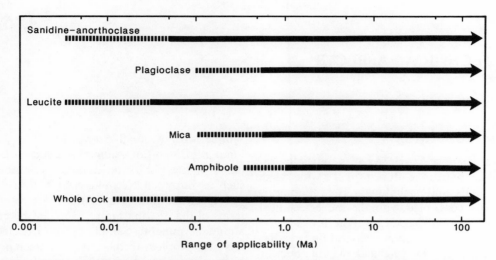

Fig. 2.7. Schematic diagram showing the approximate age range of applicability of various types of samples commonly employed in K–Ar and ^{40}Ar/^{39}Ar dating. Dashed lines extend to the younger limit achievable in favorable circumstances. Modified after Dalrymple and Lanphere (1969).

the ^{40}Ar* signal may be enhanced relative to the atmospheric argon in some steps, as the atmospheric argon may be released, at least in part, over different temperature intervals. Although relatively few ^{40}Ar/^{39}Ar age measurements have been made on very young samples, this is an area of considerable potential development for the future. Hall and York (1978) measured a whole rock basalt of age about 45 ka by means of the step heating method and obtained a relatively flat age spectrum over four of the eight steps. Age spectra on granitic xenoliths in a basalt provided evidence for eruption of the basalt at 119 ± 4 (1σ) ka ago (Gillespie et al., 1984).

Radicati di Brozolo et al. (1981) obtained essentially flat age spectra on both leucite and biotite with indicated ages of 340 ka on samples of 0.5–1 g with up to 10 steps per experiment. Similarly, McDougall (1985) obtained an excellent flat age spectrum on an anorthoclase with indicated age of 0.71 ± 0.01 Ma over 15 steps. These results serve to emphasize that the ^{40}Ar/^{39}Ar age spectrum technique can be used effectively at the younger end of the geological time scale, and as for the conventional K–Ar dating method, there is no older limit in terms of the ability to measure accurate and precise ^{40}Ar/^{39}Ar ages.

3. TECHNICAL ASPECTS

3.1 INTRODUCTION

Measurement of $^{40}Ar/^{39}Ar$ ages on geological samples requires the utilization of a wide range of methods and techniques. This chapter is primarily concerned with the more technical aspects of $^{40}Ar/^{39}Ar$ dating. The topics to be covered include sample preparation, irradiation of samples in a nuclear reactor, nuclear interferences, argon extraction procedures, and isotopic analysis of the argon. Several sections of this chapter deal with concepts drawn from the fields of nuclear physics and nuclear chemistry. In our treatment we have used the book by Friedlander et al. (1981), and also the volume by Harvey (1962), as sources for much of the background information.

3.2 SAMPLE PREPARATION

In applying both the conventional K–Ar and the $^{40}Ar/^{39}Ar$ methods, it is desirable to separate discrete mineral phases for the dating whenever possible, especially in the case of rocks that have slowly cooled or have been thermally disturbed subsequent to crystallization. An age or age spectrum measured on a single mineral phase is likely to be more readily interpretable in terms of the geological history of a rock than an age determined on an assemblage of phases, each of which will have different retention characteristics for $^{40}Ar^*$ (see Section 4.6). This approach, however, is not always feasible, particularly when dealing with fine grained samples, such as many meteorites, lunar samples, and terrestrial volcanic rocks. Whole rock samples commonly are used in these cases, as discussed in Section 2.7.7.

High-purity mineral concentrates, preferably better than 99% pure, should be obtained if possible, because the interpretation of results from impure concentrates is likely to be more equivocal than for pure samples. Thus, a sample with a small proportion of a contaminating high-potassium phase, such as biotite, in a low-potassium mineral concentrate, such as hornblende, could yield a $^{40}Ar/^{39}Ar$ age spectrum that is difficult to interpret in terms of the distribution of $^{40}Ar^*$ in the most abundant mineral. Sample inhomogeneity is less of a problem in the $^{40}Ar/^{39}Ar$ technique, however, compared with the conventional K–Ar method, at least in terms of obtaining meaningful technical measurements. This is because the determinations are made on a single aliquant of the sample in the former method. For conventional K–Ar dating, separate aliquants are required for the determination of potassium and argon; if the sample is inhomogeneous considerable difficulties can arise in obtaining representative subsamples, as demonstrated by Engels and Ingamells (1970).

We regard the examination of a thin section of a sample under a petrographic microscope as virtually an essential step prior to undertaking dating. In conjunction with field data, this enables much useful information to be obtained about the history of the rock, and what event or events the age measurements might be recording. In addition, petrographic examination of a sample enables judgments to be made as to which minerals are likely to be amenable to separation and at what grain size.

Mineral separations ideally are made at the coarsest grain size possible consistent with the high purity required, free of composites with other mineral phases. In most rocks, the target mineral or minerals will exhibit a range of grain sizes, so that it is not normally practical to prepare a pure concentrate of unbroken crystals of the size found in the samples. The main requirement is that the physical size of the mineral separate grains be greater than the distance an argon atom has to diffuse to be effectively released from the mineral structure.

Empirical evidence suggests that the effective diffusion radius for argon commonly is much less than the physical dimensions of the crystals in a rock. This is because the argon may have to diffuse only to the nearest cleavage plane, fracture, or incoherent lamella boundary within a crystal, rather than to the actual grain boundary. Clearly, however, if the grain size of the mineral separate is reduced below that of the critical diffusion distance for argon, a $^{40}Ar/^{39}Ar$ step heating experiment would be expected to yield a flat age spectrum, irrespective of whether the crystals possessed a nonuniform distribution of $^{40}Ar^*$ or not, because the argon would be outgassed from original cores and rims simultaneously.

Following crushing of the sample to an appropriate size, mineral separations are carried out by standard techniques based upon shape, density, or magnetic characteristics. These techniques involve the use of heavy liquids, magnetic separators, and other equipment. In many cases only a small amount of material, usually <1 g and often ≤ 0.1 g, is required for $^{40}Ar/^{39}Ar$ dating, so that it is possible to undertake final purification by hand picking under a binocular microscope should that be considered necessary.

3.3 MONITOR MINERALS

The $^{40}Ar/^{39}Ar$ dating method relies upon a sample of precisely known K–Ar age for flux monitoring purposes. A sample suitable as a flux monitor should meet certain criteria, clearly outlined by Alexander and Davis (1974) and followed here (see also Dalrymple et al., 1981).

1. The monitor mineral should have a uniform $^{40}Ar^*$ to ^{40}K ratio, in order that errors related to sample inhomogeneity are minimized.

2. The potassium and radiogenic argon should be homogeneously distributed in the monitor sample to ensure that a precise conventional K–Ar age can be measured. Homogeneity is required to be able to obtain representative subsamples, as the potassium and radiogenic argon have to be determined on separate aliquants.

3. The monitor mineral preferably should be of similar age and K/Ca ratio as the samples being dated by the $^{40}Ar/^{39}Ar$ method. The optimum irradiation for a sample is dependent mainly upon its age and K/Ca ratio (see Section 3.8). If there are gross differences in these parameters between monitor and sample, then difficulties may arise in providing an appropriate irradiation for one or both of the monitor and unknown samples.

4. The monitor sample should be relatively coarse grained to minimize health problems associated with handling highly radioactive material after irradiation.

5. The flux monitor sample should be available in sufficient quantity for use over an extended period to provide continuity.

Although the second criterion may appear to follow from the first, Alexander and Davis (1974) found that a sample from the meteorite St. Severin showed uniform $^{40}Ar^*/^{39}Ar_K$ values, but that it was not possible to obtain reproducible potassium and $^{40}Ar^*$ determinations on separate subsamples, preventing measurement of a precise conventional K–Ar age. This behavior probably can be explained in terms of a minor phase in the sample carrying much of the potassium and $^{40}Ar^*$.

The criteria given above are best met by choosing as a flux monitor a mineral suitable for conventional K–Ar dating from an igneous rock of simple geological history that has cooled rapidly, so that the mineral has a flat age spectrum. A mineral that has grown or has been completely isotopically reset during a metamorphic event also could be satisfactory. The mineral chosen as a flux monitor should be readily separable in reasonable quantity, and be of high purity ($>99\%$). In principle, virtually any of the minerals suitable for conventional K–Ar dating could be employed as flux monitors. In order to cover adequately the large age range of rocks and minerals and their K/Ca ratios, a minimum of two flux monitors normally is required. In our experience, a biotite (high K/Ca) and a hornblende (low K/Ca) provide the necessary compositional coverage for flux monitoring purposes. Although several flux monitors ideally are required for the proper measurement of samples over the age range encountered in geological materials, most laboratories appear

TABLE 3.1. *Potassium-argon data for a number of standards used as flux monitors*[a]

Sample number	Grain size (μm)	Mineral	K (wt%)	Radiogenic ^{40}Ar (10^{-10} mol/g)	Calculated age (Ma) ± 1σ	Reference
Hb3gr	180–500	Hornblende	1.247 ± 0.011	31.63 ± 0.19	1072 ± 11	Turner et al. (1971)
MMhb-1	180–250	Hornblende	1.556 ± 0.004	16.24 ± 0.09	519.4 ± 3.2	Alexander et al. (1978)
			1.555 ± 0.002	16.27 ± 0.05	520.4 ± 1.7	Samson and Alexander (1987)
LP-6	250–425	Biotite	8.33 ± 0.03	19.30 ± 0.20	128.9 ± 1.4	Ingamells and Engels (1976)
			8.37 ± 0.05	19.23 ± 0.10	127.9 ± 1.1	Odin et al. (1982b)
FY12a	250–600	Hornblende	0.954 ± 0.005	8.11 ± 0.04	433.8 ± 3.0	D. C. Rex in Roddick (1983)
SB-2	180–420	Biotite	7.63 ± 0.01	22.45 ± 0.28	162.1 ± 2.0	Dalrymple et al. (1981)
GA1550	250–800	Biotite	7.70 ± 0.06	13.43 ± 0.07	97.9 ± 0.9	McDougall and Roksandic (1974)
77-600	120–180	Hornblende	0.336 ± 0.003	2.712 ± 0.008	414.1 ± 3.9	Harrison (1980, 1981); this work
B4M	300–500	Muscovite	8.68 ± 0.08	2.811 ± 0.056	18.6 ± 0.4	Flisch (1982)
B4B	180–250	Biotite	7.91 ± 0.08	2.381 ± 0.023	17.3 ± 0.2	Flisch (1982)

[a] $\lambda_e + \lambda'_e = 0.581 \times 10^{-10}$/a; $\lambda_\beta = 4.962 \times 10^{-10}$/a; ^{40}K/K = 1.167 × 10^{-4} mol/mol. Errors quoted are standard deviations.

to manage satisfactorily with only one or two. Thus in the Australian National University laboratory we use a 98-Ma-old biotite (GA1550), separated from a monzonite, and a 414-Ma-old hornblende (77–600), obtained from an amphibolite, as flux monitors. Samples ranging from as old as 4 Ga (meteorite) to < 1 Ma (alkali feldspar) have been successfully dated by the ^{40}Ar/^{39}Ar step heating method using these two minerals as monitors, generally with interferences (see Section 3.8) remaining acceptably low. For samples with high K/Ca, such as biotite, muscovite, and alkali feldspar, the biotite flux monitor is most appropriate, whereas for samples with low K/Ca, for example, plagioclase, amphibole, or mafic whole rocks, the hornblende flux monitor is employed.

Most laboratories prepare flux monitors for their own use; Alexander and Davis (1974) made direct intercomparisons, mainly using the ^{40}Ar/^{39}Ar total fusion method, between five of these intralaboratory standards. For three of the standards, good agreement was found between the ^{40}Ar/^{39}Ar age and the K–Ar age given by the originating laboratory. Since that time the hornblende MMhb-1 has been prepared and made available for interlaboratory comparisons (Alexander et al., 1978; Dalrymple et al., 1981; Samson and Alexander, 1987). Roddick (1983) undertook a very careful and precise ^{40}Ar/^{39}Ar age intercalibration between MMhb-1 and three other standards. This kind of approach is to be encouraged so that comparison of results between laboratories can be more readily made. Conventional K–Ar age data for a number of the standards use for ^{40}Ar/^{39}Ar flux monitoring purposes are summarized in Table 3.1.

3.4 NUCLEAR REACTIONS

A nuclear reaction occurs when an incident particle interacts with a nucleus to produce one or more other nuclei and possibly other particles. The main nuclear reaction utilized in the ^{40}Ar/^{39}Ar dating technique involves the interaction of a neutron with the ^{39}K nucleus, which is transmuted to ^{39}Ar with emission of a proton. This reaction can be written in the form of an equation

$$^{39}_{19}K + ^{1}_{0}n = ^{39}_{18}Ar + ^{1}_{1}H + Q$$

where the superscripts denote the number of nucleons (protons plus neutrons) and the subscripts denote the atomic number (the number of protons) in each element. The symbol Q represents the kinetic energy released or absorbed in the reaction. The sum of the superscripts on one side of the reaction must balance those on the other side; similarly a balance must be preserved with respect to the subscripts.

Nuclear reactions commonly are expressed in a more abbreviated form. For example, the above reaction can be written ^{39}K(n, p)^{39}Ar, with the bombarding particle and released fragment(s) shown in parentheses between the bombarded nucleus, ^{39}K, and product nucleus, ^{39}Ar. In this notation the symbols n, p, β^-, d, α, and γ represent the neutron, proton, electron, deuteron, alpha particle, and gamma ray, respectively.

In nuclear reactions there is conservation of nucleons, charge, momentum, and energy. The energy term, Q, is positive when energy is released (an exoergic reaction) and negative when energy is absorbed (an endoergic reaction). The value of Q is calculated per nucleus transformed, and is obtained by comparison of the atomic masses of the reactants and the products. Thus, for the reaction ^{39}K(n,p)^{39}Ar the sum of the masses of the reactants is 39.972373 atomic mass units (amu), based upon the ^{12}C = 12.000000 amu scale, and the sum of the masses of the products is 39.972140, a difference of 0.000233 amu. From Einstein's special theory of relativity the equivalence of mass and energy yields the following relation

$$1 \text{ amu} = 931.5 \text{ MeV}$$

Using this relation, the value of Q for the reaction being considered is 0.22 MeV. Note that Q is positive; this is because the mass of the products is less than that of the reactants. As the reaction is exoergic it might be expected that a neutron of essentially zero energy could produce the reaction, particularly as there is no Coulomb energy barrier inhibiting the entry of a neutron into the nucleus. In practice, however, it is found that neutrons with quite high energy (>1 Mev) are required for the reaction to occur. The reason for this is that for a proton to leave the nucleus it needs to overcome an energy barrier associated with Coulomb forces. In the case of endoergic reactions, that is those with negative Q, the bombarding particle must have kinetic energy greater than the calculated Q value. The additional energy required depends upon a number of factors including the size of the effective Coulomb barrier for charged particles whether entering or leaving the nucleus.

3.5 NUCLEAR REACTORS AS A NEUTRON SOURCE

Fast neutrons are required for the reaction $^{39}K(n,p)^{39}Ar$ to proceed, and to produce sufficient ^{39}Ar within a reasonable time in a sample; irradiations are carried out in a nuclear reactor, where fast neutron fluxes in the 10^{12}–10^{14} n/cm²/s range are readily available. In this section we shall briefly discuss neutrons and their properties, the phenomenon of fission, and nuclear reactors as a source of neutrons.

Neutrons are unchanged particles of mass 1.67495×10^{-24} g, or mass 1.008665 on the atomic mass unit scale ($^{12}C = 12.000000$). They occur in the nucleus of all atoms, except that of hydrogen, which consists of a proton, the positively charged fundamental particle of mass 1.007276. The mass of a hydrogen atom, that is a proton nucleus plus an extranuclear electron, is 1.007825 amu. The discovery of the neutron by Chadwick in 1932 provided an explanation for the long-standing problem as to why nuclear masses, when expressed in terms of the mass of a proton, are about twice the mass of their nuclear charge, in units of the charge of a proton. Most nuclei consist of approximately equal numbers of protons and neutrons.

Free neutrons are unable to exist for more than short intervals as they are absorbed by nuclear matter in times of less than 1 μs, and in the absence of matter they disintegrate within minutes to a proton and an electron. Neutrons are produced only by nuclear reactions induced by α and γ radiations from radioactive nuclides, or from interactions with matter by charged particles and γ radiation produced in accelerators, or by nuclear fission or fusion processes. The most prolific source of neutrons available in a controlled manner is within a nuclear reactor, and this is the type of source that makes $^{40}Ar/^{39}Ar$ dating possible.

Neutrons are produced with a wide range of velocities, and, thus, kinetic energies. Those with energies greater than ~ 0.1 MeV generally are known as fast neutrons; those with lower energies are slow neutrons. The terms intermediate or resonance are commonly used for neutrons with energies in the 1 eV–100 keV range. As there is a continuum of possible energies, the distinctions are arbitrary, and, indeed, the terms are often employed rather loosely. Thus, the energy level for distinguishing between fast and slow neutrons appears to range from ~ 5 keV to ~ 0.5 MeV, depending upon the context or the authority. Thermal neutrons are defined as neutrons in thermal equilibrium with their surroundings; the energy distribution is similar to that of gas molecules at the same temperature. Thus, the velocities of thermal neutrons approach a Maxwellian distribution, with a most probable energy of 0.025 eV at 20°C. Epithermal is the term used for slow neutrons not in complete thermal equilibrium with the environment, and energies up to ~ 1 eV. For convenience in the present work the term slow will be used for neutrons with energies < 0.1 MeV.

Neutrons are very effective in producing nuclear reactions because, being uncharged, no electrostatic potential barrier exists for a neutron entering a nucleus. This fact was further emphasized when it was found that for most kinds of matter the probability of interaction with and subsequent transmutation of a nucleus increased as the neutron energy decreased.

Interactions of neutrons with matter are expressed in terms of cross sections, which provide a measure of the probability of a particular reaction occurring. A nuclear cross section can be thought of as the cross-sectional area presented by a nucleus to a particular particle. Consider the case of a mass with N target nuclei in the path of a neutron beam of flux ϕ, which is the number of neutrons, n, of velocity, v, passing through unit area in unit time (i.e., $\phi = nv$/cm²/s). The number of interactions, X, will be proportional to the number of target nuclei and to the neutron flux. Thus

(3.1) $$X = \sigma N \phi$$

(3.2) $$\sigma = X/N\phi$$

where the constant of proportionality, σ, is the cross section of the interaction being considered. The barn, 10^{-24} cm^2, is the unit used to express cross sections quantitatively. In Eq. (3.2), N is the number of target nuclei per unit volume.

Fission is the phenomenon whereby a heavy nucleus (atomic number >90, atomic mass >230), following capture of a particle, splits into two nuclei, usually of subequal mass. Because the total mass of the products is less than that of the heavy atom undergoing fission and the interacting particle, a great deal of energy is released in the fission process, ~ 200 MeV in the case of a uranium atom. Although spontaneous fission of ^{238}U and ^{235}U occurs in nature, the high Coulomb energy barrier to the division of the nucleus into two fragments means that fission is a low probability reaction. Nevertheless, the spontaneous fission of ^{238}U happens sufficiently frequently over geological time scales that it forms the basis of the fission track dating technique for minerals and glasses containing uranium (Fleischer et al., 1975).

A nuclear reactor is a device in which fission is constrained to occur in a controlled manner. A schematic diagram of a nuclear reactor is shown in Fig. 3.1. It consists of an assemblage of sufficient fissionable material, most commonly uranium enriched in ^{235}U, in its core, so that a self-sustaining reaction can be maintained. A minimum amount of fissionable material in the core is required for a given reactor to become "critical," that is, able to be self-sustaining. Each atom that undergoes fission produces neutrons; thus, on average, ^{235}U yields 2.4 neutrons per fission. Nuclear reactors are designed to ensure that at least one neutron from each fission is available to produce another fission so that the chain reaction is self-sustaining. In practice, a reactor is operated in such a manner that an excess of neutrons is produced by fission in order to maintain criticality as well as to ensure that sufficient neutrons are available for other purposes.

The energy of neutrons in a nuclear reactor ranges widely from ~ 0.002 eV to ~ 15 MeV. The fission process itself produces neutrons with energies dominated by fast neutrons in the 0.5–5 MeV range (Fig. 3.2), with average energy ~ 2.5 MeV, and most probable energy

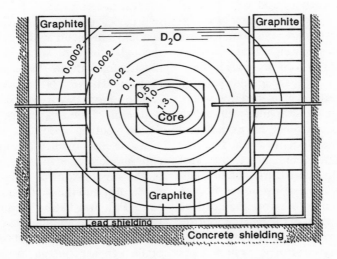

Fig. 3.1. Schematic of a nuclear reactor, based upon HIFAR of the Australian Atomic Energy Commission. The approximate fast neutron flux ($E > 0.5$ MeV) in the reactor is shown in units of 10^{14} n/cm^2/s, after Roberts (1958), in a vertical cross section through the center of the reactor. Two irradiation facilities or channels are shown; that on the right is similar to the type actually used for irradiations in this reactor in connection with ^{40}Ar/^{39}Ar dating studies.

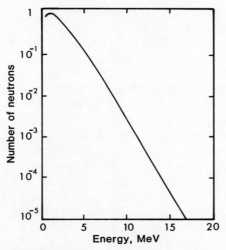

Fig. 3.2. Energy spectrum of ^{235}U fission neutrons showing relative numbers of neutrons as a function of energy. After Watt (1952).

~0.6 MeV. The distribution of neutron energies is described by the empirical expression

(3.3) $\qquad n(E) = \exp(-E)\sinh(2E)^{1/2}$

where n is the number of neutrons of energy, E (Watt, 1952). The great majority of these neutrons are too energetic to be absorbed by fissionable nuclei. However, slow neutrons, especially those with energies below a few electron volts, are readily trapped in a nucleus. The nucleus becomes unstable, as the potential energy barrier is exceeded, and will undergo fission. Thus, in a nuclear reactor the fissionable fuel within the core is surrounded by a moderator of low atomic weight that effectively slows down the fast neutrons and reflects them back into core so that the chain reaction continues. The moderator is usually water, heavy water (D_2O), or graphite, although a variety of other materials also have been used for this purpose. The moderator also acts as a coolant to facilitate removal of the vast amount of energy produced during fission, and forms the basis for generating power from a reactor. Cadmium rods or arms that can be moved at will into or out of the reactor core are used to absorb slow neutrons to provide the necessary operational control, as cadmium has a large capture cross section for slow neutrons, especially for thermalized neutrons with energies <0.4 eV.

For ^{40}Ar/^{39}Ar dating purposes, irradiations normally are done in relatively small reactors, designed and built as research devices, and for these reactors the energy liberated is dissipated rather than being used commercially. Research reactors are built with a range of facilities to enable samples to be introduced and recovered at will, so that irradiations can be performed in an appropriate flux of neutrons. For ^{40}Ar/^{39}Ar dating it is desirable that the irradiations be done in the core of the reactor where the highest fast neutron flux is available.

Reactors used for ^{40}Ar/^{39}Ar irradiation purposes differ widely in design, and range up to ~50 MW in power output. Although the external physical dimensions of these reactors are large, owing to the need to provide heavy radiation shielding, the core itself is quite small, usually somewhat less than 1 m across and in height. The core consists of an assemblage of fuel rods in the form of a cylinder, right cylinder, or cube, surrounded by the moderator. The schematic drawing of a reactor in Fig. 3.1 is based upon HIFAR, a high flux, heavy water moderated and cooled reactor, with a maximum heat output of 10 MW. We use HIFAR as an example, as we have some familiarity with it, and, although other research reactors differ considerably in design, the general principles remain the same.

In Fig. 3.1 the approximate distribution of the fast neutron flux in HIFAR is shown, and in Fig. 3.3 the theoretical distribution of the fast and slow neutron fluxes in a vertical plane through the core of a fast flux reactor is illustrated, with the distance scale adjusted to that of HIFAR. Note the decrease in fast neutron flux away from the center of the core, as a result of the progressive moderation of the fission neutron spectrum. Note also the increase in the slow neutron flux adjacent to the core–moderator boundary, as a consequence of the slowing down of fast neutrons in the moderator and their reflection back toward the core. Two irradiation facilities are shown in Fig. 3.1, one extending into the core, and the other terminating adjacent to the core. Because of the high fast neutron flux and the high fast to slow neutron flux ratio, the facility of choice for irradiation of samples in ^{40}Ar/^{39}Ar dating would be that located in the core. In the case of

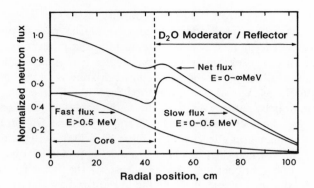

Fig. 3.3. The theoretical neutron flux distribution for a nuclear reactor, and the slow and fast neutron fluxes, in a section through the center of the reactor. Modified after Glasstone and Edlund (1952) to approximate the distributions found in HIFAR.

HIFAR a suitable irradiation facility in the core is not available, so that an irradiation channel next to the core, as illustrated in Fig. 3.1, is used.

In some reactors, for example, GSTR (TRIGA) reactor operated by the U.S. Geological Survey, it is convenient to irradiate samples within the central thimble of the core (Dalrymple et al., 1981), and this is a very satisfactory arrangement.

The distribution of fast and slow neutron fluxes in most reactors of the kind utilized for ^{40}Ar/^{39}Ar dating is expected to be quite similar to that shown in Fig. 3.3, although, no doubt, there will be differences in detail. Particularly noteworthy is that in virtually all reactors there will be significant fast (as well as slow) neutron flux gradients, which must be properly monitored if ^{40}Ar/^{39}Ar age results of high precision and accuracy are to be obtained; this question will be discussed in Section 3.9.

3.6 THE ^{39}K(n,p)^{39}Ar REACTION

Application of the ^{40}Ar/^{39}Ar method to the dating of a sample necessitates the production of enough ^{39}Ar from ^{39}K such that precise measurements can be made. The ^{39}K(n,p)^{39}Ar reaction requires fast neutrons for it to proceed, and a high neutron flux of the kind available in a nuclear reactor is necessary to produce sufficient ^{39}Ar within reasonable irradiation times.

As noted in Section 3.4, the ^{39}K(n,p)^{39}Ar reaction has a positive Q value ($+0.22$ MeV), and therefore it might be expected that the reaction would proceed with thermal neutrons, but this is not the case. The explanation as to why higher energy neutrons are required for this reaction lies in the existence of the Coulomb energy barrier, which must be overcome by charged particles leaving the nucleus, having the effect of raising the threshold for reaction to appreciably endoergic (fast neutron) levels. Turner and Cadogan (1974) derived a curve shown in Fig. 3.4 for the proportion of reactions induced by neutrons of different energy in the Herald reactor, based upon experimentally determined cross-sectional data obtained by Bass et al. (1961,

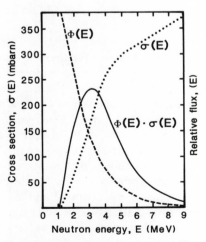

Fig. 3.4. Cross section, $\sigma(E)$, for the reaction ^{39}K(n,p)^{39}Ar plotted as a function of neutron energy, E. Neutron flux, $\varphi(E)$, from ^{235}U fission is indicated in arbitrary units. The product $\varphi(E)\cdot\sigma(E)$ shows the relative proportion of reactions induced by neutrons of energy, E. From Turner and Cadogan (1974), with permission.

1964) for the 1.46–8.7 MeV neutron energy range and at 13.7 MeV. A rather similar curve would be found for each reactor facility used for ^{40}Ar/^{39}Ar irradiation purposes. These results indicate that the threshold neutron energy for the reaction is about 1.2 MeV, and that the production of ^{39}Ar is mainly by neutrons in the 1.2–7 MeV energy range. The decrease in production of ^{39}Ar at higher neutron energies is principally because of the marked drop off in the number of neutrons of higher energy in the ^{235}U fission spectrum, despite an increasing probability of capture occurring at higher energies (Fig. 3.4).

Table 3.2 lists some of the relevant data on ^{39}Ar production and neutron fluxes for particular facilities in a number of nuclear reactors that have been utilized for irradiation of samples for ^{40}Ar/^{39}Ar dating. These data are from the sources indicated; however, the summaries previously made by Tetley et al. (1980), Dalrymple et al. (1981), and Roddick (1983) also have been extensively used. As noted by Dalrymple et al. (1981), the fast neutron fluxes and slow neutron to fast neutron flux ratios quoted for individual reactors are not directly comparable, because of different definitions, often unstated, as to the energy level used to discriminate between fast and slow neutrons. In view of this, Dalrymple et al. (1981) made a more direct comparison of ^{39}Ar production rates by means of a parameter J/h. As given in Section 2.5, Eqs. (2.14) and (2.18), J is a parameter that reflects the integrated neutron flux over the energies required for the ^{39}K(n,p)^{39}Ar reaction to proceed, and is determined from a standard sample of accurately known K–Ar age, irradiated with the unknown, whose age is to be determined. For Table 3.2 this parameter, together with the estimated ^{39}Ar production, was derived from the references given. For a given reactor location, this parameter may vary depending upon factors such as the power at which the reactor is operated and the stage at which the irradiation is performed relative to the nuclear fueling cycle. Thus variations of $\pm 50\%$, or even more in some cases, may be found in practice.

In Table 3.2 the results are listed in order of decreasing J/h parameter, which also reflects approximately the magnitude of the fast neutron flux, as might be expected. Nevertheless, because of the caveat mentioned previously regarding determination of the fast flux, the J/h parameter is a much better indication of the characteristics of a given irradiation facility used for ^{40}Ar/^{39}Ar dating purposes, as it reflects the actual production rate of ^{39}Ar. Table 3.2 shows that the J/h parameter ranges over a factor of ~ 200, with Brookhaven High Flux Beam reactor (HRBR) having the highest J/h because of the large fast neutron flux in the facility utilized for the irradiation, and HIFAR reactor exhibiting the lowest J/h. In the latter case, the low value for J/h is because irradiation is done in a facility next to the core, in the heavy water moderator rather than in the core, where the highest fast neutron fluxes occur. The only major consequence of the large variation in the J/h parameter is that irradiation times need to be varied in order that there is adequate production of ^{39}Ar. Thus, for HIFAR, irradiation times need to be increased by a factor of ~ 200 compared with HFBR, in order to obtain the same production of ^{39}Ar.

In Table 3.2, approximate ^{39}Ar$_K$ production rates are also listed. These have been derived from the relevant publications, either directly or indirectly, and can be expected to be in proportion to the J/h parameter. In fact it will be seen that differences from direct proportionality of up to $\sim 10\%$ occur. Such behavior is not unexpected because ^{39}Ar$_K$ production estimates normally are based upon manometric determinations of amounts of argon in the mass spectrometer, and, thus, are not absolute measurements. Although these ^{39}Ar$_K$ production rates are only approximate, they are included as a useful guide to the amounts of this isotope that are likely to be generated in a given facility.

Several other points also can be made about the data in Table 3.2. First, information is not always available in the relevant papers as to the neutron flux in a particular facility, and many of the figures are approximate. In a few reactors, the experimenters have used cadmium shielding around the sample to eliminate, or at least attenuate, the effect of slow neutrons. This is the reason that slow/fast neutron flux ratios are not reported for facilities utilized in reactors BR-2 and FR-2.

TABLE 3.2. Data on irradiation facilities in nuclear reactors used in $^{40}Ar/^{39}Ar$ dating[a]

Reactor and location (irradiation position)	J/h ($\times 10^{-4}$)	$^{39}Ar_K$ production (mol ^{39}Ar/g K–h $\times 10^{-10}$)	$^{37}Ar_{Ca}$ production (mol ^{37}Ar/g Ca–h $\times 10^{-10}$)	Fast flux ($n/cm^2/s \times 10^{13}$)	Slow/fast neutron flux	References
HFBR (Brookhaven, NY) (core)	35	12	6	14	0.65	Husain (1974)
RRF (Columbia, MO) (flux trap)	29	9	10	—	—	Bernatowicz et al. (1978)
BR-2 (Mol, Belgium) (core)	7.4	2.5	1.3	3.7	—	Kirsten et al. (1972); Kirsten et al. (1973b)
Herald (Aldermaston, U.K.) (core)	7.2	2.4	1.3	3.7	1.7	Turner (1970b,d; 1971b); Turner et al. (1973)
GETR (Pleasonton, CA) (shuttle tube)	4.0	1.4	0.76	1.7	7.5	Turner et al. (1971, 1972); Podosek et al. (1973); Alexander and Davis (1974)
Melusine (Grenoble, France)	3.4	0.88	0.49	—	—	Féraud et al. (1982)
Herald (Aldermaston, U.K) (core edge)	2.8	0.88	0.44	1.3	3.3	Roddick (1983)
GSTR (TRIGA) (Denver, CO) (core, centerline)	2.5	0.86	0.42	1.7	0.9	Dalrymple et al. (1981)
49-2 (Beijing, China) (H4)	1.7	0.52	—	0.6	—	Wang et al. (1986)
JMTR (Tohoku, Japan)	1.4	0.45	0.24	~0.6	—	Kaneoka et al. (1979)
McMaster University (Hamilton, Canada) (core)	1.1	0.4	—	~0.08	~19	Berger and York (1970)
JMTR (Tohoku, Japan)	0.93	0.29	0.15	0.5	—	Kaneoka (1983)
Ford (Ann Arbor, MI)(H-5)	0.78	0.24	—	—	—	Harrison and Fitz Gerald (1986)
FR-2 (Karlsrühe, Germany) (core, isotope channel)	0.68	0.22	0.12	0.34	—	Kirsten et al. (1972, 1973a); Stettler et al. (1974)
RRF (Columbia, MO) (reflector pool)	0.56	0.19	0.11	0.3	~13	Hohenberg et al. (1981); Dalrymple et al. (1981)
HIFAR (Lucas Heights, Australia) (next to core X33, X34)	0.20	0.060	0.031	~0.1	~50	McDougall (1985); Tetley et al. (1980); this work

[a] $J = (\lambda/\lambda_e + \lambda_e)(^{39}Ar_K/^{40}K)$.

Cadmium shielding also is routinely used in HIFAR, but a flux ratio is nevertheless given and refers to the ambient flux in the facility. Second, the slow to fast neutron flux ratio tends to increase with decreasing J/h. This increase probably simply indicates that the irradiations are being done in facilities progressively further from the geometric center of the core of the reactor, with progressive moderation of the fission neutron spectrum. Third, the RRF, Herald, and JMTR reactors are listed twice in Table 3.2, because two different irradiation facilities with different flux characteristics have been utilized in each nuclear reactor.

Using the published fast neutron fluxes for facilities in various reactors employed in $^{40}Ar/^{39}Ar$ dating, Roddick (1983) showed that the calculated neutron cross sections for the $^{39}K(n,p)^{39}Ar$ reaction were mainly in the 60–90 mb range, in keeping with the previous estimates of about 70 mb derived by Turner (1971a) and Turner and Cadogan (1974) for the ^{235}U neutron fission spectrum in the Herald reactor. Roddick (1983) argued that the few values that lay outside the quoted range probably simply reflected inadequate knowledge of the fast neutron flux in the particular facility, a view with which we concur.

Figure 3.5 shows the approximate production of $^{39}Ar_K$ for samples of a range of potassium contents using the parameter, J, as the measure of the fast neutron dose (fluence). For convenience, this diagram is based upon the $^{39}Ar_K$ production rate for GSTR (8.6×10^{-11} mol/g K–h) calculated from Dalrymple et al. (1981), partly because the characteristics of this reactor are very well described, but also because GSTR is in the middle of the range of reactors used for irradiations in $^{40}Ar/^{39}Ar$ dating, in terms of production rate of $^{39}Ar_K$. From the known characteristics of a given irradiation facility, especially the J/h parameter, the production of ^{39}Ar can be estimated from Fig. 3.5 probably to within about $\pm 10\%$. The irradiation time for production of the optimum amount of $^{39}Ar_K$ is dependent upon the age of the sample to be dated and the flux characteristics of the facility in which the irradiation is to be done. These aspects will be covered in some detail in Section 3.8.

It is worth noting in passing that an estimate can be made of the potassium content of an

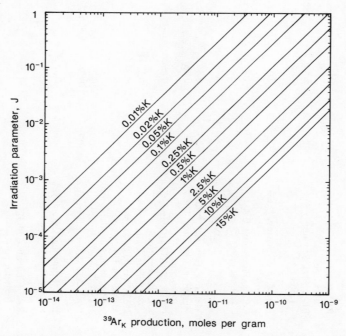

Fig. 3.5. Production of $^{39}Ar_K$ as a function of potassium content and irradiation parameter, J.

unknown sample from the size of the ^{39}Ar ion beam measured in the mass spectrometer. For this to be done the sensitivity of the mass spectrometer must be known, and the response of the machine needs to be essentially linear with pressure in the range of measurement. Calibration is effected by means of the gas released from the flux monitor, whose potassium content is available from independent measurement. Obviously the mass of each sample from which the gas is extracted also must be known. It is relatively easy to achieve a precision of $\sim 5\%$ in measuring gas amounts by this manometric technique. With greater care in monitoring the sensitivity of the mass spectrometer a precision of $\sim 1\%$ or better is possible (cf. Baksi, 1973; Roddick, 1983), in principle yielding a precision of this order in the determination of potassium, provided due allowance is made for ^{39}Ar generated from ^{42}Ca during the irradiation (see Section 3.7).

3.7 INTERFERING NUCLEAR REACTIONS AND CORRECTION FACTORS

3.7.1 General

It was recognized from the earliest stages of development of the ^{40}Ar/^{39}Ar dating method that isotopes of argon, apart from the ^{39}Ar produced in the key reaction on ^{39}K, may be formed by interaction of neutrons with isotopes of calcium, potassium, argon, and chlorine (Sigurgeirsson, 1962; Merrihue and Turner, 1966; Mitchell, 1968a; Brereton, 1970; Turner, 1971a). A wide variety of reactions may occur (Table 3.3), although, fortunately, many are of minor significance. Thus, the low abundance of argon isotopes in a sample relative to potassium and calcium, and the high negative Q values for reactions involving the most abundant isotope, ^{40}Ar (Table 3.3), mean that production of reactor-induced argon from target argon is negligible (Brereton, 1970), and some experimental evidence supporting this view was given by Tetley et al. (1980).

The most important nuclear reactions in relation to ^{40}Ar/^{39}Ar dating studies are highlighted in Table 3.3 by enclosing them in boxes. The interferences are discussed below in terms of target elements. In Section 3.8 we shall discuss how the effects of the interferences can be minimized. Correction factors for argon isotopes derived from calcium and potassium are listed in Table 3.4 for facilities in a number of reactors; their significance will be treated in the following subsections.

3.7.2 Reactions on calcium

Significant amounts of ^{36}Ar and ^{39}Ar are produced from calcium during neutron irradiation, and unless proper corrections are made, erroneous ^{40}Ar/^{39}Ar ages will be obtained. The effects are especially marked for samples with low K/Ca ratios such as hornblendes, plagioclases, pyroxenes, meteorites, and lunar rocks. For terrestrial samples, the atmospheric ^{40}Ar is corrected for by means of the ^{36}Ar. Clearly, if the ^{36}Ar produced from ^{40}Ca during irradiation is not taken into account, there will be overcorrection for atmospheric ^{40}Ar, resulting in low apparent ages. Another potential source of ^{36}Ar, from neutron interactions on ^{35}Cl (Table 3.3), in most circumstances is negligible.

Because ^{37}Ar is a quite trivial component of atmospheric argon, and because it is generated in negligible amount from ^{39}K during irradiation, owing to the high negative Q value of the reaction, production of this isotope is entirely from ^{40}Ca, and, thus, can be used to correct for both reactor-induced ^{36}Ar and ^{39}Ar derived from calcium (Brereton, 1970). The correction factors, $(^{36}$Ar$/^{37}$Ar$)_{Ca}$ and $(^{39}$Ar$/^{37}$Ar$)_{Ca}$, normally are determined by measuring the relative production rates of these isotopes in a pure calcium salt, usually CaF$_2$, after neutron irradiation. The production ratios can then be utilized to correct for neutron-induced ^{36}Ar and ^{39}Ar in the analysis of an unknown sample by means of the ^{37}Ar reference isotope. Thus, following irradiation of a pure calcium salt, argon is extracted and analyzed isotopically in the mass spectrometer. The appropriate correction factor for ^{36}Ar by reference to the ^{37}Ar is derived as follows:

$$(3.4) \quad \left[\frac{^{36}\text{Ar}}{^{37}\text{Ar}}\right]_{Ca} = \frac{^{36}\text{Ar}_m - {}^{40}\text{Ar}_A/295.5}{^{37}\text{Ar}}$$

where ^{36}Ar$_m$ refers to the measured ^{36}Ar, ^{40}Ar$_A$ is the atmospheric ^{40}Ar, and ^{37}Ar$_{Ca}$ is the total

TABLE 3.3. Reactions producing argon isotopes in neutron irradiated samples[a,b]

Argon isotope produced	Target element			
	Calcium	Potassium	Argon	Chlorine
^{36}Ar	^{40}Ca(n,nα)^{36}Ar (−7.04, 96.94)			^{35}Cl(n,γ)^{36}Cl $\xrightarrow{\beta^-}$ ^{36}Ar (+8.58, 75.77)
^{37}Ar	^{40}Ca(n,α)^{37}Ar (+1.75, 96.94)	^{39}K(n,nd)^{37}Ar (−15.99, 93.26)	^{36}Ar(n,γ)^{37}Ar (+8.79, 0.337)	
^{38}Ar	^{42}Ca(n,nα)^{38}Ar (−6.25, 0.65)	^{39}K(n,d)^{38}Ar (−4.16, 93.26) ^{41}K(n,α)^{38}Cl $\xrightarrow{\beta^-}$ ^{38}Ar (−0.12, 6.73)	^{40}Ar(n,nd)^{38}Cl $\xrightarrow{\beta^-}$ ^{38}Ar (−18.38, 99.60)	^{37}Cl(n,γ)^{38}Cl $\xrightarrow{\beta^-}$ ^{38}Ar (+6.11, 24.23)
^{39}Ar	^{42}Ca(n,α)^{39}Ar (+0.34, 0.65)	^{39}K(n,p)^{39}Ar (+0.22, 93.26)	^{38}Ar(n,γ)^{39}Ar (+6.60, 0.063) ^{40}Ar(n,d)^{39}Cl $\xrightarrow{\beta^-}$ ^{39}Ar (−10.30, 99.60)	
^{40}Ar	^{43}Ca(n,nα)^{39}Ar (−7.59, 0.14) ^{43}Ca(n,α)^{40}Ar (+2.28, 0.14) ^{44}Ca(n,nα)^{40}Ar (−8.85, 2.09)	^{40}K(n,d)^{39}Ar (−5.36, 0.01167) ^{40}K(n,p)^{40}Ar (+2.29, 0.01167) ^{41}K(n,d)^{40}Ar (−5.58, 6.73)		

[a]Sources: Brereton (1970), Turner, (1971a), and Dalrymple et al. (1981). Q values (MeV), calculated from data in Lederer and Shirley (1978), and target isotope abundance (atom%) shown in parentheses below each reaction.

[b]Half lives: ^{36}Cl $\xrightarrow{\beta^-}$ ^{36}Ar, 3.0 × 10^5 a; ^{38}Cl $\xrightarrow{\beta^-}$ ^{38}Ar, 37.3 min; ^{39}Cl $\xrightarrow{\beta^-}$ ^{39}Ar, 55.5 min; ^{37}Ar $\xrightarrow{ec}$ ^{37}Cl, 35.1 days; ^{39}Ar $\xrightarrow{\beta^-}$ ^{39}K, 269 a; ^{40}K → ^{40}Ar, ^{40}Ca, 1.25 × 10^9 a.

TABLE 3.4. *Correction factors for argon isotopes produced from calcium and potassium in various irradiation facilities in nuclear reactors used in $^{40}Ar/^{39}Ar$ dating*

Reactor (irradiation position)	Cadmium shielding	Slow/fast neutron flux	$(^{36}Ar/^{37}Ar)_{Ca}$ $\times 10^{-4}$	$(^{38}Ar/^{37}Ar)_{Ca}$ $\times 10^{-5}$	$(^{39}Ar/^{37}Ar)_{Ca}$ $\times 10^{-4}$	$(^{37}Ar/^{39}Ar)_K$ $\times 10^{-4}$	$(^{38}Ar/^{39}Ar)_K$ $\times 10^{-2}$	$(^{40}Ar/^{39}Ar)_K$ $\times 10^{-4}$	Reference
HFBR (core)	—	0.65	2.31 ± 0.01	6.2 ± 0.8	6.45 ± 0.29	<1	1.00 ± 0.01	<20	Husain (1974)
RRF (flux trap)	—	—	2.84 ± 0.02	—	7.69 ± 0.03	—	—	—	Bernatowicz et al. (1978)
BR-2 (core)	Yes	—	—	—	—	—	—	—	
Herald (core)	—	1.7	2.47 ± 0.09	13.9 ± 2.6	9.35 ± 0.17	—	—	—	Kirsten et al. (1973a)
			2.46 ± 0.11	22.4 ± 3.7	7.19 ± 0.24	—	1.14 ± 0.03	123 ± 24	Brereton (1970)
			1.1 ± 0.2	8 ± 2	7.12 ± 0.22	—	1.10 ± 0.01	164 ± 13	Brereton (1972)
			2.0 ± 0.3	13 ± 3	6.7 ± 0.3	—	—	—	Turner et al. (1973)
GETR (shuttle tube)	—	7.5	3.05 ± 0.06	—	7.32 ± 0.15	—	—	—	Turner et al. (1971)
	—	—	3.15 ± 0.09	47.7 ± 0.9	7.23 ± 0.09	—	1.46 ± 0.01	625 ± 9	Alexander and Davis (1974)
Melusine	—	—	2.39 ± 0.19	—	6.57 ± 1.11	—	—	261 ± 6	Féraud et al. (1982)
Herald (core edge)	—	3.3	2.51 ± 0.02	3.81 ± 0.21	6.44 ± 0.04	—	1.08 ± 0.02	160 ± 10	Roddick (1983)
GSTR (core)	—	0.9	2.64 ± 0.02	3.17 ± 0.02	6.73 ± 0.04	22.0 ± 0.7	1.34 ± 0.02	59 ± 7	Dalrymple et al. (1981)
								101 ± 5	
								10 ± 10	
49-2 (H4)	0.5 mm Cd	—	2.64	—	6.87	—	—	71.5	Wang et al. (1986)
49-2 (E7)	—	2.5	1.17 ± 0.05	—	7.56 ± 0.30	—	—	239 ± 12	Wang et al. (1985)
JMTR	—	—	5.5	390	7.0	—	6.7	700	Kaneoka et al. (1979)
	—	—	3.72 ± 0.06	206 ± 3	11.3 ± 0.4	—	3.47 ± 0.04	1960 ± 40	Kaneoka (1983)
McMaster	—	19	2.1 ± 0.3	—	6.6 ± 0.5	—	—	268 ± 2	Berger (1975)
	—	—	2.54 ± 0.09	—	6.51 ± 0.31	—	—	156 ± 4	Bottomley and York (1976)
Ford (H-5)	—	—	2.87	—	7.61	—	—	382	Foland (1983)
	—	—	2.21–2.26	—	8.00–8.25	—	—	250–470	Heizler and Harrison (1988)
FR-2	Yes	13	2.7 ± 0.2	6 ± 2	6.85 ± 0.20	—	1.8 ± 0.6	1300 ± 100	Stettler et al. (1973)
	—	—	2.5 ± 0.2	—	8.8 ± 0.1	—	1.7 ± 0.1	1340 ± 10	Hohenberg et al. (1981)
RRF (reflector pool)	—	—	2.13 ± 0.04	—	10.2 ± 0.3	—	1.71 ± 0.02	3008 ± 138	Honda et al. (1983)
HIFAR	No	~50?	3.06 ± 0.05	—	7.27 ± 0.08	0.15 ± 0.07	—	270 ± 20	Tetley et al. (1980); also this work
	0.2 mm Cd	—	—	—	—	—	—	30 ± 10	
	0.5 mm Cd	—	—	—	—	—	—	~20	
	1.0 mm Cd	—	3.20 ± 0.02	—	7.9 ± 0.5	—	—	—	

^{37}Ar found, after correction for its decay ($t_{1/2} = 35.1$ days), during and subsequent to the irradiation. In this calculation the assumption is made that all the ^{40}Ar present is derived from atmospheric argon; the amount of atmospheric ^{36}Ar is calculated by dividing the ^{40}Ar by 295.5, the ^{40}Ar/^{36}Ar ratio in atmospheric argon. As some ^{40}Ar may be produced by neutron interactions on other calcium isotopes (Table 3.3), the assumption is not likely to be strictly correct. Fortunately, however, no error is introduced into the calculation of a ^{40}Ar/^{39}Ar age for a terrestrial sample owing to departures from this assumption. As pointed out by Turner (1971a), and implicit in Brereton (1970), only knowledge of the composition of an arbitrary mixture of atmospheric argon and calcium-derived argon is required for an exact correction to be made in the case of terrestrial samples. Because the ^{36}Ar found during experiments on extraterrestrial samples consists of a mixture of atmospheric, trapped, and cosmogenic argon, the usual correction for atmospheric ^{40}Ar using the ^{36}Ar cannot be applied, and it becomes important to have direct information on the $(^{40}$Ar/^{37}Ar$)_{Ca}$ production ratio. Turner (1972) derived an upper limit of 4×10^{-3} for this parameter from experimental data, and, subsequently, Turner et al. (1973) provided evidence that $(^{40}$Ar/^{37}Ar$)_{Ca} \leq 6 \times 10^{-4}$, a value that is low enough that production of ^{40}Ar from calcium is unlikely to cause significant interference even when analyzing lunar anorthosites.

Production of ^{39}Ar from calcium relative to the ^{37}Ar produced from the same target element is determined readily by analysis of the argon extracted from the irradiated calcium salt. As both ^{39}Ar and ^{37}Ar are derived only from calcium isotopes in the case of a pure calcium salt, the correction factor, $(^{39}$Ar/^{37}Ar$)_{Ca}$, is obtained directly from the isotopic analysis (after corrections for ^{37}Ar and ^{39}Ar decay), without the need for simplifying assumptions.

The range of $(^{36}$Ar/^{37}Ar$)_{Ca}$ production ratios is relatively small, mainly lying between 2×10^{-4} and 3×10^{-4}, with a mean value of $2.7(\pm 0.7) \times 10^{-4}$ (Table 3.4). Similarly, the measured $(^{39}$Ar/^{37}Ar$)_{Ca}$ production ratios fall into a restricted range mainly between 6.4×10^{-4} and 9.4×10^{-4}; a mean value of $7.5(\pm 1.0) \times 10^{-4}$ is calculated from the data in Table 3.4. Although the variations of the measured production ratios ($\equiv$ correction factors) are significant, they are relatively small in comparison with the very large range of neutron fluxes and neutron flux ratios found in the irradiation facilities utilized, and virtually irrespective of whether or not cadmium is used to shield the sample from thermal neutrons. The last observation implies that fast neutrons are required for the reactions producing ^{36}Ar, ^{37}Ar, and ^{39}Ar from calcium. This similarity of these correction factors in all reactor facilities used for ^{40}Ar/^{39}Ar dating purposes is a consequence of both this finding and the expectation that the energy spectrum for fast neutrons produced from ^{235}U is likely to be similar in all reactors using this material as fuel.

Note that the Q values for the reactions producing ^{36}Ar, ^{37}Ar, and ^{39}Ar from calcium range from -7.0 to $+1.75$ MeV (Table 3.3). When the Coulomb energy barrier is taken into account, possibly as high as 2 MeV (Tetley et al., 1980), it is evident that fast neutrons are necessary for the reactions to proceed, in keeping with the empirical observations. Nevertheless, the variation in the correction factors determined for an individual irradiation facility at various times and the differences found between reactors are greater than the uncertainties associated with the measurements. Perhaps this is to be explained in terms of some variability in the fast neutron flux energy spectrum. As noted by Roddick (1983) the variability in the reported $(^{39}$Ar/^{37}Ar$)_{Ca}$ correction factor might simply reflect minor surface contamination of the calcium salt by potassium. His results showed that the small amount of gas ($< 3\%$) released from irradiated CaF$_2$ when heated at $\sim 750°$C had a variably higher ^{39}Ar/^{37}Ar ratio compared with the much more uniform ratio found for the gas released during the main extraction at higher temperatures.

A consequence of the small range of values found for the $(^{36}$Ar/^{37}Ar$)_{Ca}$ and $(^{39}$Ar/^{37}Ar$)_{Ca}$ correction factors is that in many cases it would be reasonable to assume average values, especially when measuring samples of high K/Ca ratio. Clearly, however, the prudent approach is to measure the correction factors directly for each irradiation facility used for ^{40}Ar/^{39}Ar dating purposes, and this becomes essential when age measurements are to be made on samples with low K/Ca ratios.

Production of ^{38}Ar from ^{42}Ca (Table 3.3) does not cause any interference in ^{40}Ar/^{39}Ar dating, but, especially when studying extraterrestrial samples, it is desirable to know the $(^{38}$Ar/^{37}Ar$)_{Ca}$ correction factor. This factor also can be determined from the isotopic analysis of the argon extracted from irradiated pure calcium salt, after correcting for atmospheric ^{38}Ar. Where available, the $(^{38}$Ar/^{37}Ar$)_{Ca}$ correction factor is listed in Table 3.4. Because of the large negative Q value for the ^{42}Ca(n,nα)^{38}Ar reaction (Table 3.3), fast neutrons are necessary for this reaction to proceed. Although values of the $(^{38}$Ar/^{37}Ar$)_{Ca}$ correction factor are small, there is a surprisingly large variation in reported values, also possibly explained in terms of differences in the neutron flux energy spectrum, or perhaps more simply by the presence of trace amounts of chlorine in the calcium salt.

In the same way as ^{39}Ar$_K$ can be used to provide an estimate of the actual potassium content of an unknown sample, the decay-corrected ^{37}Ar may be utilized to determine the calcium content of a sample, provided that the calcium abundance of the flux monitor has been measured accurately. Approximate production rates of ^{37}Ar from calcium are listed in Table 3.2 for the various irradiation facilities, derived from the relevant publications. Particularly noteworthy is that the production rates per unit mass of calcium average 0.52 ± 0.02 relative to the production rates per unit mass of potassium, excluding the value reported for reactor RRF (flux trap), which appears to be anomalous. The sympathetic variation of production rates of ^{39}Ar$_K$ and ^{37}Ar$_{Ca}$ indicates that the neutron capture cross section versus neutron energy curves must be remarkably similar for both the ^{39}K(n,p)^{39}Ar and ^{40}Ca(n,α)^{37}Ar reactions (cf. Fig. 3.4), or that the neutron energy spectrum in each reactor is essentially the same.

3.7.3 Reactions on potassium

The ^{39}K(n,p)^{39}Ar reaction, of paramount importance in ^{40}Ar/^{39}Ar dating, has been discussed in Section 3.6, and will not be considered further here.

The ^{39}K(n,nd)^{37}Ar reaction has a large negative Q value (Table 3.3), and therefore requires very high energy neutrons for it to occur. Thus, extremely small amounts of ^{37}Ar are expected to be produced from potassium during neutron irradiation in a reactor. This is borne out by the small value of the correction factor, $(^{37}$Ar/^{39}Ar$)_K$, listed in Table 3.4 for those few irradiation facilities for which it has been reported. The results from HFBR and HIFAR, in particular, serve to confirm that the production of ^{37}Ar from ^{39}K is indeed very low.

Some ^{38}Ar can be generated from neutron interactions on potassium (Table 3.3), but its presence has no direct effect on measurement of ^{40}Ar/^{39}Ar ages. Nevertheless, values for the correction factor $(^{38}$Ar/^{39}Ar$)_K$, where reported in the literature, are listed in Table 3.4 for completeness. Note that the values lie in a restricted range, interpreted as indicating that the reactions require fast neutrons for them to proceed, as expected from the negative Q values (Table 3.3). Knowledge of the production ratio, however, is of some importance in the case of extraterrestrial samples, in relation to calculation of exposure ages.

Neutron-induced ^{40}Ar is produced from potassium during irradiation in a reactor, and in some cases a quite significant correction needs to be made. The evidence to be discussed below, indicates that the main reaction is ^{40}K(n,p)^{40}Ar. A production ratio or correction factor, $(^{40}$Ar/^{39}Ar$)_K$, generally is determined by measurement of the isotope ratios of argon extracted from a pure potassium salt, usually K$_2$SO$_4$, that has been irradiated in the particular facility utilized for the ^{40}Ar/^{39}Ar dating studies. As no known reaction in a ^{235}U fission neutron spectrum produces significant ^{36}Ar from potassium during irradiation, all the ^{36}Ar found in the analysis is attributed to atmospheric argon, the equivalent amount of atmospheric ^{40}Ar calculated, and the balance of the ^{40}Ar accepted as neutron-induced. Thus

$$(3.5) \quad \left[\frac{^{40}\text{Ar}}{^{39}\text{Ar}}\right]_K = \frac{^{40}\text{Ar}_m - {^{36}\text{Ar}_m} \times 295.5}{^{39}\text{Ar}_m}$$

where ^{40}Ar$_m$, ^{36}Ar$_m$, and ^{39}Ar$_m$ are the measured ^{40}Ar, ^{36}Ar, and ^{39}Ar, respectively, and 295.5 is the ^{40}Ar/^{36}Ar ratio in atmospheric argon. In a pure potassium salt free of any contaminating calcium, all the ^{39}Ar will be derived from potassium so that $^{39}\text{Ar}_m = {}^{39}\text{Ar}_K$.

From the results listed in Table 3.4 it is seen that this correction factor has values that range over more than two orders of magnitude for different irradiation facilities. Obviously, therefore, it is of great importance that the correction factor be determined for each reactor facility used for ^{40}Ar/^{39}Ar dating experiments.

The reason for the large variation in the production of neutron-induced ^{40}Ar is now relatively well understood from experiments reported by Tetley et al. (1980), and from theoretical considerations. These authors showed that for HIFAR the full range of reported correction factors could be reproduced in a single irradiation facility, located in the heavy water moderator of this reactor, just outside the core. It was demonstrated that the $(^{40}\text{Ar}/^{39}\text{Ar})_K$ correction factor was ~ 0.3 when pure potassium salt was irradiated in the ambient neutron flux in this facility (X33), where the slow/fast neutron flux ratio is reported to be ~ 14 (Hickman, 1958), but we now believe the average value may be ~ 50. Thus, approximately one atom of neutron-induced ^{40}Ar is produced for every three atoms of ^{39}Ar generated from ^{39}K. Use of cadmium shielding to absorb slow neutrons reduced the production of ^{40}Ar dramatically (Table 3.4), without affecting the ^{39}Ar production. With 0.5 mm or greater thickness of cadmium shielding, the production of neutron-induced ^{40}Ar was lowered to 0.003 relative to $^{39}\text{Ar}_K$, essentially the value observed in the very hard (i.e., low thermal/fast neutron flux ratio) neutron spectrum of HFBR. These results were interpreted by Tetley et al. (1980) as indicating that neutron-induced ^{40}Ar is produced principally by the ^{40}K(n,p)^{40}Ar reaction induced mainly by slow rather than fast neutrons, in keeping with the positive Q value of 2.3 MeV. The increase in value of the $(^{40}\text{Ar}/^{39}\text{Ar})_K$ correction factor with increasing slow/fast neutron flux ratio observed in irradiation facilities in general (Table 3.4) is consistent with the findings from HIFAR. Indeed, departures from the trend probably simply indicate that the reported flux ratios for some reactors do not reflect the true values for this parameter, as was also concluded by Roddick (1983).

It will be shown subsequently that the $(^{40}\text{Ar}/^{39}\text{Ar})_K$ correction factor becomes increasingly important as the value of the ^{40}Ar/^{39}Ar in a sample decreases, for example, when measuring young samples with little ^{40}Ar*. In the case of a sample with measured ^{40}Ar/^{39}Ar = 1.0, if the $(^{40}\text{Ar}/^{39}\text{Ar})_K = 0.3$, as found for the ambient neutron flux in facility X33 of HIFAR, the correction to the measured ^{40}Ar/^{39}Ar would be $\sim 30\%$, because the correction factor essentially is simply subtracted from the measured ratio [see Eq. (3.42)]. For $(^{40}\text{Ar}/^{39}\text{Ar})_K = 0.003$, the correction to the same measured ^{40}Ar/^{39}Ar would be only $\sim 0.3\%$. Clearly, if the $(^{40}\text{Ar}/^{39}\text{Ar})_K$ correction factor is not accurately determined, major systematic error in the measured age could result, especially when measuring young samples.

If a soft spectrum (i.e., high slow/fast neutron ratio) needs to be utilized for the irradiations, the simplest solution to the problem of high neutron-induced ^{40}Ar production is to encase the samples in cadmium (≥ 0.2 mm Cd) so that the $(^{40}\text{Ar}/^{39}\text{Ar})_K$ correction factor is reduced to an acceptably low level. This has the added advantage of decreasing the overall, largely unwanted, slow neutron generated activity from other elements in the sample, thereby reducing the health risks in handling irradiated samples in the laboratory. A potential problem, but one that has not caused any serious difficulties as far as we are aware, is that absorption of thermal neutrons in the cadmium results in liberation of energy as heat, so that the sample being irradiated experiences higher temperatures than would otherwise be the case. For HIFAR, the relatively small amount of cadmium used for shielding the sample during irradiation, usually 0.2 mm thick in the form of a cylinder 34 × 11 mm, comprising ~ 3 g, does not interfere with the normal operation of the reactor. For reactor facilities with relatively hard neutron spectra, the $(^{40}\text{Ar}/^{39}\text{Ar})_K$ correction factors are sufficiently low (say < 0.02) that cadmium shielding is not required, and adjustment of irradiation times provides the best means of minimizing uncertainties in a measured age owing to application of this correction (see Section 3.8).

3.7.4 Reactions on chlorine

Mitchell (1968a) recognized that activation of the chlorine isotopes, ^{35}Cl and ^{37}Cl, during neutron irradiation, would yield ^{36}Cl and ^{38}Cl, respectively, with subsequent β^- decay to produce ^{36}Ar and ^{38}Ar, respectively (Table 3.3). Because of the slow decay of ^{36}Cl to ^{36}Ar ($t_{1/2} = 3 \times 10^5$ a), the amount of ^{36}Ar produced from neutron interactions on ^{35}Cl would not normally cause an increase in ^{36}Ar sufficient to introduce an error in the application of the atmospheric ^{40}Ar correction based upon ^{36}Ar. Nevertheless, Bernatowicz et al. (1978) and Roddick (1983) showed that for samples with relatively high $Cl/^{40}Ar^*$, as found in some hornblendes, for example, delays between irradiation of a sample and extraction and analysis of the argon in the order of a year can result in sufficient ^{36}Ar being produced from decay of ^{36}Cl, that errors of a few percent in the reported age can be expected, unless due allowance is made.

The (n,γ) reactions on the chlorine isotopes have high positive Q values (Table 3.3), so that they are exoergic, and require only thermal neutrons to proceed. Roddick (1983) calculated from the natural abundance of the chlorine isotopes and the measured thermal neutron capture cross sections that the production of ^{36}Cl relative to ^{38}Cl is about 316:1. Because of the rapid decay of ^{38}Cl to ^{38}Ar ($t_{1/2} = 37.3$ min), this isotope normally will have completely decayed to ^{38}Ar by the time of analysis. Thus measurement of ^{38}Ar can provide a useful method of determining the chlorine content of a sample, especially in terrestrial materials, but also, in principle allows corrections to be calculated for ^{36}Ar produced from the decay of ^{36}Cl. For KCl irradiated in the Herald reactor, facility G7N, Roddick (1983) found $^{38}Ar_{Cl}/^{39}Ar_K = 3.21$ with production rates of 3.1×10^{-10} mol $^{38}Ar/g$ Cl-h and 8.8×10^{-11} mol $^{39}Ar/g$ K-h, so that ^{38}Ar is produced at more than three times the rate of ^{39}Ar per unit mass of parent element in this irradiation facility. Roddick (1983) also showed that the ^{38}Ar production from ^{37}Cl is proportional to the thermal neutron fluence, in an analogous manner to that found for reactor-induced ^{40}Ar production from ^{40}K by Tetley et al. (1980). Thus, as in the case of neutron-induced ^{40}Ar production, the yield of ^{38}Ar and potential ^{36}Ar from chlorine can be minimized by use of cadmium shielding to absorb thermal neutrons. In terms of $^{40}Ar/^{39}Ar$ dating, this provides a further reason for using cadmium shielding where practicable. With appropriate shielding, the interference from ^{36}Ar produced from ^{35}Cl via ^{36}Cl would be reduced to insignificant levels, even with protracted delays (≥ 1 year) between irradiation and analysis. However, should it be necessary to determine chlorine abundances accurately, irradiations might well be done without cadmium shielding to provide a very high sensitivity. In any case, the desirability of determining production rates for argon isotopes from chlorine is indicated by the work of Roddick (1983), particularly as it has been shown that each irradiation facility will have different production rates, related to the thermal neutron flux.

For extraterrestrial samples, such as meteorites and lunar rocks, exposed to cosmic rays, a significant amount of ^{38}Ar also can be produced by spallation, and its measurement can yield cosmic ray exposure ages in those cases in which the chlorine content is demonstrated to be low, and provided trapped and atmospheric ^{38}Ar can be properly allowed for.

3.8 OPTIMIZATION OF IRRADIATION PARAMETERS

3.8.1 Introduction

Turner (1971a) argued that, in principle, the $^{40}Ar/^{39}Ar$ dating method can be applied to samples covering essentially the same age range as the conventional K–Ar method. For this to be achieved in practice a number of requirements need to be met. First, the amount of sample chosen for an experiment obviously must be large enough to provide sufficient $^{40}Ar^*$ for accurate measurement. Second, during irradiation of the sample, adequate ^{39}Ar must be produced to enable its measurement relative to ^{40}Ar at an appropriate precision. Third, the various interferences, and especially the production of reactor-induced ^{40}Ar and ^{36}Ar, must be minimized by judicious choice of the integrated neutron flux (fluence) to which the sample is subjected. Following Turner (1971a) we shall discuss these and other requirements for optimizing the irradiation in the following sections. These considerations show that the optimum fluence is directly related to the age of the sample, but independent of the potassium content, with further constraints imposed by the K/Ca ratio of the sample.

3.8.2 Sample size

The amount of sample for an experiment should be chosen so that there is sufficient ^{40}Ar* available to enable a ^{40}Ar/^{39}Ar age spectrum to be measured with appropriate resolution, as would be the case, for example, if the argon was extracted in a minimum of 10 approximately equal steps. Using an argon extraction line which has a relatively low blank (say ^{40}Ar $\simeq 10^{-13}$ mol), and a mass spectrometer of good sensitivity (say $\sim 3 \times 10^{-15}$ mol/mV), it is possible to measure a 10-step age spectrum on $\sim 1 \times 10^{-11}$ mol ($\sim 2 \times 10^{-7}$ cm^3 STP) total ^{40}Ar*. The amount of sample required to yield this volume of ^{40}Ar* can be read off the curves given in Fig. 3.6. Clearly other minimum amounts of ^{40}Ar* can be calculated for mass spectrometers of different sensitivity. Note, however, that handling of an order of magnitude greater amount of ^{40}Ar* permits an age spectrum to be measured much more readily and with greater resolution, should that be required.

Samples ranging up to ~ 1 g can be conveniently irradiated, and subsequently run in an argon extraction system; larger masses may be used but difficulties can arise in maintaining both uniform neutron doses during irradiation and temperatures over the sample during step heating.

It can be seen from Fig. 3.6 that for samples older than 10^8 years and potassium content greater than 0.06%, 1 g of material contains more than sufficient ^{40}Ar* for age spectrum measurement, and indeed with increasing age, progressively smaller amounts of sample are required. For a sample of age 10^7 years there is sufficient ^{40}Ar* in a 1 g aliquant to measure an age spectrum, provided the potassium content is greater than $\sim 0.6\%$. For younger samples, say of age $\sim 10^6$ years, only high potassium minerals ($>6\%$ K) yield enough ^{40}Ar* from 1 g to be able to obtain an age spectrum readily (Fig. 3.6). For samples of lower potassium

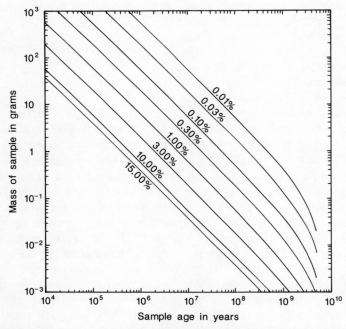

Fig. 3.6. Mass of sample required to yield 1×10^{-11} mol of radiogenic argon for a given age. Potassium content (weight percent) is indicated on the curves.

content and/or younger age, it may be possible to irradiate a greater amount of sample, but, with current gas extraction systems and procedures, it is difficult to handle more than a few grams of material. A more promising avenue is to increase the sensitivity of the mass spectrometer by using a more sensitive ion source, or by employing an electron multiplier at the collector. However, background in the machine and the magnitude of the blank from the extraction line provide limits as to how far this enhancement of sensitivity can be usefully taken. As in the conventional K–Ar dating method, the ultimate limiting factor relates to the detection of a small $^{40}Ar^*$ component above a background of atmospheric argon derived from the sample itself and from the blank of the extraction system.

3.8.3 Production of sufficient ^{39}Ar

The requirement that there be enough ^{40}Ar present for analysis commonly can be met by choosing an appropriate sample size. The condition that there be sufficient ^{39}Ar produced is best realized by setting an upper limit for the

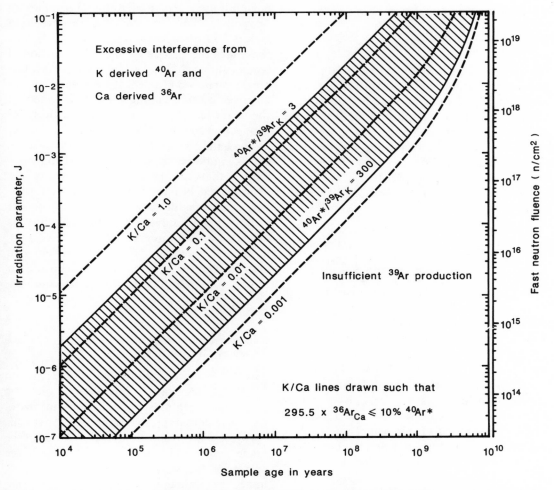

Fig. 3.7. Diagram showing the irradiation parameter, J, proportional to the fast neutron fluence, required to optimize ^{39}Ar production from ^{39}K for differing age of sample (shaded area), while minimizing ^{40}Ar and ^{36}Ar interference effects from neutron interactions on potassium and calcium. Insufficient ^{39}Ar production occurs for irradiations to the right of the shaded area, and excessive interference from $^{40}Ar_K$ occurs for irradiations to the left of the shaded area. Dashed curves show the maximum J for the indicated K/Ca ratio for minimization of $^{36}Ar_{Ca}$ interference. For convenience the fast neutron fluence also is given on the right-hand vertical axis, based upon the measured fluence of neutrons with energies >0.6 MeV in the GSTR reactor (Dalrymple et al., 1981). For other reactors the correlation between J and measured fast neutron fluence is likely to differ slightly. Diagram modified from that of Turner (1971a).

^{40}Ar*/^{39}Ar$_K$ ratio, and irradiating accordingly. Turner (1971a) suggested a ratio of 300 as a reasonable maximum value, and this remains a sensible choice. Although higher ratios, up to ~1000, can be measured on a gas sample with acceptable precision (say, ~1%), provided that the mass spectrometer has adequate resolution, low background (especially at $m/e = 39$), and the ion beam size is appropriate, it is desirable to aim for a maximum value of ~300.

Thus, from Eq. (2.15), we obtain

$$(3.6) \qquad \frac{^{40}\text{Ar}^*}{^{39}\text{Ar}_K} = \frac{(\exp \lambda t) - 1}{J} \leq 300$$

Rearranging we find

$$(3.7) \qquad J \geq \frac{(\exp \lambda t) - 1}{300}$$

On this basis, the lower full curve in Fig. 3.7 is derived, separating fields of insufficient and sufficient ^{39}Ar production as a function of the age of a sample. It is important that this same condition be satisfied for the flux monitor as well as for the unknown being measured as they normally will differ in age.

3.8.4 Minimization of reactor-induced ^{40}Ar interference

The previous sections establish the criteria for adequate production of ^{39}Ar for measurement relative to the expected ^{40}Ar*. It is also necessary to limit the fluence the sample receives so that the correction for reactor-derived ^{40}Ar from the ^{40}K(n,p)^{40}Ar reaction does not inhibit the precise measurement of age.

This requirement is related to the magnitude of the (^{40}Ar/^{39}Ar)$_K$ correction factor, determined from measurement of an irradiated pure potassium salt. The correction factor appears to vary by as much as 50% for a given irradiation facility, even when similar irradiation conditions are utilized (Table 3.4). This problem can be minimized by measuring the correction factor in each irradiation undertaken, but a more practical solution is to ensure that the correction remains less than some specified small proportion of the ^{40}Ar*, say 1% (Turner, 1971a). This condition is satisfied if

$$(3.8) \qquad \frac{^{40}\text{Ar}^*}{^{39}\text{Ar}_K} = \frac{(\exp \lambda t) - 1}{J} \geq 100 \left[\frac{^{40}\text{Ar}}{^{39}\text{Ar}} \right]_K$$

or

$$(3.9) \qquad J \leq \frac{(\exp \lambda t) - 1}{100(^{40}\text{Ar}/^{39}\text{Ar})_K}$$

Expressed in terms of the irradiation parameter, J, and the age, this is shown as the upper solid curve on Fig. 3.7, separating the field of excessive neutron-induced ^{40}Ar production from that of acceptably small interference from reactor-produced ^{40}Ar. The curve has been drawn for a (^{40}Ar/^{39}Ar)$_K$ correction factor of 300×10^{-4}, and readily can be adjusted for other values for this parameter. The basis for this choice is that the majority of reactor facilities used for irradiation of samples in relation to ^{40}Ar/^{39}Ar dating have correction factors less than 300×10^{-4} (Table 3.4). For those facilities that show higher production of reactor-induced ^{40}Ar, owing to a slow/fast flux greater than ~5 (Table 3.4), consideration can be given to using cadmium shielding to reduce the incidence of slow neutrons and, thus, the production rate for ^{40}Ar.

The hatched area between the two curves in Fig. 3.7 indicates the range of irradiation parameters that can be chosen relative to the expected age of the sample to be dated for the two conditions discussed to be satisfied. The range of acceptable irradiation parameters extends over two orders of magnitude, so that, in general, there should be little difficulty in ensuring choice of an appropriate fluence, even if the sample and flux monitor differ considerably in age.

3.8.5 Minimization of interference from ^{40}Ca(n,nα)^{36}Ar reaction

The ^{36}Ar produced from the reaction ^{40}Ca(n,nα)^{36}Ar causes interference with the correction for atmospheric ^{40}Ar by means of the ^{36}Ar found in the sample. Clearly this correction for reactor-induced ^{36}Ar must be made as precisely as possible. Although the correction factor, (^{36}Ar/^{37}Ar)$_{Ca}$, can be determined to within a few percent for a given irradiation facility (Table 3.4), it would seem sensible to accept a maximum desirable interference of, say, 10% of the ^{40}Ar*/^{39}Ar ratio, after the atmospheric ^{40}Ar correction has been applied. In this we follow Dalrymple et al.

(1981) rather than the stricter condition suggested by Turner (1971a). As the $(^{40}Ar/^{36}Ar)_A = 295.5$, then this condition requires that

(3.10) $\qquad ^{40}Ar^*/^{36}Ar_{Ca} \geqq 2955$

Following Turner (1971a) we can modify Eq. (2.11) to read

(3.11) $\qquad ^{40}Ar^* = K \dfrac{^{40}K}{K} \dfrac{\lambda_e + \lambda'_e}{\lambda} [(\exp \lambda t) - 1]$

where K refers to total potassium, and the other symbols remain as before.

Now

(3.12) $\qquad ^{36}Ar_{Ca} = S_{36Ca} Ca \, \phi$

where S_{36Ca} is an effective cross section for the production of ^{36}Ar from calcium (Ca) for neutrons with a ^{235}U fission spectrum, and ϕ is the integrated fast neutron flux (fluence).

Dividing Eq. (3.11) by Eq. (3.12) we obtain

$\dfrac{^{40}Ar^*}{^{36}Ar_{Ca}} = \dfrac{1}{S_{36Ca}\phi} \dfrac{K}{Ca} \dfrac{^{40}K}{K} \dfrac{\lambda_e + \lambda'_e}{\lambda} [(\exp \lambda t) - 1]$

(3.13) $\qquad \geqq 2955$

Rearranging

$\phi \leqq \dfrac{1}{2955 \, S_{36Ca}} \dfrac{K}{Ca} \dfrac{^{40}K}{K} \dfrac{\lambda_e + \lambda'_e}{\lambda}$

(3.14) $\qquad \times [(\exp \lambda t) - 1] \quad n/cm^2$

Because J, the irradiation parameter, is proportional to ϕ, we may replace ϕ by J/C, where C is a constant for a given reactor facility. Thus,

$J \leqq \dfrac{C}{2955 \, S_{36Ca}} \dfrac{K}{Ca} \dfrac{^{40}K}{K} \dfrac{\lambda_e + \lambda'_e}{\lambda} [(\exp \lambda t) - 1]$

(3.15)

Note that this equation is of the same form as Eq. (3.9), but that there is the additional dependence upon the K/Ca ratio of the sample.

Turner (1971a) has reported an effective cross section for ^{37}Ar production from calcium of 28 ± 5 mb, based upon measurements in the Herald reactor. Calculations for other reactor facilities yield similar results, as might be anticipated, so that this figure can be applied universally. We can obtain an effective cross section for ^{36}Ar production (S_{36Ca}) from the relative production rates of ^{36}Ar and ^{37}Ar listed in Table 3.4. Although there is a significant variation of the $(^{36}Ar/^{37}Ar)_{Ca}$ between irradiation facilities (Table 3.4), for the purposes of calculation we shall choose a value of 3.0×10^{-4}, near the upper limit of the observed range, in order to account for the higher production rates. On this basis we derive a value for $S_{36Ca} = 8.4 \times 10^{-3}$ mb. As $^{40}K/K = 1.167 \times 10^{-4}$, $\lambda_e + \lambda'_e = 0.581 \times 10^{-10}/a$, $\lambda = 5.543 \times 10^{-10}/a$, and $S_{36Ca} = 8.4 \times 10^{-30} \, cm^2$, we find from Eq. (3.15)

(3.16) $\quad J \leqq 4.9 \times 10^{20} \, C(K/Ca)[(\exp \lambda t) - 1]$

In Fig. 3.7 curves (dashed lines) are shown for the maximum irradiation permitted, in terms of the J parameter and age, to meet this condition for the K/Ca ratios indicated. These curves are based upon the value for the constant, C, of 0.4×10^{-20} for the GSTR reactor, as reported by Dalrymple et al. (1981). Although it might be expected that this constant would be similar for irradiation facilities in all reactors used for $^{40}Ar/^{39}Ar$ dating, values ranging over a factor of ~ 5 have been given in the literature. This probably arises mainly because of different criteria for measuring or defining the fast neutron flux, rather than because there are real differences in the value of C. The GSTR value is used because it is well documented and lies near the middle of the reported range. Nevertheless, the curves can be recalculated readily for other values of C, if required, but any changes will cause the curves to move vertically only by minor amounts.

It can be seen from Fig. 3.7 that for $K/Ca \geqq 0.002$, an irradiation can be chosen that satisfies this condition as well as that required to produce sufficient ^{39}Ar, as discussed previously. For samples with K/Ca ratios <0.002, if the ^{39}Ar production is to be sufficient to meet the condition set out previously, then a correction of $>10\%$ for calcium-derived ^{36}Ar has to be accepted. As the $(^{36}Ar/^{37}Ar)_{Ca}$ correction factor can be determined quite precisely, then, in principle, it is possible to make the necessary corrections, even if the $^{36}Ar_{Ca}$ production is larger than the desirable upper limit discussed above. The alternative is to tolerate a $^{40}Ar^*/^{39}Ar_K$ ratio of >300. In such circumstances a judgment needs to be made as to the best compromise.

The problem of $^{36}Ar_{Ca}$ interference becomes particularly severe when attempting to date young samples that have low K/Ca ratios. In such cases the low proportion of $^{40}Ar^*$ relative to the total ^{40}Ar can result in small errors in applying the correction for $^{36}Ar_{Ca}$ causing large systematic errors in the calculated age, exacerbated by marked error magnification effects. These combined effects constitute a major limitation in the application of the $^{40}Ar/^{39}Ar$ dating method to young samples of plagioclase, hornblende, and calcium-rich whole rocks, say < 10 Ma old. For calcium-rich plagioclase from potassium-poor rocks, as found, for example, in ophiolitic complexes, the K/Ca ratio may be < 0.001, and the potassium content < 0.01%. In such an extreme case, the presence of little $^{40}Ar^*$ relative to atmospheric argon, combined with high $^{36}Ar_{Ca}$ production, may preclude reliable measurement of $^{40}Ar/^{39}Ar$ ages, even on samples as old as 50 Ma. For much older samples of low K/Ca ratio, the greater amount of $^{40}Ar^*$ present, and the higher proportion of $^{40}Ar^*$ relative to total ^{40}Ar, removes one of these limitations, although requiring a longer irradiation to produce sufficient ^{39}Ar for precise measurement.

3.8.6 Interference from the $^{42}Ca(n,\alpha)^{39}Ar$ reaction

Some ^{39}Ar is produced from calcium during irradiation so it is important to determine the magnitude of its effect on the potassium-derived ^{39}Ar. We have seen already that the relative production rates for $^{39}Ar_K$ and $^{37}Ar_{Ca}$ are similar for all irradiation facilities used for $^{40}Ar/^{39}Ar$ dating purposes. From Section 3.7.2

$$(3.17) \qquad \frac{^{39}Ar_K}{^{37}Ar_{Ca}} = \frac{1}{0.52(\pm 0.02)} \frac{K}{Ca}$$

Similarly, the range of production rates of ^{39}Ar and ^{37}Ar from calcium was found to be relatively small, with a mean value of $7.5(\pm 1.0) \times 10^{-4}$ for $(^{39}Ar/^{37}Ar)_{Ca}$, derived from Table 3.4. Thus,

$$(3.18) \qquad ^{37}Ar_{Ca} = \frac{^{39}Ar_{Ca}}{7.5(\pm 1.0) \times 10^{-4}}$$

Substitution in Eq. (3.17) gives

$$(3.19) \qquad \frac{^{39}Ar_K}{^{39}Ar_{Ca}} = [2.6(\pm 0.4) \times 10^3] \frac{K}{Ca}$$

This indicates that the relative contribution of $^{39}Ar_{Ca}$ is independent of the fluence, depending only upon the K/Ca ratio of the sample. If we wish to keep the $^{39}Ar_{Ca}$ production to less than 1% of the $^{39}Ar_K$ production (i.e., $^{39}Ar_K/^{39}Ar_{Ca} \geqq 100$), then from Eq. (3.19) we obtain

$$(3.20) \qquad K/Ca \geqq 0.039 \pm 0.005$$

Thus, for K/Ca ratios < 0.04, we must accept a correction > 1% to the ^{39}Ar from $^{39}Ar_{Ca}$. For example, for K/Ca = 0.004, a correction of ~10% to the measured ^{39}Ar is required. However, as the correction factor $(^{39}Ar/^{37}Ar)_{Ca}$ can be measured quite precisely for a given irradiation facility, often to within 1%, the correction for $^{39}Ar_{Ca}$ can be made without difficulty through the measured ^{37}Ar (decay corrected) in the sample. Even a 10% uncertainty in the correction factor would cause only a 1% uncertainty in the calculated $^{40}Ar^*/^{39}Ar_K$, and, hence, age, when the K/Ca of the sample is 0.004. In general, interference from $^{39}Ar_{Ca}$ is not a serious problem, although it can become significant when measuring $^{40}Ar/^{39}Ar$ ages on low potassium, low K/Ca samples, such as calcic plagioclases, especially if they are relatively young.

3.8.7 Other interferences

Interference from ^{40}Ar produced by neutron interactions on calcium is effectively corrected for in the case of terrestrial samples, as discussed previously, and generally is not a serious problem when dating extraterrestrial samples. Turner (1971a) showed that production of ^{36}Ar is likely to be quite negligible from neutron interactions on potassium. Production of ^{36}Ar from Cl can be significant in some circumstances (see Section 3.7.4) and should be corrected for if necessary (cf. Roddick, 1983). Finally, Turner (1971a) argued that the effect of neutron interactions in removing argon isotopes from a sample during irradiation is negligible, even for fluences as great as 10^{19} n/cm^2.

3.8.8 Conclusions

Interferences from unwanted argon isotopes, produced by neutron interactions on calcium, potassium, and chlorine can be minimized by judicious choice of irradiation parameter, J. For most samples, it is possible to adequately correct for these interferences. The optimum fluence for a given sample is related to its age and also to its K/Ca ratio, as shown in Fig. 3.7, which provides a guide to the appropriate irradiation parameter, J, to be aimed for, and the approximate fast neutron fluence required. Difficulties can arise, however, in dating samples of low K/Ca ratio, say <0.001, especially if the samples are young. In principle, the ^{40}Ar/^{39}Ar dating method can be applied over the same age range as conventional K–Ar dating; there is no older limit, but the detection of a small amount of ^{40}Ar* in a much larger amount of ^{40}Ar, mainly atmospheric, ultimately determines the younger limit of application.

3.9 NEUTRON FLUX GRADIENTS

It is expected that any irradiation facility in a nuclear reactor is likely to exhibit a neutron flux gradient across it. Reported gradients of the fast neutron flux component range from as low as 0.5%/cm in a facility in the Karlsrühe FR-2 reactor (Kirsten et al., 1972) to ~13%/cm in the irradiation facility employed for ^{40}Ar/^{39}Ar dating purposes in the HIFAR reactor (McDougall, 1974). Such gradients obviously must be properly monitored if accurate ^{40}Ar/^{39}Ar ages are to be obtained. Most workers go to considerable trouble to ensure that flux variations are suitably measured and taken into account.

The most commonly utilized procedure is simply to irradiate monitor minerals in known geometric relation to the unknown samples, so that it is possible to interpolate and obtain an appropriate J (irradiation parameter) value for each unknown. This approach is very satisfactory because monitoring is done using the identical nuclear reaction as that of interest in the sample whose age is to be measured.

An alternative method, used in some laboratories in conjunction with a mineral flux monitor, is to employ flux wire monitors, usually nickel wire, to map the spatial variation of the neutron flux. Wire monitors of this kind have the considerable advantage that they occupy a very small space, and, thus, can be placed within or in close proximity to samples. After irradiation the nickel wires are counted to determine the activity produced in the reaction ^{58}Ni(n,p)^{58}Co, from which a measure of the fast neutron fluence ($E > 3.0$ MeV) is obtained for each wire. The measured fluence is not necessarily equivalent to that which produces the ^{39}K(n,p)^{39}Ar reaction, because of differences in the excitation functions, although it is likely to be proportional to the fluence that causes the reaction of interest. Thus, provided a flux wire is in close proximity to a standard flux monitor mineral so that calibration can be made, the fluence variations observed from flux wires placed in other parts of the irradiation package can be used to derive an appropriate J value at these other locations. In practice it is possible to measure J using a flux monitor mineral to a precision of ~0.2%. In monitoring the fluence by means of flux wires it is obviously desirable, therefore, to aim for a similar counting precision.

To obtain a better understanding of, and solutions to, the problems associated with flux gradients, the fast neutron flux characteristics in the GSTR and HIFAR reactor facilities utilized for irradiations for ^{40}Ar/^{39}Ar dating purposes will be described briefly.

Dalrymple et al. (1981) documented, in considerable detail, the neutron flux gradients found within the central thimble of GSTR, the location where ^{40}Ar/^{39}Ar irradiations are done. This reactor is of the TRIGA type, with uranium–zirconium hydride fuel moderator elements arranged in a cylindrical core ~0.5 m in diameter, the fuel extending vertically for ~0.4 m. It is a water-cooled and reflected reactor that is able to operate continuously at 1 MW. The central thimble, located in the geometric center of the core, has a high fast to thermal neutron flux ratio of ~1.2 and a high fast neutron flux ($E > 0.6$ MeV) of ~1.7×10^{13} n/cm^2/s. The main fast neutron flux gradient in this facility is vertical, where changes as great as 3.5% cm are found (Fig. 3.8). Samples for irradiation are placed in an aluminum sample holder contained within a sealed aluminum vessel or tube of dimensions as great as 30 × 3.2 cm, which is lowered dir-

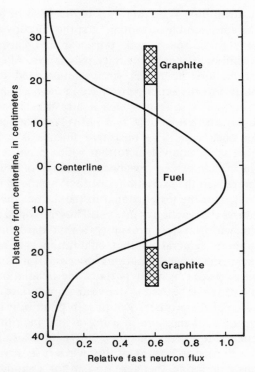

Fig. 3.8. Fast neutron flux gradient in the GSTR (TRIGA) reactor operated by U.S. Geological Survey. After Dalrymple et al. (1981), with permission.

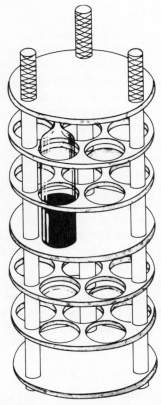

Fig. 3.9. Sample holder employed in the GSTR reactor. After Dalrymple et al. (1981), with permission. One sample in its sealed quartz vial is shown on the second level of the holder.

ectly into the central thimble of GSTR. The sample holder may have up to four levels in which individual samples, both flux monitors and unknowns, are placed in sealed flat-bottomed vials made of pure silica glass (Fig. 3.9). Vials ranging from as small as 3 mm to as large as 12 mm in diameter can be used with appropriate configurations of the sample holder. This provides considerable flexibility as to the amount of sample or flux monitor irradiated, as well as to the number of samples included at each level. The most common configuration for terrestrial samples consists of a ring of six 8-mm vials surrounding a seventh vial centrally located. To minimize the effects of the vertical flux gradient, the vials within a given level of the sample holder are filled with sample to the same height, usually 10 mm, and at least one flux monitor is included at each level. Dalrymple et al. (1981) showed that with this geometry, the fluence can be monitored routinely to $\sim 0.5\%$ (σ) for samples within a given horizontal level. Their experiments indicated that the fast neutron flux in the center of each level can be $\sim 0.5\%$ less than that in the outer ring of the sample holder, indicating the presence of small horizontal flux gradients. Experiments also were reported showing that nickel flux wire monitors provide a relative measure of the fast neutron flux to within uncertainties of $\sim 1\%$.

The irradiation facilities used for $^{40}Ar/^{39}Ar$ dating purposes in HIFAR are located just outside the core of the reactor in the heavy water moderator (Fig. 3.1), in a region where the gradient of the fast neutron flux is large (Fig. 3.3). The fast neutron flux decreases $\sim 13\%$/cm along the horizontal facility, away from the reactor core (Fig. 3.10). The seemingly rather difficult problem of flux monitoring has been successfully overcome simply by providing each unknown sample with its own flux monitor, centrally located within the sample (Fig. 3.11). The aluminum reactor vessel has usable internal dimensions of 34 × 11 mm, and

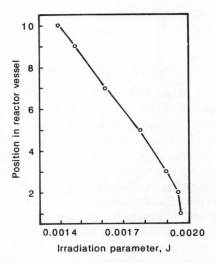

Fig. 3.10. Plot illustrating the large fast neutron flux gradient, reflected in the irradiation parameter, J, in facility X33 of the HIFAR reactor. After McDougall (1974). Position 1, at the base of the reactor vessel, was closest to the core of the reactor during the 60-h irradiation. Note that this flux gradient can be virtually eliminated by inverting the reactor vessel exactly half way through the irradiation.

a small indentation machined in its bottom to facilitate positive location. The outer container is then completely filled with the unknown sample and sealed with a lid. Thus, the sample to be dated concentrically surrounds the flux monitor, so that the geometry is such that the monitor might be expected to accurately reflect the fluence the unknown sample has received, irrespective of the magnitude or direction of the neutron flux gradient.

Using this configuration, and with the same standard sample in both inner and outer cans, Tetley (1978) showed that after irradiation the argon extracted from the samples in the two locations yielded $^{40}Ar^*/^{39}Ar_K$ ratios that agreed, on average, to better than 0.2%. This indicated that the geometry adopted provides very satisfactory flux monitoring, even in the extreme case such as that found in HIFAR where particularly large neutron flux gradients are observed. An obvious further improvement in operating procedure is to invert the reactor vessel, end for end, exactly half way through the irradiation. Using this approach in the HIFAR facilities, we find that the overall gradient recorded in the samples in the vessel is reduced to only 1–2%, instead of ~40%, as the major component of the neutron flux gradient is parallel to the length of the reactor can in the horizontal facilities employed. Thus, with this concentric configuration of sample and flux monitor, the neutron flux gradient is accounted for irrespective of the actual direction and magnitude of the gradient. No doubt this kind of geometry also could be used, with ad-

is shown in Fig. 3.11 with three sample cans each ~10 mm long. Each of these sample cans in turn has a smaller diameter, cylindrical container located centrally within it, in which the flux monitor normally is loaded. All of these sample containers are accurately machined out of aluminum. The small internal can is completely filled with the flux monitor, which is carefully tamped down. After sealing with its own aluminum lid, the inner can is placed in the center of the outer can, which has

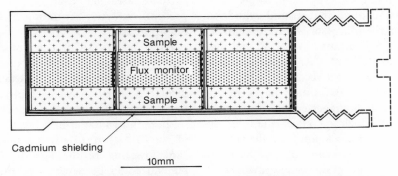

Fig. 3.11. Reactor vessel used in the HIFAR irradiation facilities X33 and X34. The flux monitor is placed in a machined container located centrally within a larger diameter container that is filled with the sample to be dated. A three can configuration is shown, but larger or smaller numbers of sample and flux monitor containers may be used. Cadmium, usually 0.2 mm thick, normally is employed to shield the samples and flux monitors from slow neutrons.

vantage, in irradiation facilities in other reactors, especially in those cases in which flux gradients are large. The system described has considerable flexibility in that longer or shorter containers can be manufactured in order that different amounts of material can be irradiated conveniently. Another alternative that can be utilized, where appropriate, is to place the unknown sample in the inner container and surround it by the flux monitor. As an approximate guide, for the geometry shown in Fig. 3.11, the inner can accommodates ~ 0.14 g of sample, and the outer can ~ 0.9 g.

3.10 SELF-SHIELDING

Mitchell (1968a) recognized that neutron absorption could mean that the inner parts of a sample might experience a lower neutron dose than the outer parts. To test whether self-shielding effects were of importance, Mitchell irradiated a sphere of muscovite 10 mm in diameter. Subsequent measurement of the $^{40}Ar^*/^{39}Ar_K$ ratio in gas extracted from material taken from the inner and outer parts of the sphere yielded results that agreed to within 2%. Mitchell (1968a) therefore concluded that self-shielding was not a serious source of error. Tetley (1978) irradiated a mineral sample in the form of small grains in an inner (~ 4 mm diameter) and outer can (~ 10 mm diameter), as described in Section 3.9, and found that the average difference in the measured $^{40}Ar^*/^{39}Ar_K$ ratio was $<0.2\%$ in three separate experiments. With this geometry it is evident that self-shielding is negligible, at least for granular material. Dalrymple et al. (1981), however, found that a solid cylindrical sample of dolerite (diabase) 24 mm in diameter and 25 mm high, irradiated in GSTR, and then analyzed, yielded $^{40}Ar^*/^{39}Ar_K$ results indicating that the inner parts received $\sim 3\%$ fewer fast neutrons than the outer parts of the cylinder. Their data suggested that for a solid sample the self-shielding effect may be detectable for distances >4 mm from the surface, but they believed the effect is likely to be less in samples of mineral grains and powders, in keeping with the observations made by Tetley (1978). Overall it appears that for granular samples the effect of self-shielding of neutrons is quite small and generally can be disregarded.

3.11 TEMPERATURE EFFECTS

Most reactors used for $^{40}Ar/^{39}Ar$ dating purposes have water as a moderator, reflector, and coolant. Thus, the temperature of a sample during irradiation normally can be expected to remain below 100°C, although recognizing that local heating within the sample is likely to occur owing to absorption of neutrons and γ radiation. If cadmium shielding is utilized, the temperature experienced by the sample during irradiation undoubtedly will increase, owing to the heat released by absorption of the thermal neutrons in the cadmium. For most minerals, temperatures as great as 200°C should not cause serious problems. Temperature-sensitive indicators can be used to monitor the temperature if required. Employing this approach, Mitchell (1968a) demonstrated that some mica samples he irradiated in the Herald reactor reached temperatures of $\sim 200°C$. In keeping with experimental evidence for lack of diffusional loss of argon from micas at temperatures in this order, no evidence for argon loss was found by Mitchell (1968a).

3.12 LATTICE DAMAGE

Neutron interactions with atoms of the elements comprising the sample being irradiated will cause radiation damage to the mineral lattices. Such damage can change the argon retention properties of minerals, and, in extreme cases, may even vitiate the underlying precepts of $^{40}Ar/^{39}Ar$ dating by the step heating approach.

Horn et al. (1975) calculated that a fast neutron fluence of $\sim 10^{19}$ n/cm^2 is likely to cause several percent of the lattice atoms to be displaced. They showed, empirically, that for fine grained lunar plagioclase crystals (<15 μm), irradiated in a fast neutron fluence of 2×10^{19} n/cm^2 ($J \sim 0.1$), a small, but significant lowering of the age of the $^{40}Ar/^{39}Ar$ plateau could be detected relative to the age found for a coarser fraction (>35 μm) of the same mineral irradiated at the same time. Horn et al. (1975) concluded that a fluence of $\sim 2 \times 10^{18}$ n/cm² ($J \sim 0.01$) does not cause such effects.

From these observations, it would seem sensible to restrict irradiations to a fluence of fast neutrons $\leq 5 \times 10^{18}$ n/cm² ($J \sim 0.03$). It should be noted, however, that irradiations of this magnitude are required only for rather old samples, say ≥ 3.5 Ga (Fig. 3.7). For younger samples, there is little difficulty in choosing a fluence less than the upper limit proposed, and for samples younger than ~ 1 Ga the fast neutron dose can be kept well below 10^{18} n/cm².

The question of recoil of argon isotopes produced by neutron interactions will be reserved for discussion in Section 4.9.

3.13 DECAY FACTORS

As both ^{37}Ar and ^{39}Ar are radioactive, it is important to correct for their decay during and subsequent to irradiation. As the half life for ^{39}Ar is 269 ± 3 years (Stoenner et al., 1965), the correction always will be small; a delay of 1 year between irradiation and analysis of the gas extracted from a sample requires a correction of only $\sim 0.3\%$ to the measured ^{39}Ar. In contrast, the half life of ^{37}Ar is 35.1 ± 0.1 days (Stoenner et al., 1965), so that significant decay will occur during irradiation, and in the interval between the end of the irradiation and analysis. Proper allowance for its decay must be made, because ^{37}Ar is used as the reference isotope to enable corrections to be made for ^{36}Ar and ^{39}Ar produced from calcium during irradiation. Following Brereton (1972) and Dalrymple et al. (1981) the relevant general equation is

$$(3.21) \quad N_0 = \frac{N \lambda t \exp(\lambda t')}{1 - \exp(-\lambda t)}$$

where N_0 is the total number of atoms formed during irradiation of duration t, N is the number of atoms measured in the analysis at time t' after completion of the irradiation, and λ is the decay constant. With the appropriate decay constants, this equation can be used to correct for the decay of ^{37}Ar and ^{39}Ar.

Because of operational requirements of the reactor, or for other reasons, irradiations may be done in a number of discrete intervals or increments. In addition, the power level, P, of the reactor may be varied. As given by Wijbrans (1985), the generalized form of Eq. (3.21) for n irradiation segments of length t_i is

$$(3.22) \quad N_0 = N \sum_{i=1}^{n} P_i t_i \bigg/ \sum_{i=1}^{n} P_i \left(\frac{1 - \exp(-\lambda t_i)}{\lambda \exp(\lambda t'_i)} \right)$$

where P_i is the reactor power level for the segment, and t'_i corresponds to the time between the end of the irradiation and analysis. This equation differs somewhat from that given by Dalrymple et al. (1981); the two equations yield similar decay factors provided the intervals between irradiation segments are short in comparison with the half life. For intervals between irradiations that approach or exceed the half life of the decaying species, Eq. (3.22) must be used, as the Dalrymple et al. (1981) equation gives incorrect results in these circumstances.

3.14 TOTAL SAMPLE ACTIVITY

During irradiation in the nuclear reactor, many of the elements in a sample will undergo activation by neutrons to produce nuclei that are radioactive. Samples irradiated for ^{40}Ar/^{39}Ar dating purposes generally become highly radioactive, and it is often desirable to have an estimate of the activity.

Following Rakovic (1970), the activity, A_t, of a radioactive nuclide produced during an irradiation of duration, t, is

$$(3.23) \quad A_t = \phi \sigma N [1 - \exp(\lambda t)]$$

where ϕ is the neutron flux, σ is the nuclear reaction cross section, N is the number of atoms of the parent element present, λ is the decay constant for the radioactive nuclide produced, and A_t is given in disintegrations per second. The activity in curies is found by dividing by 3.7×10^{10}, the number of disintegrations per second in 1 Ci. The activity at any time, t', subsequent to the end of the irradiation is given by

(3.24) $\quad A_{t'} = \phi\sigma N[1 - \exp(\lambda t)][\exp(-\lambda t')]$

Summation of the activity produced in each reaction allows a total activity to be estimated.

Dalrymple et al. (1981) provided a compilation of likely reactions induced by neutrons, and showed that of more than 100 potential nuclear reactions, only 26 result in significant activity. In addition, Dalrymple et al. (1981) listed a computer program that enables calculation of the total activity immediately after irradiation and at subsequent times. For these calculations to be made, knowledge of the neutron flux and its energy distribution in the particular reactor facility used for the irradiations is required. Because the ^{39}K(n,p)^{39}Ar reaction requires fast neutrons to proceed, the slow neutrons in the flux are not needed. A significant reduction in unwanted activity, therefore, readily can be achieved by using cadmium shielding around the sample being irradiated to absorb slow neutrons.

Depending upon sample composition, mass, irradiation time, and neutron flux, activities as great as 1 Ci can be expected at the end of an irradiation. However, as most of the nuclides produced have short half lives, the activity of a geological sample typically decreases by two or three orders of magnitude within a week or so, to levels that can be handled safely in the laboratory.

3.15 SAFETY ASPECTS

Samples irradiated for ^{40}Ar/^{39}Ar dating become highly radioactive, and as radiation is hazardous, it is essential that appropriate precautions be taken in handling them. For details of procedures and practices in working with radioactive samples see, for example, Morgan and Turner (1967). In addition, most countries or states have their own safety guidelines and regulations to be adhered to. Nevertheless, a few general comments usefully can be made.

It is highly desirable to set aside a small laboratory especially for safe storage and handling of radioactive samples. Following irradiation, samples normally will not be sent to the laboratory from the reactor until the activity has decreased below some designated level, perhaps 10 mCi or less, depending in part upon local regulations. Samples often are shipped in lead containers, and conveniently can be kept therein or in a lead castle until required for analysis. It is common practice to store samples for several months after irradiation, to allow further decay of the activity. However, because the half life of ^{37}Ar is only 35 days, progressively longer delays between irradiation and analysis increasingly lead to difficulties in measuring the ^{37}Ar with sufficient precision to make corrections for calcium-derived ^{36}Ar and ^{39}Ar satisfactorily. A delay of 1 year, more than 10 half lives, causes the ^{37}Ar to be reduced to <0.001 of its original amount, too small to measure in many cases.

Geological samples used for dating normally are in the form of crushed material, either as mineral separates or whole rocks. Samples for irradiation should be as coarse as possible; fine powders with grain size $<100\,\mu\mathrm{m}$ should be avoided, as the potential for inhalation or ingestion of such fine material is considerable should spillage occur.

The handling of samples after irradiation preferably is carried out in a tray within a fume hood that has an adequate air flow. To minimize exposure to radiation, shielding in the form of lead bricks and lead glass ought to be placed between the person and the samples, also remembering that the radiation from a source decreases inversely as the square of the distance. Samples should not be handled with bare hands; gloves are required, together with appropriate tongs and other tools to assist in safe manipulation. A suitable instrument must be available to determine the levels of radioactivity being handled, and to check that no spillage has occurred. It is usually required that personal dosimeters are worn when working with radioactive samples to monitor the radiation received, and to ensure that the dose remains below the limits set by health authorities. After the argon from a sample has been extracted in the vacuum system, the melted material must be stored properly, and returned to appropriate authorities subsequently for disposal.

3.16 ARGON EXTRACTION SYSTEMS

3.16.1 Introduction

The small amounts of $^{40}Ar^*$ that are generated in geological samples, and the presence of ^{40}Ar in the atmosphere, require that extraction and handling of argon from a sample be undertaken in a vacuum system, isolated from its pumps. In this section a brief description will be given of the techniques employed in argon extraction and purification, noting that the approach used is similar for both the conventional K–Ar and the $^{40}Ar/^{39}Ar$ dating methods. The major differences are that in K–Ar dating the sample is fused directly, and in most laboratories a tracer of ^{38}Ar is added to the extracted gases so that the amount of argon can be determined by isotope dilution.

An argon extraction system is illustrated schematically in Fig. 3.12. It consists essentially of a high vacuum line with appropriate pumping facilities, a furnace assembly in which a sample can be heated in a controlled manner to release argon, getter systems for purification of the released noble gases, interconnecting tubulation and isolation valves, and activated charcoal traps for moving gas from one part of the line to another by cooling with liquid nitrogen. Although argon extraction systems all have the same general features, in practice there is great variability in design and mode of building, which must be to ultrahigh vacuum standards, generally using stainless-steel or borosilicate glass as the main construction material.

It is advantageous to have the extraction system connected through to the mass spectrometer so that a gas sample can be analyzed directly after its extraction and purification.

3.16.2 Pumps and pressure measurement

The high vacuum required in an argon extraction system normally is achieved by means of a liquid nitrogen trapped diffusion pump, backed by a mechanical rotary pump. Mercury traditionally has been used as the fluid in diffusion pumps in systems of this kind, but special very low vapor pressure oils now are quite commonly employed. A turbomolecular pump can be utilized as the primary means of pulling a vacuum instead of a diffusion pump. Because ion pumps are not well suited to cycling from high pressure, generally they are used in extraction systems, if at all, only after a high vacuum has been produced by other means.

To achieve a high vacuum relatively quickly, baking of the extraction system to a temperature of about 300°C is most advantageous. Often this is conveniently done overnight. Provision commonly is made to bake the sample itself at a lower temperature compared with the rest of the extraction system. This is of importance when dealing with samples that begin to release argon at low temperatures; for example, some clays or claylike materials may commence to break down in the vacuum at little more than 100°C. Baking the whole system helps to achieve an appropriately low argon blank.

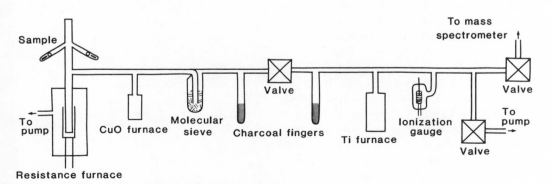

Fig. 3.12. Schematic diagram of a typical argon extraction line.

Prior to undertaking an argon extraction it is necessary to establish that a high vacuum has been achieved, to better than 10^{-6} torr, and that the system is completely free of leaks. An ionization gauge installed in the vacuum line enables low pressures to be effectively monitored and together with judicious operation of valves allows the system to be leak tested. Note that an ionization gauge, when operating, acts as a small pump, so that it is imperative that gas extracted from a sample not be exposed to the gauge when it is on. A nonpumping vacuum measuring device, for example a Pirani gauge, can be helpful in an extraction system as a means of monitoring the pressure during and subsequent to the heating of a sample, and to ensure that gettering is successful.

3.16.3 Furnaces

Complete extraction of argon from a sample is achieved by fusing it within the vacuum system. For most minerals used for $^{40}Ar/^{39}Ar$ dating a maximum temperature of $\sim 1400°C$ is adequate, although in some special cases (diamond, tektite glass) temperatures of $\sim 2000°C$ are required. The sample normally is placed within a crucible constructed of a very high melting point ($>2600°C$) metal such as molybdenum or tantalum. In the $^{40}Ar/^{39}Ar$ step heating technique the sample is heated at successively higher temperatures for discrete intervals, so that effective control of temperature also is required.

Controlled heating of a sample can be achieved by means of a radiofrequency generator or by some form of resistance heating. The work coil of a radiofrequency generator is positioned around the metal crucible, and the eddy currents induced in the crucible cause it, and the sample contained therein, to heat up. An advantage of this type of heating is that no electrical connections need to be made inside the vacuum. Obviously, however, the wall of the vacuum system between the generator work coil and the crucible must be an insulator, normally borosilicate or silica glass, often with an outer jacket to allow the vacuum wall to be water cooled. Care must be taken that glow discharges do not occur in the gas phase during heating of the sample, as this may result in isotopic fractionation (Bernatowicz and Fahey, 1986). Temperature control and monitoring may be by means of a thermocouple or optical pyrometer; these methods will be discussed in Section 3.16.4.

As an example of a resistance heated furnace, that described by Staudacher et al. (1978) is particularly satisfactory because of the excellent temperature control and the very low blanks that can be achieved. This type of all-metal furnace, illustrated in Fig. 3.13, has two separately pumped sections. An inner vessel or crucible is directly joined to the rest of the argon extraction system, and consists of

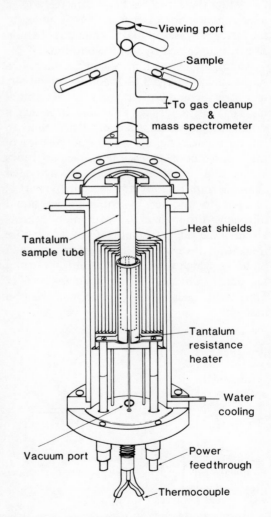

Fig. 3.13. Schematic diagram of a double vacuum resistance heating furnace similar to the kind described by Staudacher et al. (1978).

a tantalum tube (~17 mm diameter), with a tantalum plug forming its base, into which a thermocouple is inserted. A sample can be dropped into this tube from a multiple sample holder above, after the crucible has been thoroughly outgassed. The sample tube projects into the much larger stainless-steel vacuum chamber, separately pumped, in which a split ring resistance heater, built of a thin tantalum sheet, surrounds the lower part of the crucible (sample tube) and in turn is surrounded by tantalum heat shields. The inner vessel is isolated from the furnace chamber via a vacuum seal utilizing the soft tantalum crucible top as the gasketing material or by welding the crucible directly to a standard stainless steel flange (Harrison and Fitz Gerald, 1986). Power for the resistance heater, derived from a low-voltage, high-current transformer, is delivered by means of copper feedthroughs, which connect to molybdenum supports for the heater. Temperature control may be achieved electronically by means of a feedback loop between the thermocouple and the current controller for the resistance heater. Temperatures exceeding 1500°C can be obtained with a power input of ~1000 W provided the geometry and electrical resistance of the heating elements are appropriate and the heat shielding adequate. Temperatures in the 1600°–1800°C range are achievable. As the walls of the crucible actually form the furnace wall, particularly thorough outgassing is possible.

For handling samples as small as 1 mg, Huneke (1978) developed a technique for differential thermal release of argon by placing the sample within a special heating filament in a vacuum system attached to a sensitive mass spectrometer. Huneke was able to obtain excellent age spectra on some lunar samples using this approach. Hohenberg (1980) also described a low blank extraction system for small samples.

On an even smaller scale, heating of a sample to release gas for analysis may be performed using a laser microprobe technique (Megrue, 1967; 1973; Schaeffer, 1982). For $^{40}Ar/^{39}Ar$ dating a previously prepared and irradiated sample is placed in a small ultrahigh vacuum chamber attached to a suitable sensitive mass spectrometer. The sample can be viewed by microscope, and the laser system is arranged in such a way that its beam can be focused on any selected location on the surface of the sample. Typically a laser beam spot diameter of 25–100 μm is employed. The laser is operated most commonly in the pulsed mode, when up to ~0.5 μg of material is excavated from the melt pit formed in silicate material. Following purification the argon is analyzed in the mass spectrometer, usually measuring the ion beams on an electron multiplier owing to their small size. Although very low system blanks are achievable ($<10^{-15}$ mol ^{40}Ar), it is often found necessary to accumulate gas released from a number of laser pulses to provide sufficient gas for precise analysis.

This laser microprobe technique of heating initially was applied to meteorites and lunar samples in relation to noble gas geochemical studies (Megrue, 1967, 1971). Megrue (1973) subsequently used the laser microprobe for $^{40}Ar/^{39}Ar$ geochronology on a lunar breccia; Schaeffer and colleagues then undertook extensive $^{40}Ar/^{39}Ar$ dating on lunar rocks (Plieninger and Schaeffer, 1976; Müller et al., 1977; Schaeffer et al., 1977; Eichhorn et al., 1979). Reconnaissance age measurements on terrestrial samples, using the laser microprobe as the means of releasing gas, were reported by York et al. (1981), Maluski and Schaeffer (1982), and Sutter and Hartung (1984).

The major advantage of the laser microprobe is its ability to focus on very small areas so that age measurements may be made on individual mineral grains or parts thereof *in situ* without mineral separation. In complex or heterogeneous samples this spatial resolution can be extremely useful (cf. McConville et al., 1985). In addition, smaller samples can be dated compared with the requirements for the usual $^{40}Ar/^{39}Ar$ step heating technique. A very important limitation, however, is that an age measured on gas released by melting using a laser is equivalent to a total fusion $^{40}Ar/^{39}Ar$ or conventional K–Ar age, and thus suffers from the same difficulties of interpretation. By utilizing the pulsed laser technique it is not possible to produce an age spectrum of the kind obtained during differential thermal release of gas from a sample in a resistance or radiofrequency furnace. Nevertheless, York et al. (1981) used a defocused beam, about 600 μm in diameter, from a continuous laser to heat a small area of ~2.5-Ga-old slate for 5-min intervals at progressively higher power settings to produce a 4-step age spectrum indistinguishable from that derived by normal step heating procedures. Subsequently, Layer et al. (1987) have shown that detailed age spectra can be generated from single grains of hornblende and biotite using a defocused continuous laser beam. Up to the present time, there have been surprisingly few detailed applications of these microtechniques to terrestrial samples, but it is anticipated that increasing use of such approaches is likely in the future.

3.16.4 Temperature monitoring

Monitoring of the temperature of a sample during a ^{40}Ar/^{39}Ar incremental heating experiment is important as it provides information necessary for the good control of the temperature increments between successive steps, as well as permitting any predetermined heating schedule to be followed. Knowledge of temperature is not required, of course, for the formal calculation of a ^{40}Ar/^{39}Ar age. However, if results are to be used to examine diffusion characteristics of argon from the phase being analyzed, then accurate measurement of the temperature of the sample is essential.

Temperature monitoring can be undertaken by optical pyrometry or by means of one or more thermocouples. Optical pyrometry cannot be used satisfactorily below about 700°C, and at high temperatures coating of the walls of the vacuum system, by evaporation from the metal crucible or the sample, may lead to underestimates of the temperature. Despite these and other difficulties, optical pyrometer readings can be extremely useful as a means of checking the temperature of a sample, as well as temperature gradients in a crucible or furnace assembly. Thus, in our experience, a molybdenum crucible of height 30 mm, diameter 15 mm, and wall thickness 0.6 mm, centrally located within the work coil of a radio-frequency generator, commonly shows a gradient of ~100°C from the hottest part inside the crucible near its top, to the coolest part at the base of the crucible where the sample is located.

A thermocouple can be a very satisfactory way of monitoring the temperature. In addition, in conjunction with appropriate electronic circuitry, a thermocouple can be utilized for controlling the power input to a resistively heated furnace. Platinum versus platinum–10% rhodium (Pt–Rh) and tungsten–3% rhenium versus tungsten–25% rhenium (W–Re) are commonly used thermocouples for the range of temperatures to be measured, although the factor of four greater emf produced by W–Re compared with Pt–Rh makes it the better choice. Unless there is good contact between the thermocouple hot junction and the crucible, underestimates of the temperature will be obtained. It is also important to locate the thermocouple immediately adjacent to the sample, owing to the problem of marked thermal gradients in most furnace systems. If actual, rather than relative temperatures are required, it is desirable to check, by means of optical pyrometry or by observing the melting of metals or other substances whose melting points are accurately known, whether the thermocouple is recording the temperature experienced by the sample itself. In ideal circumstances, the temperature of a sample may be known to better than ~10°C. Temperature increments between steps often may be controlled to better than 1°C. Note, however, that the reported temperatures for steps of a ^{40}Ar/^{39}Ar incremental heating experiment commonly are nominal, and the actual temperatures experienced by the sample may differ considerably.

In a ^{40}Ar/^{39}Ar step heating experiment, the time a sample is maintained at a particular temperature is arbitrary, but a standard duration is used in most laboratories, commonly in the 10–45 min range.

3.16.5 Gas cleanup

Subsequent to extraction of argon from a sample it is essential that the bulk of the other (active) gases be removed prior to transfer for isotopic analysis in order that pressure scattering effects in the mass spectrometer are minimized, and to prevent oxidation of the hot filament. Gas cleanup is most readily accomplished by reaction of the active gases with one or more getters in the extraction system, noting that the other noble gases also will remain with the argon. Thus, gases released from a sample normally are exposed to hot ($\sim 800°C$) titanium metal sponge, or other suitable metal or alloy getter, for at least 5 min, when most of the active gases will either chemically combine with, or be absorbed by, the getter. Because of the relatively low solubility of hydrogen in most getters at high temperature, and its much greater solubility at low temperature, the getter generally is cooled to $\leq 400°C$ to reduce the hydrogen partial pressure, which may be quite high. It is common practice also to have a copper oxide furnace, operated at $\sim 600°C$, in the extraction line to assist in oxidizing hydrogen to water, which can be absorbed onto molecular sieve (chabazite) at room temperature (Fig. 3.12). The getters are located within suitable stainless-steel or silica glass containers attached to the vacuum system. An alternative approach is to retain the argon on a liquid nitrogen cooled charcoal trap while pumping away the less effectively sorbed hydrogen (and helium).

It is increasingly common practice to include one or more nonevaporable Zr–Al getter pumps of the type manufactured by S.A.E.S. Getters S.p.A., Milan, Italy, in extraction systems or installed on mass spectrometers to provide further gas cleanup facilities. In some cases they are used for primary gas purification.

3.16.6 Extraction system blanks

The blank of an extraction system is the amount of ^{40}Ar found, generally of atmospheric argon composition, by following the normal argon extraction procedure, including heating of the crucible, but in the absence of sample. Clearly it is desirable to reduce the blank to as low a value as possible, because its size is an important factor in determining the minimum amount of argon from a sample that can be measured adequately.

In borosilicate glass extraction systems, which usually have some metal parts such as valves, and with heating of the crucible by a radiofrequency generator, it is possible to obtain blanks in the 10^{-12}–10^{-13} mol ^{40}Ar range ($\sim 2 \times 10^{-8}$–2×10^{-9} cm^3 STP ^{40}Ar). This is for a 30-min heating interval, with gas cleanup and transfer to the mass spectrometer taking ~ 60–90 min, depending upon the procedures adopted. To achieve such blanks it is necessary for the crucible to be degassed at high temperature ($\sim 1500°C$) prior to blank measurement, and that the getters be outgassed for at least 1 day under vacuum. Note that it is advantageous to keep the getters under vacuum when the extraction system is vented to load new samples, as this helps to maintain the blank at a low and reproducible level.

A double vacuum furnace assembly and associated extraction system of the kind described by Staudacher et al. (1978) can yield blanks as low as 3×10^{-15} mol ($\sim 7 \times 10^{-11}$ cm^3 STP) ^{40}Ar (Harrison et al., 1986), and are found to be very reproducible. Hohenberg (1980) reported a blank of 2×10^{-15} mol ^{40}Ar (4.5×10^{-11} cm^3 STP) for an extraction system described by him.

3.17 MASS SPECTROMETRY

3.17.1 Introduction

Following extraction and purification of the argon, the gas sample is analyzed isotopically in a mass spectrometer and the relative abundances of the argon isotopes determined, providing data necessary for the formal calculation of an age. Although it is not our intention to present a comprehensive discussion of mass spectrometry, certain basic information is relevant in the present context. The interested reader is referred to texts on this topic by Barnard (1953), Duckworth (1958), Roboz (1968), Duckworth et al. (1986), and White and Wood (1986) for greater detail.

A mass spectrometer is a device in which it is possible to measure isotope abundance ratios on extremely small samples. An instrument of this type is essential for measurement of isotope ratios of argon extracted from geological samples for age determination by both the conventional K–Ar and the $^{40}Ar/^{39}Ar$ methods. Most of the mass spectrometers used for this purpose are developments from the gas source, magnetic sector, first order direction focusing machines described by Nier (1940, 1947). Noteworthy improvements in mass spectrometer design by Reynolds (1956) lead directly to greatly enhanced sensitivity for measurement of all the noble gases, including argon.

A schematic diagram of a mass spectrometer of the type most commonly used for argon isotope analysis is shown in Fig. 3.14. The main components of the machine include the flight tube, the electron bombardment ion source, the magnetic analyzer, and the collector, with appropriate electronics and a suitable pumping system. The latter is required to produce a high vacuum in the machine so that travel of the ions along the flight tube is not hampered by collisions with extraneous gas molecules. The beam of positive ions from the gas sample being analyzed, produced as essentially monoenergetic ions in the source, is collimated and accelerated along the flight tube. The curved portion of the mass spectrometer tube lies within the gap between the pole pieces of the sector magnet, which produces a homogeneous magnetic field perpendicular to the trajectory of the ion beam. In traversing the magnetic field the forces experienced by the ions cause them to describe circular paths, the radii of which depend upon the mass to charge (m/e) ratio of the ions. The ion beam is arranged to enter and leave the mass analyzer at a particular angle, most commonly at right angles to the pole faces of the magnet. In this configuration, a monoenergetic ion beam is not only dispersed into its component beams corresponding to different m/e values, but is refocused at a position located on the extension of the line joining the final exit slit of the ion source and the apex of the sector magnet. The collector is placed in this position, so that the ion beams of discrete m/e value can be successively focused on it and the current measured. In recent years, designs embodying non-normal entrance and exit of the ion beam relative to the sector magnet pole faces have been found to produce enhanced resolution and sensitivity for a given radius of curvature of the magnet sector. The turning angle of the sector magnet in most mass spectrometers of the type used for this work is 60° or 90°, but other angles can be utilized.

3.17.2 Basic concepts

If e is the charge of an ion produced in the source of a mass spectrometer and accelerated through a potential V volts, then the energy will be equal to the kinetic energy, so that

$$(3.25) \quad eV = (1/2)mv^2$$

where m and v are the mass and velocity, respectively, of the ion.

The magnetic field of strength H, applied in the plane normal to the trajectory of an ion, causes the latter to describe a circular path of radius r, such that the force equation is

$$(3.26) \quad mv^2/r = Hev$$

These two equations (3.25 and 3.26) can be combined to yield the basic mass spectrometer equation

$$(3.27) \quad \frac{m}{e} = \frac{H^2 r^2}{2V}$$

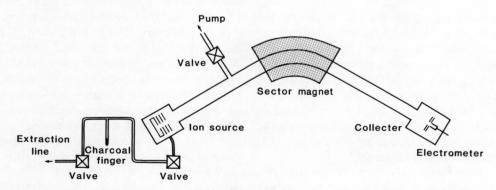

Fig. 3.14. Schematic diagram of a gas source mass spectrometer of the kind used for isotopic analysis of argon.

This equation indicates that ions of a particular mass to charge ratio will have a circular path of radius r in the mass analyzer for a given combination of magnetic field and accelerating voltage. Rearranging, and with m in atomic mass units (amu) on the $^{12}C = 12.0000$ scale, H in gauss, V in volts, and e the number of unit electronic charges, the radius r is given in millimetres

(3.28) $$r = \frac{1439}{H}\left[\frac{m}{e}V\right]^{1/2}$$

Thus, the radius of the path described by ions of the same energy but different mass to charge ratio varies according to the square root of that ratio in a homogeneous magnetic field. For example, a $^{36}Ar^+$ ion accelerated to 2000 V will be deflected by a magnetic field of 3000 G into a path of radius 128.7 mm, whereas a $^{40}Ar^+$ ion, under the same conditions, will travel on a path of radius 135.7 mm.

If the magnetic sector radius, a, is substituted in Eqs. (3.27) and (3.28) for r, then the magnetic field for focusing a particular ion species on the collector can be calculated for a given accelerating potential. Ions of different mass to charge ratio will have different trajectories, and, thus, will not enter the collector, but instead will collide with the walls of the flight tube or other internal (earthed) parts of the machine. In passing it is worth noting that most gas source mass spectrometers used for isotope abundance ratio measurements are operated at a standard accelerating potential, commonly in the range 2000–4000 V.

To consider further the focusing properties of a homogeneous sector magnetic field, the trajectory is shown in Fig. 3.15 of a beam of monoenergetic ions emerging from the beam-defining slit of the ion source at A, the object point, with a small angular divergence, α, relative to the central ray. The geometry is arranged in such a way that the ion beam enters the magnetic field essentially at right angles to the pole face, and the ion beam for a particular m/e value will leave the field normally as shown. Ion beams of higher mass to charge ratio will be deflected less in the magnetic field, and those with lower m/e values will show greater deflection. Ions of the particular

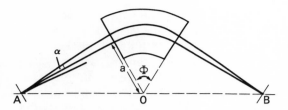

Fig. 3.15. Diagram showing the trajectory of an ion beam in a mass spectrometer.

m/e ratio will be refocused at point B, the image point, where the slit in front of the collector is located. Refocusing occurs at the image point where r, the radius of curvature of the particular ion beam, determined from the mass spectrometer Eq. (3.27), coincides with the magnet sector radius, a. The object point, the image point, and the center of curvature of the ion beam at the apex of the magnet sector will lie on a straight line, according to Barber's rule, for the case of normal entry and exit of the ion beam, irrespective of the angle of deflection in the magnetic field.

Herzog (1934) demonstrated that the magnet field acts both as a prism and as a lens, and that equations analogous to that for a thick optical lens can be set up to describe the focusing properties. Thus, the first-order direction focusing equation for an ion beam of particular mass to charge ratio that enters and leaves the magnetic field at right angles is

(3.29) $$f^2 = (l' - g)(l'' - g)$$

where l' and l'' are the distances of the object and image points, respectively, from the edge of the magnetic field. The focal length, f, is defined by

(3.30) $$f = a/\sin \phi$$

and g is defined by

(3.31) $$g = a/\tan \phi$$

where a is the radius of curvature of the path of the ions in the magnetic field, and ϕ is the angle through which the beam is deflected. Angles of deflection commonly used are 180°, 90°, and 60°. In many cases it is arranged that $l' = l''$ so that the analyzer is symmetrical. For 90° deflection, the object and image distances are equal to a, and for 60° deflection these distances are 1.73a. Note that the above examples are special cases of the more general theory of focusing in sector magnetic fields derived by Herzog (1934) in which non-normal entry into the magnetic field can be accommodated, always providing that the angular divergence of the ion beam and its energy spread are small.

Most of the mass spectrometers used for noble gas isotopic analysis are variants of the symmetrical, normal beam entry and exit designs, and for this type of geometry the magnification is unity. Ion beams of different m/e ratio are focused on the collector in turn by varying the magnetic field or accelerating potential, usually the former so that optimum focusing of the ion beam emerging from the source is maintained and mass discrimination minimized.

More recently, mass spectrometer designs incorporating non-normal entry of the ion beam into the magnetic field according to the geometry described by Cross (1951) have been utilized, and these have the advantage of providing greater dispersion of the ion beam for a given magnetic sector radius.

For precise measurement of isotopic ratios in a mass spectrometer, it is necessary to ensure that there is sufficient separation or dispersion so that interference between ion beams of adjacent masses is essentially zero, or at least small enough to be readily allowed for. The resolving power of a single focusing magnetic sector mass spectrometer is dependent upon the mass dispersion of the magnetic analyzer and the width of the ion beams at the point of focus, and is determined by a number of factors including the radius of curvature of the sector magnet, the widths of the source and collector slits, and the general ion optics of the instrument. For symmetrical analyzers with normal entry of a monoenergetic ion beam into the magnetic field, the resolution, R, is given by

$$(3.32) \qquad R = \frac{M}{\Delta M} = \frac{r}{w_b + w_c}$$

where $M/\Delta M$ is the inverse of the fractional mass difference that can be separated, w_b is the actual beam width at the collector, and w_c is the width of the collector slit. The beam width at the collector invariably is greater than the width of the beam emerging from the source slit owing to aberrations of various kinds related to the effect of the fringing field adjacent to the magnet, small variations in the energy of the ions leaving the source, and other factors.

In principle, the resolution can be calculated from the ion optics of the mass spectrometer, but it needs to be measured directly. Several different conventions have been adopted for measuring and reporting the resolution; see Roboz (1968) for a full discussion. A widely used definition gives resolution as the mass at which peaks of singly charged ions of one atomic mass unit separation are just resolved. In practical terms this means the highest mass at which two adjacent peaks of the same amplitude, differing by unit mass, can be separated with a valley between the peaks of some designated percentage, often 10% of the peak amplitude. Usually it is more convenient to determine resolution from a single peak by scanning across it slowly by varying either the magnetic field or the accelerating potential in a systematic manner. The width of the peak at say 5% peak amplitude, also expressed in terms of atomic mass units, is measured and the resolution calculated from Eq. (3.32) (Fig. 3.16). This is essentially equivalent to the 10% valley definition previously discussed.

For argon isotopic analyses in relation to ^{40}Ar/^{39}Ar age determinations, only a modest resolution is required, enough to allow good unit mass separation at $m/e = 40$. In practice, a minimum resolution of ~ 100 is required to ensure that it is possible to measure small ion beams at mass 39 and 36 relative to the commonly very much larger peaks at adjacent masses of 40 and 37, respectively. At lower resolution the tailing of the larger ion beams beyond their nominal mass number—because of elastic scattering collisions of the ions with gas molecules in the flight tube and other factors—makes precise measurement of small ion beams difficult.

The abundance sensitivity is a measure of the ability to measure small peaks adjacent to much larger ones, and can be defined as the ratio of the ion current at mass M relative to that found above the true baseline at mass $M \pm 1$ in the absence of a peak at that mass (Roboz, 1968). Commonly, however, it is convenient to express the abundance sensitivity as the inverse of the defined value, that is,

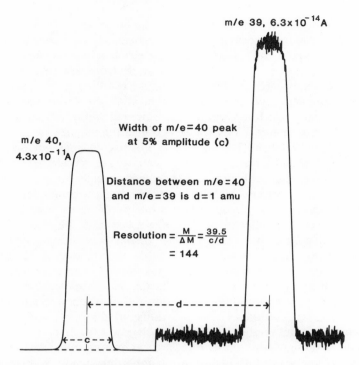

Fig. 3.16. Diagram illustrating measurement of resolution of a mass spectrometer. Taken from an actual chart record of a portion of the argon spectrum obtained by changing the magnetic field at a uniform rate. Amount of ^{40}Ar $\sim 1.5 \times 10^{-11}$ mol. Measured in a VG Isotopes Limited 12 cm radius of curvature mass spectrometer, accelerating potential = 2 kV, source trap current = 110 μA.

the magnitude of the tail from the large peak at an adjacent mass in terms of the size of the large peak.

By way of example, an AEI MS10 mass spectrometer in a typical configuration for argon isotopic analyses has a resolution of ~ 100 (definition based on 5% peak width), and an abundance sensitivity of ~ 30 ppm at $m/e = 39$ at $\sim 2 \times 10^{-6}$ torr, corresponding to a ^{40}Ar ion beam of $\sim 1 \times 10^{-10}$ A. The MS10 is a 180° machine of 50 mm radius of curvature, and is found to be adequate for argon isotopic measurements. In cases in which the ^{40}Ar/^{39}Ar ratio is ≥ 200, tailing effects of the $m/e = 40$ peak under m/e 39 become noticeable and need to be allowed for. A 120 mm radius of curvature, 60° sector mass spectrometer, like the VG Isotopes Limited MM1200, can be readily operated at a resolution of 150, and has an abundance sensitivity of better than 10 ppm at $m/e = 39$ for a ^{40}Ar ion beam of 1×10^{-10} A and a pressure in the mass spectrometer of $\sim 2 \times 10^{-7}$ torr. Under these operating conditions, tail corrections are negligible, even for

^{40}Ar/^{39}Ar ratios > 500 (see, for example, Fig. 3.16). These comparative results illustrate the value of employing a larger radius of curvature mass spectrometer with greater inherent resolution and abundance sensitivity for isotopic abundance measurements. Note also that the abundance sensitivity deteriorates rapidly as pressure in the mass spectrometer increases to the level at which gas scattering becomes significant.

An additional point is that it is important to ensure that the collector slit of the mass spectrometer is sufficiently wide so that the focused ion beam is completely accommodated within it, so that a flat-topped peak results. This is required for high sensitivity, and to permit the magnetic field or accelerating potential to be altered in a stepwise manner without difficulty for peak hopping, thus allowing data to be obtained in a minimum time. For the AEI MS10 mass spectrometer it is only just possible to obtain a flat-topped peak as well as maintaining the appropriate resolution. In larger radius of curvature machines, say > 100 mm, there usually is no problem achieving flat-topped peaks, high abundance sensitivity, and high sensitivity without the need for compromise.

A radically different approach to obtaining isotope ratio measurements of rare gases, utilizing a Fourier Transform-Ion Cyclotron Resonance Mass Spectrometer (FT-ICR), is described by DeLong et al. (1988). The underlying principle is that radiofrequency excitation of ions in a magnetic field results in resonant ions achieving cyclotron orbit with a frequency specific to the mass to charge ratio. Conveniently, this effect is manifested in Eq. (3.26). By dividing both sides of the equation by v, it can be seen that the angular velocity (v/r), or frequency, is proportional to the magnetic field strength divided by the mass to charge ratio. Thus, detection of the various frequency components in the ICR signal allows isotope ratios to be determined. DeLong et al. (1988) demonstrated that, although other oscillation modes in an ICR cell complicate interpretation of the acquired signal, precise isotope ratio measurement is possible.

3.17.3 Ion sources

Electron bombardment ion sources invariably are used in mass spectrometers in which isotope abundance measurements on gaseous samples are made. Such sources can be readily regulated and can produce very stable positive ion beams exhibiting only a small energy spread, usually less than a few electron volts. Ion sources most widely used for this purpose can be considered to have their genesis in designs put forward by Nier (1940, 1947).

Electrons produced by thermionic emission from a hot filament, most commonly made of tungsten, are accelerated across an ionization chamber by a small potential difference. Positive ions are formed when electrons collide with the gas molecules, causing removal of one or more extranuclear electrons. Most gases, including argon (Bleakney, 1930), have maximum efficiency of ionization when the bombarding electrons have energies in the 50–100 eV range, and electrons of 70 eV are frequently employed. At these energies, the ions formed are dominantly singly charged. It is common practice to place a small permanent magnet, yielding a field of ~ 100 G, parallel to the electron beam to improve its collimation. The electrons are caused to travel in helical paths, resulting in an increase in efficiency of ionization by about an order of magnitude compared with the same source without a collimating magnet.

Positive ions drawn out of the ionization chamber are accelerated down a potential difference, usually of several kilovolts, applied across an electrostatic lens system, to produce a finely collimated beam that travels along the flight tube into the magnetic mass analyzer.

In this conventional type of electron bombardment ion source, the electron beam is at right angles to the trajectory of the positive ion beam. The proportion of ions formed in the ionization chamber that are drawn out and transmitted along the flight tube is quite low, probably less than 10% (Roboz, 1968). In recent years a much higher efficiency electron bombardment ion source, the Baur–Signer source, has become increasingly widely used (Baur, 1980; see also Hohenberg, 1980). In this source, the filament is a small circular ring; it and the focusing electrodes are symmetrical about the trajectory of the ion beam, resulting in almost total extraction of the ions formed and thus higher sensitivity. Mass discrimination effects generally are small in a machine fitted with a Baur–Signer source. For the VG Isotopes Limited MM1200 mass spectrometer fitted with a Nier-type ion source a sensitivity for argon of 3×10^{-4} A/torr is readily obtained (accelerating potential 2000 V, trap current 110 μA, total emission 750 μA), and the modified version of this machine, the MM1200S, fitted with the Baur–Signer source, is said to have a sensitivity at least a factor of three greater than the standard MM1200, with 250 μA total emission.

3.17.4 Ion detection and collection

Positive ion beams of different m/e ratio separated in the magnetic analyzer are individually focused on the collector where the ion currents are detected and measured. Collector assemblies consist of a defining or resolving slit located at the point of focus, behind which is located at least one plate held at a modest negative potential, usually in the 50–100 V range, for suppression of secondary electrons produced when high-energy ions collide with the collector. In some collectors an additional slit is included, together with guard rings, to suppress the passage of ions that have lost energy owing to collision during their travel along the flight tube. The ion beam passing through the slit assembly may be collected on a Faraday cage or cup or on the conversion electrode of an electron multiplier. If the beam is collected on a Faraday cup, the ion current effectively is measured by means of the potential drop observed across a high value resistor, normally in the 10^{10}–10^{11} Ω range, through which the collector is grounded. The signal is amplified through a dc amplifier or a vibrating reed electrometer, displayed on a chart recorder, and usually also fed to an integrating digital voltmeter to provide output suitable for computer processing.

For most argon isotopic analyses, measurement of the signal on a Faraday cup is adequate provided the largest ion beam, generally ^{40}Ar, is greater than about 5×10^{-12} A. In many cases there is little difficulty in ensuring the major ion beam is $\sim 10^{-10}$ A, close to the optimum for highly precise measurement. In argon isotopic analyses, ratios exceeding several thousand commonly need to be measured. With Faraday cup detection, and measurement of the ion current by electrometer, the smallest beam detectable is $\sim 10^{-15}$ A. This can be measured only very imprecisely as it corresponds approximately to the thermal noise level of the resistor.

If the major ion beam is less than $\sim 5 \times 10^{-12}$ A, it becomes essential to detect and measure the beams using an electron multiplier. Such multipliers can be operated at gains of up to $\sim 10^7$, and are capable of extremely high sensitivity as the inherent noise level can be as low as 10^{-19} A (Roboz, 1968). An electron multiplier often is used as a preamplifier for the conventional type of electrometer already discussed, although at low currents ion counting can be employed with advantage.

In the larger radius of curvature sector mass spectrometers it is common for both an electron multiplier and a Faraday cup to be fitted, so that the most appropriate detector can be utilized for a particular isotopic analysis. With the trend toward measuring smaller samples, the detection and measurement of argon isotopes by electron multiplier will increase. A practical problem with electron multipliers is that the output signal is dependent upon the mass and energy of the impinging ions, although this effect is obviated by use of ion counting methods. Proper calibration, using atmospheric argon as the standard in the case of argon isotope analyses, can effectively compensate for these discrimination effects. Another problem in the use of electron multipliers in ^{40}Ar/^{39}Ar dating is that ^{39}Ar and ^{37}Ar are radioactive. These ions collected in an electron multiplier will continue to decay, and may well produce an increase in background or dark current of the multiplier.

3.17.5 Static mode operation

Measurement of the isotopic composition of argon extracted from a sample almost invariably is done by operating the mass spectrometer in the static mode, that is, with the flight tube isolated from the pumping system. Generally the purified gas sample is adsorbed onto an activated charcoal finger using liquid nitrogen in a small volume separated from the mass spectrometer by an ultrahigh vacuum valve and which can be isolated from the argon extraction system by another valve. The gas is then released from the charcoal and admitted wholly to the machine through the valve, and after pressure equilibration, which normally takes less than 1 min, the analysis can be commenced. If the amount of gas is too large, it can be partly let into the machine through a leak valve, and then analyzed. Isotopic fractionation will take place during this process, but this can be accurately corrected for (see Section 3.17.7).

Static operation provides sensitivity of one or two orders of magnitude greater than the earlier dynamic mode of analysis whereby the gas was leaked into the source region of the mass spectrometer and pumped away continuously. Isotopic analysis of noble gases by the static method was a most important innovation introduced by Reynolds (1956) at the time he described a mass spectrometer designed and built for this kind of study. To operate statically, the mass spectrometer must be quite free of leaks, and have a low background and outgassing rate, in practice possible only if the machine is bakeable to ~300°C. All these features were achieved in the mass spectrometer designed by Reynolds (1956), which set new standards in noble gas analysis. Reynolds constructed the flight tube of 11.4 cm radius of curvature sector machine using borosilicate glass, because at that time it was not possible to attain a completely leak-tight, ultrahigh vacuum system by working with metal. Because of great improvements in vacuum technology, including readily available flange sealing systems and superior welding techniques, today most mass spectrometers for noble gas studies are fabricated from stainless steel and are routinely operated in the static mode. Indeed the development of the Reynolds machine and its operation statically were major factors in the successful application of the K–Ar method to a wide range of dating problems not previously tractable, especially the precise and accurate dating of young igneous rocks, say those less than 10 Ma old. Interesting historical aspects of these developments by Reynolds have been recorded by Glen (1982). The increase in sensitivity achieved by this approach applies with even greater force in dating by the ^{40}Ar/^{39}Ar dating method, as in many cases very small amounts of gas must be analyzed.

It should be stressed that the static mode of operation can be applied to the dating of old as well as young samples, resulting in much smaller amounts of material being required for argon extraction than previously.

Operation in the static mode introduces an undesirable side effect, which, however, can normally be allowed for. This is the so-called memory effect, whereby the peak heights of the individual isotopes change progressively with time in the mass spectrometer owing to some form of exchange with atoms embedded in the machine, especially in the source region, from previous analyses. Generally there is a decrease of peak height with time owing to the pumping action of the machine. In the case of an isotope of low abundance being analyzed in the mass spectrometer, where previously a sample of high abundance in that isotope was analyzed, it is not uncommon for the peak height to increase as the analysis progresses. Typically during a 10-min analysis, peak height changes of several percent are found, although much larger changes may occur. To overcome this problem, the peak heights or ratios are extrapolated to the time at which the gas sample was admitted to the mass spectrometer. This is done routinely by application of regression analysis to the data, generally using a straight line extrapolation, although a polynomial or asymptotic extrapolation often is found to provide a better fit to the results.

3.17.6 Machine calibration

It is a truism that mass spectrometers in general do not yield absolute abundance ratios of the isotopes being analyzed, so that a machine must be calibrated and appropriate corrections applied to the measured ratios. Departures from the absolute values commonly are mass dependent, and, thus, are included under the general term mass discrimination. Effects of this kind may be caused by a variety of factors. For example, stray magnetic fields in the source region of the mass spectrometer are likely to result in the extracted ion beam not reflecting exactly the composition of the gas in the machine. The difference in ionization efficiency of the various argon isotopes also is a source of mass discrimination. Additionally, ion currents measured by an electron multiplier vary approximately inversely as the square root of mass. Irrespective of the causes of these effects, calibration of the mass spectrometer for argon analysis is readily achieved by analyzing atmospheric argon, often introduced directly to the machine from a gas pipette attached to it.

The procedure normally adopted is to measure the ^{40}Ar/^{36}Ar ratio in an aliquot of atmospheric argon and then to calculate the discrimination by comparing the value measured with that given by Nier (1950), 295.5. The mass discrimination is calculated from the ^{40}Ar/^{36}Ar measurements and applied to the measured ratios on the assumption that it is linear with mass. In most mass spectrometers used for argon analysis, the mass discrimination rarely exceeds 1% per mass unit. Depending upon the type of machine, the heavier or lighter isotopes may be favored. It is important to determine the range of gas quantities over which the discrimination remains essentially constant, and then to restrict the size of gas sample analyzed to within this range; this ensures that any pressure effects or nonlinearities of the measurement system are minimized.

In $^{40}Ar/^{39}Ar$ analyses, it is common practice to use the mass spectrometer as a manometric device to determine the amount of gas being measured. This kind of information is required for plotting an age spectrum, and also to allow calculation of an incremental total fusion age by combining results from each step in proportion to the amount of gas present. Thus, it is of some importance to determine the relation between the size of the ion beam and the partial pressure of argon in the mass spectrometer. This can be done several ways, but perhaps most conveniently by measuring the peak heights for ^{40}Ar when one to several aliquots of atmospheric argon of known amount are admitted to the machine. At the gas pressures employed in mass spectrometers for argon analysis, usually $< 10^{-6}$ torr, there is normally a linear relationship between ion beam and amount of argon present. Should it be necessary to analyze at higher pressures, for example, owing to the incomplete removal of other gases, then it can be expected that the sensitivity will decrease because of gas scattering effects.

3.17.7 Orifice correction

If it is anticipated that a gas sample is too large to be analyzed statically in the mass spectrometer by equilibration between the sample reservoir and the machine, it is common practice to make isotope ratio measurements on some smaller portion of the gas admitted to the mass spectrometer through a leak valve. In argon dating experiments, gas pressures almost invariably are low enough that molecular flow conditions prevail, resulting in isotopic fractionation as the gas enters the mass spectrometer. If pressure equilibration is allowed across the valve between the reservoir and mass spectrometer volumes, then back diffusion ensures that isotopic equilibration also occurs. However, if only part of the gas is leaked into the mass spectrometer, kinetic gas theory predicts that the lighter isotopes will be enriched relative to the heavier isotopes in the gas fraction admitted to the machine, compared with the original isotopic composition of the bulk gas; the gas remaining in the reservoir will be fractionated in the opposite sense. Thus the measured isotopic composition of a gas sample analyzed in this way has to be corrected for fractionation in order to obtain a proper estimate for the isotopic composition of the original, unfractionated gas. Many laboratories used an orifice correction formula derived by J. H. Reynolds in 1958, but this was never formally published. Baksi and Farrar (1973) derived and experimentally documented a rather similar orifice correction formula:

$$(3.33) \quad \left[\frac{x}{y}\right]_R = \left[\frac{x}{y}\right]_D \frac{f_y}{1 - (1 - f_y)^k}$$

where $(x/y)_D$ is the measured ratio for isotopes x and y as determined in the mass spectrometer, $(x/y)_R$ is the corrected isotopic ratio for the gas in the reservoir prior to fractionation, f_y is the fraction of the isotope y admitted to the mass spectrometer for analysis, and $k = (M_y/M_x)^{1/2}$, the square root of the ratio of the masses of molecules y and x. The fraction f_y is measured by allowing the gas from the reservoir to equilibrate with that in the mass spectrometer after completion of the isotopic analysis. Note that for $f_y = 1.0$, that is, for complete equilibration of pressure between the reservoir and mass spectrometer volumes, the measured isotopic ratios are identical to those in the original gas sample, as would be expected intuitively. For small gas fractions, say $f_y \leq 0.01$, the correction factor applied to the measured ratio is the square root of the ratio of the masses ($= 1/k$).

Our own extensive, empirical, experience confirms that the orifice correction works extremely well in practice. We find that argon isotopic ratios near unity are reproducible to within a part or two in a thousand when duplicate analyses are made, the first involving analysis of an equilibrated gas fraction from a portion of the gas reservoir, requiring no orifice correction, the second analysis being made on a fraction leaked in from the reservoir, and, thus, needing an orifice correction. Valves appropriately located within the gas reservoir system attached to the mass spectrometer often provide sufficient flexibility that an unfractionated portion of the gas sample can be equilibrated with the mass spectrometer volume, eliminating the need to apply an orifice correction. Indeed for most ^{40}Ar/^{39}Ar step heating experiments, it is possible to arrange for unfractionated gas fractions to be measured.

3.17.8 Data acquisition

In ^{40}Ar/^{39}Ar dating experiments, the gas extracted from a sample is analyzed isotopically in the mass spectrometer to determine the relative abundances of ^{40}Ar, ^{39}Ar, ^{37}Ar, and ^{36}Ar. In addition, ^{38}Ar may also need to be measured in cases in which the presence of significant chlorine is suspected, with potential production of some ^{36}Ar from ^{36}Cl (see Section 3.7.4).

Detailed procedures for data acquisition vary widely, but the general approach used in most laboratories is similar. Following isolation of the mass spectrometer from its pumps, the purified gas sample is admitted to the mass spectrometer. Isotopic analysis commences as soon as possible, and data are collected over a period of the order of 10 min. After completion of data gathering, the peak heights of the isotopes of interest, or interpolated ratios between peaks, are extrapolated to the time of admission of the gas to the mass spectrometer to allow for the effects of memory and the pumping action of the machine. Corrections for machine mass discrimination and for fractionation, where necessary, are applied, and the decay factors for ^{37}Ar and ^{39}Ar are calculated, to finally arrive at a set of ratios that are used to determine the ^{40}Ar*/^{39}Ar$_K$ ratio in the gas sample.

Most mass spectrometers utilized for argon isotopic analysis have a single Faraday cup collector, so that ion beams corresponding to the different isotopes to be measured need to be focused in turn on the collector, where the signal is detected by an electrometer. Focusing generally is done by changing the magnetic field, but in some cases the accelerating potential is varied instead; this is mandatory if the mass spectrometer is fitted with a permanent magnet (e.g., AEI MS10 machine). Adequate data may be obtained by systematic scanning of the magnetic field or accelerating potential to focus the ions beams of interest on the collector, with the output from the electrometer being displayed on a chart recorder, from which data can be scaled. Generally, however, data are acquired in digital form and stored and processed in a computer. Data obtained in this way are more precise, and there is a great saving in time and labor compared with the earlier analogue methods.

Instead of continuous mass scanning, it is now common practice to arrange for the magnetic field or accelerating voltage to be changed to preset values in order to focus individual ion beams on the collector together with other intermediate values for baseline measurements. This is done under computer control or through a peak hopping device, and allows much more efficient data collection compared with the earlier scanning technique. In the case of magnetic field changes, this is best done by using a Hall-effect probe in the gap of the sector magnet linked in a feedback loop with the power supply, permitting rapid changes of field to be made. A minimum of about 5, and usually 7 data sets is required to enable proper extrapolation of the peak heights or interpolated ratios to be made to the time of letting the gas into the mass spectrometer. The signal from the electrometer normally is fed to an integrating digital voltmeter. For the smaller ion beams, for example, that for ^{36}Ar which is usually $<10^{-13}$ A, it is desirable to integrate for up to ~ 10 s in order to obtain a good signal-to-noise ratio. The thermal noise associated with the high value feedback resistor, commonly $10^{11} \Omega$, in the electrometer, and other sources of noise correspond to a background signal of the order of 2×10^{-15} A. For the larger ion beams such as ^{40}Ar, which may be in the 10^{-10}–10^{-11} A range, integration times as small as 1 s are adequate to obtain good statistics. These data in digital form are readily accumulated in a computer, followed by on-line processing of the results, so that isotope ratios and calculations leading to an age often can be made within minutes of completion of an analysis.

As noted previously, when operating a mass spectrometer in the static mode, peak heights change progressively with time, necessitating extrapolation of the data to zero time, the time of gas admittance to the machine. Simple linear regression of peak heights versus time is satisfactory in many cases, but it is not uncommon to observe distinct curvature of the plots. Thus, extrapolations involving some form of curve fitting is often more appropriate. Our own experience shows that an asymptotic fit (Stevens, 1951) is generally satisfactory, but other types of curve fitting are commonly used. For the larger ion beams, say $> 10^{-12}$ A, extrapolated values of the peak heights commonly have uncertainties of $<0.1\%$ standard deviation.

Although most mass spectrometers used for argon isotopic analyses have single collectors, machines with multiple collectors have been built. Thus a five collector mass spectrometer, specifically designed for argon analysis, was constructed by Stacey et al. (1981). The major advantage of this approach is that data can be obtained much more rapidly than with a conventional machine, so that in principle extrapolations to zero time can be made more precisely. In practice, data appear to be comparable to those obtained with a single collector machine, as the accurate intercalibration of the electrometers poses some problems.

3.17.9 Calculation of $^{40}Ar^*/^{39}Ar_K$

Calculation of a $^{40}Ar/^{39}Ar$ age requires that the ratio of radiogenic argon ($^{40}Ar^*$) to ^{39}Ar produced by neutron interactions on ^{39}K be obtained, as discussed in Section 2.5. This ratio, $^{40}Ar^*/^{39}Ar_K$, must be determined on the argon extracted both from the flux monitor mineral and from the sample whose age is to be measured. Because we have considered neutron interference reactions (Section 3.7), and measurements of argon isotope ratios by mass spectrometry (Section 3.17), we are now in a position to derive the expression for $^{40}Ar^*/^{39}Ar_K$.

In general, argon extracted from a geological sample containing some potassium and calcium, and which has been irradiated for $^{40}Ar/^{39}Ar$ dating purposes, will consist of a mixture of the isotopes ^{36}Ar, ^{37}Ar, ^{39}Ar, and ^{40}Ar. Some ^{38}Ar also will be present, but as this isotope is not normally involved in the calculation of age (see Section 3.7.4), it will not be considered further here. We will assume that the amount of chlorine-derived ^{36}Ar is negligible.

Consistent with previous usage, the following subscripts identify the source of each isotope:

m = measured (corrected for radioactive decay where necessary);
Ca = neutron-induced on calcium;
K = neutron-induced on potassium;
A = atmospheric, noting that $(^{40}Ar/^{36}Ar)_A = 295.5$.

Following irradiation, the total ^{40}Ar present in a geological sample will be

$$^{40}Ar_m = {}^{40}Ar^* + {}^{40}Ar_A + {}^{40}Ar_K + {}^{40}Ar_{Ca} \quad (3.34)$$

As noted in Section 3.7.2, for terrestrial samples the $^{40}Ar_{Ca}$ is effectively taken into account through the atmospheric argon correction so that Eq. (3.34) can be rewritten in the form

$$(3.35) \quad {}^{40}Ar^* = {}^{40}Ar_m - {}^{40}Ar_A - {}^{40}Ar_K$$

Now

$$(3.36) \quad {}^{36}Ar_m = {}^{36}Ar_A + {}^{36}Ar_{Ca}$$

and as

$$\left[\frac{{}^{40}Ar}{{}^{36}Ar}\right]_A = 295.5$$

then

$$(3.37) \quad {}^{40}Ar_A = 295.5[{}^{36}Ar_m - {}^{36}Ar_{Ca}]$$

Substitution of Eq. (3.37) into Eq. (3.35) yields

$$\begin{aligned}{}^{40}Ar^* = {}^{40}Ar_m &- 295.5\,{}^{36}Ar_m \\ &+ 295.5\,{}^{36}Ar_{Ca} - {}^{40}Ar_K\end{aligned} \quad (3.38)$$

The total ^{39}Ar generated during irradiation is given by

$$(3.39) \quad {}^{39}Ar_m = {}^{39}Ar_K + {}^{39}Ar_{Ca}$$

so that

(3.40) $\quad ^{39}\text{Ar}_K = {}^{39}\text{Ar}_m - {}^{39}\text{Ar}_{Ca}$

Dividing Eq. (3.38) by Eq. (3.40), and dividing both numerator and denominator on the right-hand side by $^{39}\text{Ar}_m$, we obtain

(3.41) $\quad \dfrac{^{40}\text{Ar}^*}{^{39}\text{Ar}_K} = \dfrac{[^{40}\text{Ar}/^{39}\text{Ar}]_m - 295.5[^{36}\text{Ar}/^{39}\text{Ar}]_m + 295.5(^{36}\text{Ar}_{Ca}/^{39}\text{Ar}_m) - (^{40}\text{Ar}_K/^{39}\text{Ar}_m)}{1 - (^{39}\text{Ar}_{Ca}/^{39}\text{Ar}_m)}$

As

$$\dfrac{^{36}\text{Ar}_{Ca}}{^{39}\text{Ar}_m} = \dfrac{^{36}\text{Ar}_{Ca}}{^{37}\text{Ar}_{Ca}} \dfrac{^{37}\text{Ar}_m}{^{39}\text{Ar}_m}$$

$$= \left[\dfrac{^{36}\text{Ar}}{^{37}\text{Ar}}\right]_{Ca} \left[\dfrac{^{37}\text{Ar}}{^{39}\text{Ar}}\right]_m$$

because $^{37}\text{Ar}_{Ca} = {}^{37}\text{Ar}_m$, and

$$\dfrac{^{39}\text{Ar}_{Ca}}{^{39}\text{Ar}_m} = \left[\dfrac{^{39}\text{Ar}}{^{37}\text{Ar}}\right]_{Ca} \left[\dfrac{^{37}\text{Ar}}{^{39}\text{Ar}}\right]_m$$

for the same reason, and

$$\dfrac{^{40}\text{Ar}_K}{^{39}\text{Ar}_m} = \left[\dfrac{^{40}\text{Ar}}{^{39}\text{Ar}}\right]_K \dfrac{^{39}\text{Ar}_K}{^{39}\text{Ar}_m}$$

$$= \left[\dfrac{^{40}\text{Ar}}{^{39}\text{Ar}}\right]_K \left[1 - \dfrac{^{39}\text{Ar}_{Ca}}{^{39}\text{Ar}_m}\right]$$

[see Eq. (3.40)] also noting that the last term is the same as the denominator of Eq. (3.41), which can be rewritten as follows:

and ^{39}Ar and for mass discrimination of the mass spectrometer prior to insertion in Eq. (3.42). The $^{40}\text{Ar}^*/^{39}\text{Ar}_K$ value derived from Eq. (3.42) is then used in Eqs. (2.16) or (2.17) for calculation of age or the irradiation parameter, J.

The formula given in Eq. (3.42) is identical to those derived by Brereton (1970) and Mak et al. (1976), although presented in a somewhat different form. The expression derived by Dalrymple and Lanphere (1971) and Dalrymple et al. (1981), also given in McDougall and Roksandic (1974), differs slightly from Eq. (3.42). However, both expressions yield essentially indistinguishable results.

It is evident from Eq. (3.42) that to derive the $^{40}\text{Ar}^*/^{39}\text{Ar}_K$ for argon extracted from a terrestrial sample, the measured $^{40}\text{Ar}/^{39}\text{Ar}$ ratio is corrected for atmospheric argon, with due allowance being made for the presence of neutron-induced ^{36}Ar and ^{39}Ar generated from ^{40}Ca and ^{42}Ca, respectively, and neutron-induced ^{40}Ar generated from ^{40}K.

For extraterrestrial samples, such as lunar rocks and meteorites, calculation of $^{40}\text{Ar}^*/^{39}\text{Ar}_K$ from the isotopic data is somewhat more complex, because of the number of

(3.42) $\quad \dfrac{^{40}\text{Ar}^*}{^{39}\text{Ar}_K} = \dfrac{[^{40}\text{Ar}/^{39}\text{Ar}]_m - 295.5[^{36}\text{Ar}/^{39}\text{Ar}]_m + 295.5[^{36}\text{Ar}/^{37}\text{Ar}]_{Ca}[^{37}\text{Ar}/^{39}\text{Ar}]_m}{1 - [^{39}\text{Ar}/^{37}\text{Ar}]_{Ca}[^{37}\text{Ar}/^{39}\text{Ar}]_m}$

$\qquad \qquad - \left[\dfrac{^{40}\text{Ar}}{^{39}\text{Ar}}\right]_K$

All of the quantities on the right-hand side of Eq. (3.42) are known from measurement of the argon isotope ratios in the gas extracted from the sample, or are separately determined correction factors derived from measurements on irradiated pure salts (see Section 3.7) for the particular nuclear reactor facility being used. It should be emphasized, however, that the measured ratios are corrected for decay of ^{37}Ar

possible sources of argon isotopes. The various components include radiogenic, atmospheric, cosmogenic, and trapped as well as argon produced from calcium and potassium during irradiation in the reactor. The usual procedure (Turner, 1970d, 1971b; Podosek and Huneke, 1973) is to correct for the neutron-induced interferences, and then subtract an estimate of the atmospheric argon contributed from the extraction system based upon blanks measured in the same system under similar conditions, before and after the experiment. Provided that chlorine-derived ^{38}Ar is absent, the measured ^{38}Ar/^{36}Ar ratio can be useful in distinguishing between cosmogenic and trapped argon. The presence of significant chlorine-derived ^{38}Ar interferes with this approach. In such circumstances the ^{36}Ar remaining after corrections for neutron-induced interferences and atmospheric argon is assumed to have some nominal ^{40}Ar/^{36}Ar ratio, commonly 1 ± 1. In many analyses on extraterrestrial materials, corrections for cosmogenic and trapped ^{40}Ar are quite minor, having a very small effect on the calculated ages. Even in cases in which the trapped component is significant it may be possible to interpret the data by using a three isotope correlation plot (Turner, 1971b).

3.17.10 Error estimates

From the mass spectrometric measurements on the argon extracted from a sample, errors can be assigned to the ratios used in Eq. (3.42) to calculate the ^{40}Ar*/^{39}Ar$_K$ ratio. Extrapolation of peak heights or ratios measured in the mass spectrometer to the time of admittance of the gas into the machine, using least-squares procedures, provides a quantitative measure of the errors (see Section 3.17.8), which may be combined quadratically with other identified errors, such as those associated with the determination of the mass discrimination of the mass spectrometer, to yield an overall error estimate for a given ratio. The error in ^{40}Ar*/^{39}Ar$_K$ obviously will be a complex function of the errors in all the terms in Eq. (3.42), and there also will be error magnification as the proportion of radiogenic ^{40}Ar to total ^{40}Ar decreases. Dalrymple and Lanphere (1971) and Dalrymple et al. (1981) derived an error formula by differentiation of an equation very similar to (3.42). They obtained the following expression for the variance of ^{40}Ar*/^{39}Ar$_K$, incorporating some justifiable simplifications:

$$
(3.43) \quad \begin{aligned} \sigma_F^2 &= \sigma_G^2 + C_1^2 \sigma_B^2 \\ &+ [C_4 G - C_1 C_4 B + C_1 C_2]^2 \sigma_D^2 \end{aligned}
$$

where the errors, σ, are absolute errors, usually expressed as a standard deviation, and where
$F = {}^{40}\text{Ar}*/{}^{39}\text{Ar}_K$
$G = ({}^{40}\text{Ar}/{}^{39}\text{Ar})_m$
$B = ({}^{36}\text{Ar}/{}^{39}\text{Ar})_m$
$D = ({}^{37}\text{Ar}/{}^{39}\text{Ar})_m$
$C_1 = ({}^{40}\text{Ar}/{}^{36}\text{Ar})_A = 295.5$
$C_2 = ({}^{36}\text{Ar}/{}^{37}\text{Ar})_{Ca}$
$C_3 = ({}^{40}\text{Ar}/{}^{39}\text{Ar})_K$
$C_4 = ({}^{39}\text{Ar}/{}^{37}\text{Ar})_{Ca}.$

The expression can be converted readily to handle percentage errors, a form that often is found to be more convenient for calculation purposes.

An expression similarly can be derived for the error in the calculated age, t, as shown by Dalrymple et al. (1981) [see also Berger and York (1970) and Dalrymple and Lanphere (1971)]. Thus,

$$
(3.44) \quad \sigma_t^2 = \frac{J^2 \sigma_F^2 + F^2 \sigma_J^2}{\lambda^2 (1 + FJ)^2}
$$

where J is the irradiation parameter, and the other terms are as previously defined.

An alternative, and indeed a more exact approach to error calculation is to use a form of numerical differentiation, a procedure advocated by Roddick (1987). For example, in calculating the error in ^{40}Ar*/^{39}Ar$_K$, each of the measured isotopic ratios in turn is deviated by a given amount, say 0.1%, and a new notional value for ^{40}Ar*/^{39}Ar$_K$ calculated by Eq. (3.42). This new value is compared with the actual calculated value to derive an error magnification term. The overall error for the ^{40}Ar*/^{39}Ar$_K$ ratio is obtained by summing quadratically the errors derived by multiplying the error magnification term by the determined analytical error for each ratio. Thus, by this method, the error associated with each ratio is taken into account by directly calculating its effect on the ^{40}Ar*/^{39}Ar$_K$ ratio, removing the need to make simplifying assumptions or approximations.

4. INTERPRETATION OF RESULTS: AGE SPECTRUM AND ISOCHRON APPROACHES

4.1 INTRODUCTION

The ^{40}Ar/^{39}Ar step heating approach provides a wealth of information on the internal distribution of argon in a sample. Early workers immediately saw the need for an appropriate format to present their results. Although Merrihue and Turner (1966) recognized the value of isotope correlation diagrams for undisturbed samples (see Section 4.13), those studying both disturbed extraterrestrial (Turner et al., 1966) and terrestrial samples (e.g., Fitch et al., 1969) proposed plotting apparent age of a gas fraction against some measure of the state of completion of the experiment. Turner et al. (1966) chose as this parameter the cumulative ^{39}Ar release. Because ^{39}Ar is derived from potassium this plot has the advantage of representing the age of each step in proportion to the total parent isotope present. Fitch et al. (1969) proposed a more qualitative scheme, giving each gas fraction equivalent status regardless of the proportion of gas contained in the step. This approach reflected their more descriptive model of argon behavior. Thus, Fitch et al. (1969) suggested that argon release is dominated by variable activation energy sites rather than by its spatial distribution within the mineral grain as proposed by Turner et al. (1966). A significant disadvantage of their method is that the form of the release pattern can change dramatically with variations in the heating schedule of the sample. The approach advocated by Turner et al. (1966), in which the ^{40}Ar/^{39}Ar apparent age is plotted against cumulative percentage ^{39}Ar released, proved very popular, and has become the standard way of displaying step heating results. The diagram is known as an age spectrum or ^{40}Ar/^{39}Ar release pattern.

The use of the stepwise degassing technique to resolve the internal distribution of ^{40}Ar* within crystals utilizes some aspect of volume diffusion, be it controlled by grain dimension (e.g., Turner et al., 1966) or a thermal activation barrier (e.g., Fitch et al., 1969). In this chapter current models for the interpretation of age spectra based upon diffusion theory are covered. Many release patterns, however, display behavior apparently not attributable to this theory, and possible explanations for these departures are given where possible.

Before embarking on a detailed discussion of age spectra and their interpretation, it is important to point out the fragility of systems dated in this manner. To be successful, the age spectrum approach requires that the original diffusion surfaces are preserved together with the transport mechanism—requirements perhaps unique in geochronology. Thus, interpretation of data from the literature may be difficult because details of sample preparation, not generally reported, are often vital for the unambiguous interpretation of certain argon release patterns.

4.2 SINGLE-SITE DIFFUSION (TURNER MODEL)

Turner et al. (1966), with the publication of the first age spectrum on a sample of the disturbed Bruderheim chondritic meteorite, presented an interpretative model, based upon diffusion theory, that corresponded well with empirical results (Fig. 4.1). This model predicted that the first infinitesimal increment of gas extracted in the laboratory represents the age at which a relatively brief episodic thermal disturbance of the chondrite terminated. Subsequent gas extractions in effect sampled regions of the crystals remote from surfaces of gas loss during the thermal event. In this interpretation, the steps of an age spectrum are a coarse representation of the internal distribution of

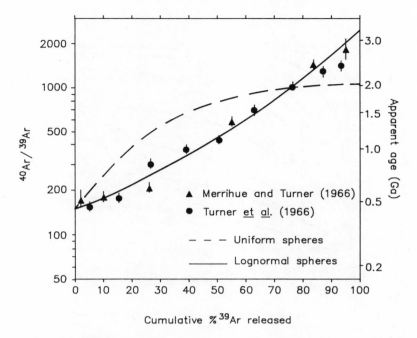

Fig. 4.1. The first age spectrum diagram. After Turner et al. (1966). Data from two separate irradiations of the Bruderheim chondrite are shown together with model age spectra for spheres of uniform radius and spheres with a lognormal distribution of radii that have undergone 90% ^{40}Ar* loss 0.5 Ga ago.

^{40}Ar* resulting from volume diffusion within mineral grains. This theoretical model, elaborated by Turner (1968, 1969), predicted the form of age spectra for both spheres of uniform radii and spheres with a lognormal distribution of radii that have undergone a postcrystallization heating (Fig. 4.2). The modelling assumed a homogeneous distribution of ^{39}Ar, a single thermally activated transport mechanism, a zero ^{40}Ar* boundary concentration, and similar diffusion rates for all argon isotopes. Turner's method was to generate expressions for both the instantaneous release of ^{40}Ar and ^{39}Ar and cumulative loss of ^{39}Ar, and then solve them parametrically for a variable he called x, which is a function of temperature, time, and grain radius. Using this approach a family of curves was generated, corresponding to zero, intermediate, or complete outgassing (Fig. 4.2) of spheres during a younger heating event.

Turner's theoretical spectra for uniform spheres were characterized by convex curves (Fig. 4.2, solid lines), rising, from the age of outgassing, asymptotically toward the original age. Curves for the lognormal distribution of grain radii (Fig. 4.2, dashed lines) were similar to the uniform spheres for samples that had lost less than ~60% of their ^{40}Ar* during the younger heating, but for greater proportions of outgassing the curves became concave.

Huneke (1976), in deriving an identical single-site diffusion model for uniform grain radii, pointed out several implications of the Turner model. One useful contribution of this work was the plotting of theoretical age spectra on a linear age scale. In contrast, Turner's original diagram (Turner et al., 1966; Turner, 1968, 1969) with a logarithmic age axis did not allow easy recognition of pattern similarity between his model results and data presented in a conventional age spectrum diagram. Huneke (1976) made the important point that a sample conforming to the assumptions of the Turner model theoretically would yield a meaningful original age only if no ^{40}Ar* had been lost from the sample. As a result of the decrease in the flux of ^{40}Ar* at the surface of the sample that had experienced ^{40}Ar* loss in nature, the increasing ^{40}Ar*/^{39}Ar$_K$ ratio would asymptotically approach, but never reach the original value. In practice, a sample with

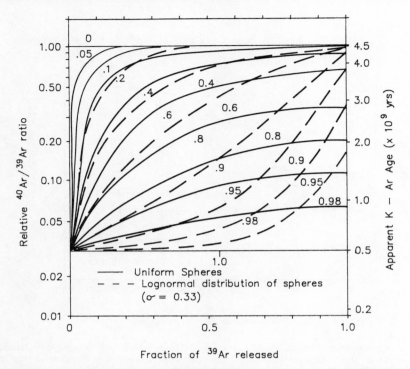

Fig. 4.2. Theoretical age spectra for both aggregates of uniform spheres (solid lines) and a lognormal distribution of sphere radii (dashed lines). After Turner (1968), with permission of author and publisher. In this example, a 4.5-Ga-old sample has experienced varying degrees of $^{40}Ar^*$ loss during an event 0.5 Ga ago.

<10% gas loss could be expected to yield maximum ages in the gas fractions released at the higher temperatures that were within measurement precision of the original age. For samples outgassed by more than ~10%, assessment of $^{40}Ar^*$ loss from the form of the age spectrum can be used, together with the predictive model, to correct for the lowering of maximum ages (Harrison and McDougall, 1980b).

The observation that a number of steps, representing a significant proportion of the gas release, yielded essentially identical ages in undisturbed and some disturbed samples led earlier workers (Dalrymple and Lanphere, 1974; Fleck et al., 1977; Berger and York, 1981a) to propose that meaningful ages of geological events or conditions could be obtained from such age spectra. The broad, flat portions of age spectra of this kind enable a plateau age to be calculated. The Turner model essentially rules out the possibility of uniform ages in all but undisturbed or completely reset samples; this concept, however, has been used by many authors, who require the existence of a plateau before meaningful interpretations can be made. There are also special cases in which experimental artifacts may cause homogeneous $^{40}Ar/^{39}Ar$ release even though the distribution of $^{40}Ar^*$ is nonuniform (see Sections 4.6, 4.9, 4.10, and 4.11).

Despite these caveats, there are samples that can be expected to yield essentially ideal flat age spectra with well-developed plateau segments. For example, measurements on a mineral from a rapidly cooled igneous rock, not subsequently reheated, should give an age spectrum dominated by a plateau accurately reflecting the time that has elapsed since crystallization and cooling of the rock. A good example of an apparently undisturbed age spectrum of this kind is shown in Fig. 4.3, measured on biotite GA1550 from a rapidly cooled monzonite stock or small pluton emplaced at a high level in the continental crust of eastern Australia. This biotite has a K–Ar age of 97.9 ± 0.7 Ma, taken to be the age of crystallization and cooling, supported by much additional data (McDougall and Roksandic, 1974; Williams et al., 1982). The age spectrum

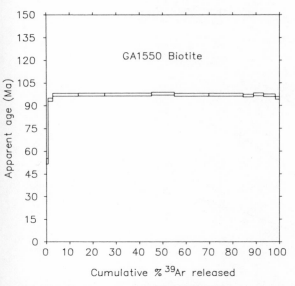

Fig. 4.3. Age spectrum for GA1550 biotite from the Mount Dromedary complex, New South Wales, Australia. After Tetley (1978). This spectrum conforms well with that expected from a sample that has remained thermally undisturbed since crystallization and rapid cooling.

measured on biotite GA1550 (Fig. 4.3) has a well-defined, essentially ideal flat release pattern with an age of 97.2 ± 0.3 Ma over 96% of the ^{39}Ar release (Tetley, 1978).

Several workers have proposed criteria for the identification of such a plateau in an age spectrum, in order to distinguish between undisturbed and disturbed systems (Dalrymple and Lanphere, 1974; Fleck et al., 1977; Lanphere and Dalrymple, 1978; Berger and York, 1981a). There is general agreement that a minimum of at least three contiguous steps with concordant ages is required for a plateau to be defined, often with the additional specification that a significant proportion of the ^{39}Ar release be represented in these same steps, which should also yield a well-defined isochron. Ages are concordant if they do not differ at the 95% confidence level, and comparisons should be made based upon assessment of internal errors only. Thus, any allowance for uncertainties associated with the neutron dosimetry should be excluded in the comparisons, as such factors are common to all steps within a single heating experiment. When comparisons of age are made between results from different experiments, however, the uncertainties in J, the irradiation parameter, must be taken into account. A point worth noting is that, especially for samples that yield good plateaus, the error in the incremental total fusion ^{40}Ar/^{39}Ar age, that calculated by summing results from all the steps in a single heating experiment, may be smaller than the error for an individual step (Berger and York, 1981a).

Harrison (1983) presented a variation of the single-site diffusion model by considering the kind of age spectrum that would be obtained on a sample that had undergone two episodes of gas loss separated by a significant time interval. He found that for samples that had experienced a small loss of ^{40}Ar* followed by a larger degree of outgassing, the form of the resulting age spectrum could not be distinguished from a single heating event (Fig. 4.4). For the other case of a large loss followed by a small degree of outgassing, the ideal model spectrum was distinct, but unlikely to be sufficiently so in natural samples not conforming perfectly to the assumptions, to allow resolution of the two-stage history (Fig. 4.5). However, Harrison (1983) observed that coexisting phases with differing retentivities of ^{40}Ar* might provide the basis for distinguishing a single episodic heating from a multistage history.

Calculation of single-site diffusion models is based on solutions of Fick's Second Law (see Section 5.4) for one of three plausible diffusion geometries, spherical, cylindrical, or infinite slab. In the case of radial diffusion from a sphere, the theoretical ^{40}Ar/^{39}Ar release pattern can be determined from the following calculation using the parameters indicated:

M_0 = initial mass of ^{40}Ar*
M_t = mass of ^{40}Ar* after time, $t > 0$
C = concentration of ^{40}Ar*
C_1 = concentration of ^{40}Ar* produced prior to first outgassing
C_2 = concentration of ^{40}Ar* produced after first outgassing
a = effective spherical diffusion radius, r
D = diffusion coefficient
Δt_1 = duration of first geological outgassing
Δt_2 = duration of second geological outgassing
$\Delta t'$ = duration of laboratory outgassing
$\Delta t_3 = \Delta t_2 + \Delta t'$
C_{39} = unit value of ^{39}Ar$_K$ concentration

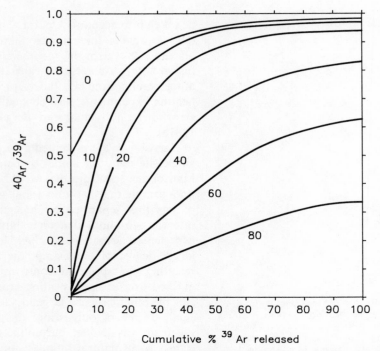

Fig. 4.4. Theoretical $^{40}Ar/^{39}Ar$ age spectra for a sample aggregate of grains of uniform radius that experienced two outgassings separated in time. After Harrison (1983). The sample, with an original age corresponding to $^{40}Ar/^{39}Ar = 1.0$, is 20% outgassed at the time corresponding to $^{40}Ar/^{39}Ar = 0.5$ and then variously outgassed by 0, 10, 20, 40, 60, and 80% of the gas present today (i.e., $^{40}Ar/^{39}Ar = 0$).

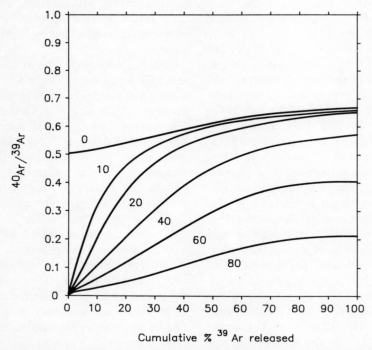

Fig. 4.5. Theoretical $^{40}Ar/^{39}Ar$ age spectra for a sample aggregate of grains of uniform radius that experienced two outgassings separated in time. After Harrison (1983). The sample, with an original age corresponding to $^{40}Ar/^{39}Ar = 1.0$, is 80% outgassed at the time corresponding to $^{40}Ar/^{39}Ar = 0.5$ and then variously outgassed by 0, 10, 20, 40, 60, and 80% of the gas present today (i.e., $^{40}Ar/^{39}Ar = 0$).

Although in our subsequent discussion we assume D to be constant, it in fact varies with temperature (T) in a manner described by $D = D_0 \exp(-E/RT)$, where E and D_0 are the Arrhenius parameters, activation energy and frequency factor, respectively, and R is the gas constant (see also Section 5.9.1).

The fractional loss of ^{40}Ar from a sphere can be derived as follows:

(4.1) $$\delta M = 4\pi r^2 \, dr C(r)$$

Here δM is the differential mass of ^{40}Ar* contained in an infinitesimal radial shell, dr. Integrating from the center to the rim, the initial mass of ^{40}Ar* is the product of the concentration and volume:

(4.2) $$M_0 = \int_0^a 4\pi r^2 C_1 \, dr = (4/3)\pi a^3 C_1$$

The equation for the initially uniform radial distribution of a diffusant in a sphere [Crank, 1975; see Chapter 5, Eq. (A.5.3.14)] describes how the ^{40}Ar* concentration within the crystal changes as a function of the dimensionless parameter, Dt/a^2, which incorporates the cumulative effects of both temperature and duration of heating, as well as defining the dimension of the system.

$$M_t = \int_0^a 4\pi r^2 (-C_1) 2a(\pi r)^{-1} \sum_{n=1}^{\infty} n^{-1}(-1)^n$$
$$\times \exp(-Dn^2\pi^2 t/a^2) \sin(n\pi r/a) \, dr$$
$$= 8\pi^{-1} a^3 C_1 \sum_{n=1}^{\infty} n^{-2} \exp(-Dn^2\pi^2 t/a^2)$$

(4.3)

An expression for the fractional loss, f, of ^{40}Ar* from the sample can be obtained by subtracting the mass of ^{40}Ar* remaining in the sphere after heating from the starting concentration, and dividing by the initial concentration. This gives

$$f = (M_0 - M_t)/M_0$$
$$= 1 - 6\pi^{-2} \sum_{n=1}^{\infty} n^{-2} \exp(-Dn^2\pi^2 t/a^2)$$

(4.4)

The instantaneous flux of gas from a sphere is the product of the concentration gradient at the surface and the surface area, or

(4.5) $$(\text{flux}) = 4\pi a^2 (dC/dr)$$

The concentration gradient at the surface is obtained from the expression describing the ^{40}Ar* distribution in the solid, that is (Crank, 1975)

(4.6) $$C = -2\pi^{-1} a C_1 \sum_{n=1}^{\infty} n^{-1}(-1)^n$$
$$\times \exp(-Dn^2\pi^2 t/a^2) \sin(n\pi r/a) r^{-1}$$

By taking the first derivative with respect to r and evaluating at $r = a$, the surface

(4.7) $$dC/dr|_{r=a} = (d/dr)\left\{\left[-2\pi^{-1} a C_1 \sum_{n=1}^{\infty} n^{-1}(-1)^n \right.\right.$$
$$\left.\left.\times \exp(-Dn^2\pi^2 t/a^2) \sin(n\pi r/a) r^{-1} \right]\right\}$$

Thus, from Eq. (4.5), the instantaneous flux is

(4.8) $$(\text{flux}) = 8\pi a C_1 \sum_{n=1}^{\infty} \exp(-Dn^2\pi^2 t/a^2)$$

Then for a single outgassing in nature of duration Δt_1, the ^{40}Ar*/^{39}Ar$_K$ ratio observed in the laboratory for any fractional loss of ^{39}Ar$_K$ brought about by a period of $\Delta t'$ laboratory heating can be obtained by dividing the function that describes the ^{40}Ar* flux from the sphere [Eq. (4.8) using $t = \Delta t_1 + \Delta t'$] by the ^{39}Ar flux equation [Eq. (4.8) using $t = \Delta t'$]. After cancelling the constant terms we obtain

(4.9) $$\frac{^{40}\text{Ar}^*}{^{39}\text{Ar}_K} = \frac{C_1}{C_{39}} \left\{ \left(\sum_{n=1}^{\infty} \exp[-n^2\pi^2 D(\Delta t_1 + \Delta t')/a^2] \right) \Big/ \left[\sum_{n=1}^{\infty} \exp(-n^2\pi^2 D \Delta t'/a^2) \right] \right\}$$

This expression is relevant only if the laboratory experiment is performed immediately following the natural thermal event. If time elapses between the termination of the geological outgassing and the analysis, the ^{40}Ar*/^{39}Ar$_K$ ratio equivalent to that duration is added to Eq. (4.9).

Equation (4.9) is equivalent to the expressions derived by Turner (1968) and

Huneke (1976). For two natural outgassings of duration Δt_1 and Δt_2, separated in time, the theoretical age spectra can be constructed using the following expressions, derived in a manner analogous to the single outgassing (see Harrison, 1983).

The instantaneous flux of gas from the surface of the sphere following the second geological outgassing is again given by Eq. (4.5) except dC/dr is now defined as

$$\left.\frac{dC}{dr}\right|_{r=a} = \frac{d}{dr}\left\{\left(-2\pi^{-1}aC_1 \sum_{n=1}^{\infty} n^{-1}(-1)^n\right.\right.$$
$$\times \exp[-n^2\pi^2(D\Delta t_1 + D\Delta t_3)/a^2]$$
$$\left.\times \sin(n\pi r/a)r^{-1}\right)$$
$$+ \left[2\pi^{-1}aC_2 \sum_{n=1}^{\infty} n^{-1}(-1)^n\right.$$
$$\times \exp(-r^2\pi^2 D\Delta t_3/a^2)$$
$$\left.\left.\times \sin(n\pi r/a)r^{-1}\right]\right\}$$
$$= 2a^{-1}C_1 \sum_{n=1}^{\infty} \exp[-n^2\pi^2(D\Delta t_1 + D\Delta t_3)/a^2]$$
(4.10)
$$+ 2a^{-1}C_2 \sum_{n=1}^{\infty} \exp(-n^2\pi^2 D\Delta t_3/a^2)$$

In a similar fashion to the single-stage outgassing, the instantaneous flux from a crystal affected by two stages of outgassing is given by

(flux)

(4.11)
$$= 8\pi aC_1 \left\{\sum_{n=1}^{\infty} \exp[-n^2\pi^2(D\Delta t_1 + D\Delta t_3)/a^2]\right.$$
$$\left. + C_2 \sum_{n=1}^{\infty} \exp(-n^2\pi^2 D\Delta t_3/a^2)\right\}$$

For a sample that has experienced two outgassings in nature separated by some time, the $^{40}Ar^*/^{39}Ar_K$ ratio corresponding to a fractional loss of $^{39}Ar_K$ at the time of laboratory outgassing is given by

(4.12)
$$\frac{^{40}Ar^*}{^{39}Ar_K} = \frac{C_1 \sum_{n=1}^{\infty} \exp[-n^2\pi^2(D\Delta t_1 + D\Delta t_3)/a^2]}{C_{39} \sum_{n=1}^{\infty} \exp(-n^2\pi^2 D\Delta t'/a^2)}$$
$$+ \frac{C_2 \sum_{n=1}^{\infty} \exp[-n^2\pi^2 D\Delta t_3/a^2]}{C_{39} \sum_{n=1}^{\infty} \exp(-n^2\pi^2 D\Delta t'/a^2)}$$

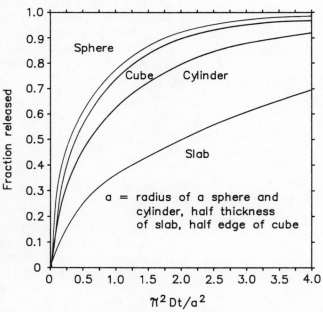

Fig. 4.6. Relationship between fractional loss and the dimensionless parameter $\pi^2 Dt/a^2$ for spherical, cylindrical, plane sheet, and cubic geometries.

The physical significance of this calculation perhaps can best be appreciated from the graphical representation in Fig. 4.6, a plot of cumulative fractional loss against the dimensionless parameter $\pi^2 Dt/a^2$, calculated for the sphere from Eq. (4.4). (For other solutions, see Table 5.1.) The flux of ^{40}Ar* or ^{39}Ar$_K$ from the crystal for any value of $\pi^2 Dt/a^2$ is given by the tangent to the curve. By way of illustration let us consider the curves for spherical and slab geometry. Because of the pronounced curvature of the line calculated for the sphere, which results from the center of mass of the gas being closer to the crystal surface compared with the slab model, a sample that has experienced even minor ^{40}Ar* loss will not reach a uniform ^{40}Ar*/^{39}Ar$_K$ value until very late in the degassing. Why? The small offset in $\pi^2 Dt/a^2$ values between the ^{40}Ar* and ^{39}Ar$_K$ distributions, the magnitude of which corresponds to the intensity of the heating event, places the ^{40}Ar* gradient well up the steeply convex curve and prevents the two gradients from overlapping (i.e., having the same slope) until the sample is essentially outgassed. In contrast, the more linear nature of the slab geometry curve allows the ^{40}Ar* and ^{39}Ar$_K$ gradients to reach a common value (i.e., the slope of the curve above a $\pi^2 Dt/a^2$ value of ~ 0.5) more rapidly than is the case for the sphere. For comparison, theoretical age spectra for a slab outgassed to various degrees are shown in Fig. 4.7 together with age spectra for a sphere outgassed 60 and 80%. It is clearly seen that for equivalent losses the slab geometry reaches a "plateau" much sooner than an equivalent sphere.

4.3 AGE SPECTRA CONFORMING TO A SINGLE-SITE DIFFUSION MODEL

The first published age spectrum, that for the Bruderheim chondrite (Turner et al., 1966), corresponded very well to a theoretical age spectrum of an initially 4.5-Ga-old aggregate with a lognormal distribution of grain radii [for which Turner (1968) later produced independent evidence] that had experienced $\sim 90\%$ outgassing ~ 500 Ma ago. This is the age of a well-documented extraterrestrial fragmenta-

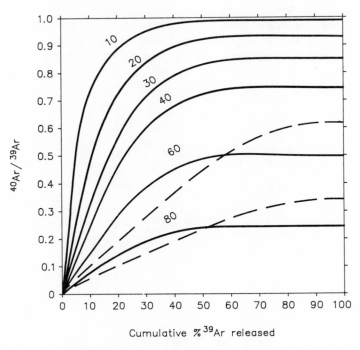

Fig. 4.7. Theoretical ^{40}Ar/^{39}Ar age spectra for a sample of plane sheet geometry (solid lines) that has been outgassed today by 10, 20, 30, 40, 60, and 80% of ^{40}Ar* present. For comparison, curves for 60 and 80% loss from spheres of uniform radius are shown as dashed lines.

tion event (Heymann, 1967). Subsequently, many excellent examples of age gradients in lunar and meteoritical materials were published (e.g., Turner, 1970a, 1972; Bogard et al., 1976), but as these were primarily on whole rock samples, most with complex grain size and potassium distributions, the usefulness of the original model was diminished. Turner (1972), however, produced an age spectrum on a lunar anorthosite, which consisted of essentially homogeneous anorthite. This sample revealed an ^{40}Ar* gradient that could be explained very well by the simple diffusion model (Fig. 4.8). Subsequently, Albarède (1978) published an analysis enabling the transformation of an age spectrum back to the original concentration profile using inverse theory methods (e.g., Parker, 1977). A critical requirement in this treatment is that gas loss be effected by a diffusion mechanism with known Arrhenius parameters and that one argon isotope be uniformly distributed. Albarède argued that the good correlation of diffusion coefficients derived from the release of ^{37}Ar in the first five steps of Turner's experiment on the anorthosite, when plotted against the extraction temperature in an Arrhenius diagram, provided an accurate measure of the general transport rate of argon in the crystals. The calculations resulted in apparent concentration distributions for ^{40}Ar*, ^{39}Ar$_K$, ^{38}Ar, and ^{36}Ar as well as estimates of the error. Although this would seem to be a powerful approach in reconstructing the concentration distribution of argon isotopes in crystals, potentially providing a basis for estimating the original age of a sample subsequently affected by gas loss, the absence of subsequent studies attests to both the complexity of the analysis as well as the scarcity of samples to which the diffusion-related assumptions hold during heating *in vacuo* (see Section 4.11). However, having identified an age spectrum that corresponds to a single-site, single-radius episodic-loss model, a kind of inverse experiment can be effected by observing the profile directly from Fig. 4.9, taken from Crank (1975, p. 92), for the appropriate value of Dt/a^2.

Terrestrial samples were slow to reveal unambiguous examples of age spectra that could be interpreted in terms of a Turner-type model, for reasons discussed in this chapter. Nevertheless some good examples of well-behaved terrestrial age spectra were documented by Harrison and McDougall (1980b) in a study of the response of igneous hornblendes to a later heating episode by intrusion of a nearby granitic pluton. Although most of the hornblendes also revealed excess ^{40}Ar uptake profiles (see Section 4.8), three age spectra conformed well

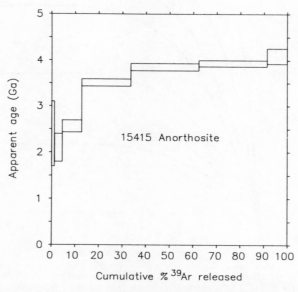

Fig. 4.8. Age spectrum of lunar anorthosite 15415 showing an apparent diffusion loss profile. After Turner (1972).

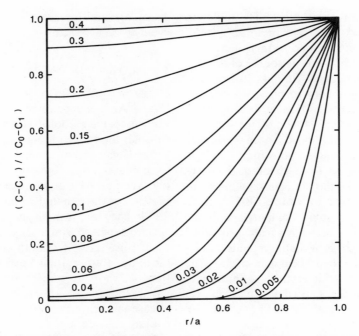

Fig. 4.9. Concentration of ^{40}Ar (C) within a spherical mineral grain of radius a, following recent thermal disturbance of varying intensity Dt/a^2 (values shown on curves), assuming an initially uniform concentration (C_1) and a boundary concentration of C_0. After Crank (1975), with permission, from *The Mathematics of Diffusion*, Clarendon Press, Oxford.

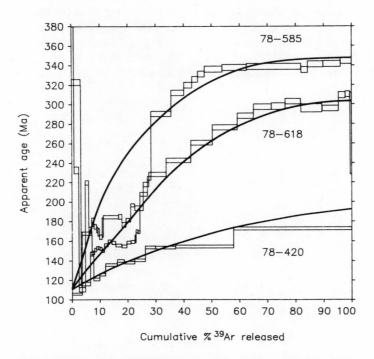

Fig. 4.10. Age spectra of hornblendes from samples at varying distances away from the contact with a 114-Ma-old granodiorite. From Harrison and McDougall (1980b). The hornblendes, originally 367 Ma old, lost 31, 57, and 78% of their ^{40}Ar* present during reheating. Also shown as solid lines are model age spectra for the corresponding degree of ^{40}Ar* loss. The excellent fit gives confidence in the simple single-site diffusion model.

to model spectra (Fig. 4.10). The different amounts of $^{40}Ar^*$ loss from these samples of 78, 57, and 31% correspond to their increasing distance away from the heat source of 0.3, 1.0, and 2.5 km, respectively. In fact, using conductive heat flow theory to provide estimates of temperature and duration of heating, Harrison and McDougall (1980a,b) showed that these data were well aligned on an Arrhenius diagram for a single activation energy. Although deviations from theoretical curves occur, for example, the plateaulike segment at 10–20% ^{39}Ar release in sample 78-585, these perturbations seem small compared with the likely differences between real geological samples and the kinds of simple crystals modelled in deriving the theoretical age spectra. The results on all three samples allow estimation of the age of reheating, caused by the emplacement of the nearby granitic body 114 Ma ago, as having occurred ~110 Ma ago. The two least affected spectra from samples 78-618 and 78-585 permit estimation of the original age of crystallization of the host gabbro at ~370 Ma [keeping in mind Huneke's (1976) advice, see Section 4.2] very close to the established age of 367 Ma. In summary, three samples of hornblende with chronologically meaningless conventional K–Ar ages yielded $^{40}Ar/^{39}Ar$ age spectra that allowed both the original crystallization age and the age of the younger thermal event to be assessed.

4.4 SLOW COOLING MODELS

Harrison and McDougall (1982) presented $^{40}Ar/^{39}Ar$ age spectra on microcline samples known to have cooled slowly (at ~5°C/Ma) through the range of temperatures over which $^{40}Ar^*$ began to be retained. These age spectra were characterized by a linear increase in age over the first 65% of gas release, followed by a plateau for the remainder of gas evolution (Fig. 4.11). They are similar in some respects to spectra expected from episodic loss of $^{40}Ar^*$ by reheating (cf. Fig. 4.7), but have a very different meaning. Rather than corresponding to the time of a short-lived thermal event, the ages for gas extracted at low temperature in these release spectra reflect the cessation of a long cooling episode.

In his development of the theory of closure temperatures, Dodson (1973) produced expressions for the limiting isotope distribution within crystals of different geometry following slow cooling. The problem is to determine the net radial accumulation of a diffusant that is being uniformly produced at a constant rate in a system that is losing the substance from the boundary, but at a rate that is decreasing

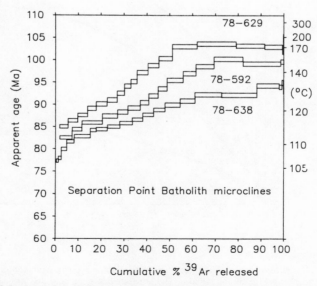

Fig. 4.11. Age spectra of slowly cooled microclines from the Separation Point Batholith, New Zealand. After Harrison and McDougall (1982). The cooling history for 78-592 is shown on the right vertical axis.

geometrically with time. The apparent closure temperature of some radial position in the crystal, which corresponds to a concentration deficit, can be calculated from these expressions and translated into an apparent age, assuming a cooling history linear in $1/T$. Harrison and McDougall (1982) evaluated the equation for a plane sheet employing parameters appropriate for their samples, and demonstrated that the form of the concentration distribution was similar enough to that produced by episodic loss to preclude using this information alone as a diagnostic tool. Dodson (1982, personal communication, 1986) has calculated age spectra resulting from slow cooling in spherical, cylindrical, and plane sheet geometries (Fig. 4.12). These results show that the magnitude of the preserved age gradient resulting from slow cooling is strongly dependent upon the diffusion geometry. For a given cooling history, the age spectrum obtained from a sphere is characterized by virtually no age plateau and a steeper and larger age gradient than is the case for either the cylinder or slab. The equivalently affected cylinder is intermediate in its plateau development and age profile, whereas the slab has by far the least pronounced age gradient, with a well-developed plateau over the last 60% of ^{39}Ar release. Without sufficient geochronological information, it is possible to misinterpret the significance of apparent age gradients in release patterns for samples that have cooled slowly through the temperature region of partial ^{40}Ar* retention. Detailed mathematical derivation of Dodson's closure temperature theory is given in Chapter 5.

4.5 NUMERICAL SOLUTIONS

The similarity between single-site diffusion models of episodic loss and slow cooling models, documented in the preceding section, often precludes obtaining a unique solution from a single age spectrum experiment. These models, apart from not describing actual mineral geometries realistically, are highly idealized representations of thermal histories, restricted by the need to make mathematical simplifications in order for the analytical solutions to be tractable. An attractive adjunct to these interpretative models are numerical solutions employing a wide range of initial ^{40}Ar* distributions in minerals, boundary concentrations, and thermal histories. Zeitler (1988, personal communication) has described a numerical model of diffusion in a plane sheet that can accommodate an arbitrary thermal history. The model employs a finite-difference solution to Fick's Second Law (see Chapter 5), using the Crank–Nicolson implicit method (Crank, 1975), and describes both the concentration distribution within a slab and the resulting age spectrum, assuming either a zero argon surface concentration or uptake of argon by the crystal from an ambient reservoir (see Section 4.8). Several of Zeitler's examples illustrate both the power of the method in accommodating realistic temperature histories and the ambiguities in interpretation. Figure 4.13 shows the age spectrum for a sample that experienced slow cooling at a constant rate through the interval corresponding with the closure temperature. As mentioned in the preceding section, the similarity of this spectrum to that expected in the case of episodic loss requires additional geological information to infer this interpretation from the ^{40}Ar/^{39}Ar data alone. Figure 4.14 illustrates a case in

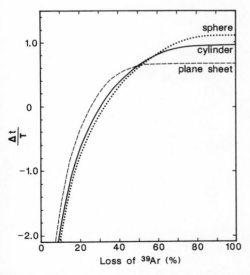

Fig. 4.12. Theoretical age spectra for slowly cooled minerals of spherical, cylindrical, or plane slab geometry, as calculated by M. H. Dodson (personal communication). The vertical axis ($\Delta t/\tau$) is the age difference between the bulk sample and age at any radial position.

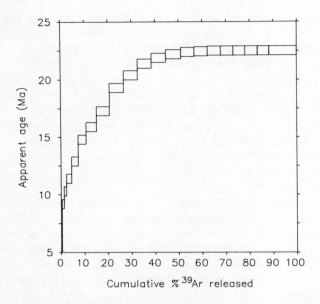

Fig. 4.13. Theoretical age spectrum for a sample that has experienced slow cooling at a constant rate through the closure temperature interval. After Zeitler (1988). Calculated for a cooling rate of 12°C/Ma for an aggregate of uniform thickness slabs ($a = 1\ \mu$m), with activation energy of 50 kcal/mol and pre-exponential factor $D_0 = 2.7 \times 10^{-4}$ cm^2/s for argon diffusion.

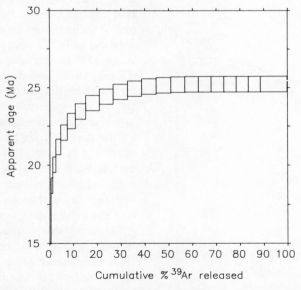

Fig. 4.14. Identical theoretical age spectra generated from two different thermal histories. After Zeitler (1988). Calculations based upon slab geometry and with the same parameters for size, activation energy, and D_0 as used in Fig. 4.13. One case is for slow cooling from 400°C at 100 Ma to 275°C at 11 Ma, followed by more rapid cooling to 10°C by the present time. The second case is for a sample undergoing episodic heating, starting from 11°C prior to 11 Ma, heated to 300°C from 10.5 to 9.6 Ma ago, and returning to 11°C by 9 Ma.

which a simple two-stage cooling history, with slow cooling at a constant, linear rate followed by rapid cooling, yields an age spectrum identical to that found for a single episodic loss. These examples demonstrate the need for independent geological and geochronological evidence when interpreting age spectra.

Careful sample characterization and a sound understanding of the geological context, coupled with numerical solutions should prove to be very useful in assessing subtle aspects of temperature histories; for example, deviation from linear or hyperbolic cooling curves in cases in which episodic loss is ruled out. Future developments could include more realistic mineral geometries, describing crystallographic dependent diffusion rates, but will have to await advances in understanding of argon diffusion in silicates, which is still at a fairly primitive stage (see Chapter 5).

4.6 MIXED PHASES

The interpretation of disturbed spectra is least ambiguous in samples that may be described by the simplest model, that is, one activation energy for argon diffusion and a uniform grain shape and size. However, a great many samples of geological importance are more complex and analysis of multiphase aggregates is necessary. A few examples are fine-grained rocks that make mineral separation unattractive (e.g., shale), small sample size (some lunar and meteoritic materials), and minerals with complex exsolution features (e.g., intermediate plagioclase).

Gillespie et al. (1982) presented a theoretical analysis of age spectra resulting from partially outgassed samples with two different grain sizes and activation energies, which were the result of either the presence of two separate phases or one mineral with two distinctive lattice sites for argon (Fig. 4.15). Where the contrast in activation energy is strong, $^{40}Ar/^{39}Ar$ analysis effectively separates argon from the two domains and relatively unambiguous temporal information related to both outgassing and crystallization can be obtained from the age spectrum. Gillespie et al. (1982) illustrated their model with an age spectrum analysis of a granitoid xenolith (Fig. 4.16), which was partially degassed upon being incorporated into a basalt flow. The first fractions of argon released, from sites of low activation energy, gave the time of outgassing in the hot magma. The subsequent argon fractions, from sites of high activation energy, yielded dates which rose precipitously toward the original age of the granitoid. Extension of the treatment of Gillespie et al. (1982) from

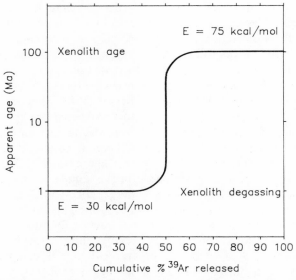

Fig. 4.15. Theoretical age spectrum for a 100-Ma-old sample, degassed 1 Ma ago, consisting of grains with two different activation energies (E). After Gillespie et al. (1982).

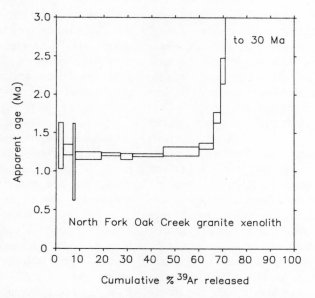

Fig. 4.16. Age spectrum for a granitoid xenolith from a Cretaceous quartz monzonite that was found in an early Pleistocene basalt (Gillespie et al., 1982). The reheating by the basalt caused a portion of the K-feldspar to degas. This is revealed in the first 60% of gas release, allowing assessment of the age of the basalt as 1.19 ± 0.03 Ma. Older ages found for gas released at higher temperatures in the experiments reflect more retentive siting of the argon.

binary models to assemblages of arbitrary numbers and types of argon-bearing phases is relatively straightforward.

Analysis of the Kirin (Jilin) meteorite by the $^{40}Ar/^{39}Ar$ technique provides an excellent example of the amount of chronological information that can be recovered from a thermally disturbed polymineralic aggregate with a nonuniform grain distribution. Wang et al. (1980) and Harrison and Wang (1981) measured age spectra on three different samples of the Kirin meteorite (see also Müller and Jessberger, 1985). All three spectra are characterized by apparent ages that sequentially rise, fall, and then rise monotonically to establish a plateau (Fig. 4.17). In essence, these age spectra result from the superimposition of $^{40}Ar^*$ diffusion profiles from both feldspars and pyroxenes. Fortunately, the contrast in release kinetics between the feldspars and pyroxenes produces an offset that is sufficient to resolve the feldspar diffusion profile, recorded early in the release pattern, from the age spectrum in each case. The age gradients corresponded well with the model of Turner (1968) for partially degassed samples with a lognormal distribution of grain radii, an expected condition for a whole rock meteorite sample.

It is possible to produce apparent diffusion-type age spectra in bimineralic samples if both phases have a uniform $^{40}Ar^*$ distribution, and if as a result of their contrasting retentivities, the two minerals have different ages. This effect is often observed in hornblende release patterns which contain a small (∼1%) amount of composite biotite. Biotite typically contains an order of magnitude more potassium than hornblende, and is known to record younger K–Ar ages if cooling were slow. The mixing of argon from these two reservoirs can produce artifacts in age spectra that in most respects resemble those caused by partial diffusive loss. Berger (1975) pointed out a good example of this problem in the age spectra for hornblendes from the contact aureole of the Eldora Stock, Colorado. Although the spectra revealed apparent age gradients similar to those predicted for the samples by the model of Turner (1968), Berger (1975) noted that the samples contained 1–2% fine-grained biotite composites, and that subtraction of the biotite component would remove the gradient from the hornblende spectra. Although it would seem useful to monitor the apparent K/Ca ratio to assess this effect, the first fractions of argon from pure amphibole separates commonly show high K/Ca

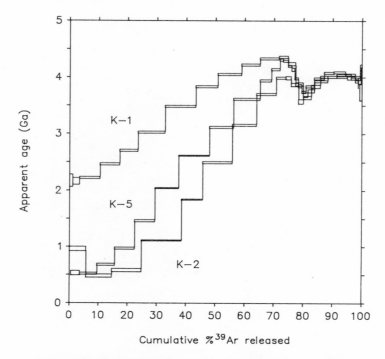

Fig. 4.17. Three age spectra from different samples of the Kirin meteorite (Wang et al., 1980; Harrison and Wang, 1981). The first 70% of gas release corresponds to feldspar and the subsequent degassing reflects argon loss from pyroxene.

ratios, possibly due to exsolution features, making this test somewhat ambiguous.

A more cryptic example of this effect was described by Wijbrans and McDougall (1986) for white mica separates from metamorphic rocks on the island of Naxos in the Greek Cyclades. Two metamorphic events are recognized in the rocks of Naxos; the first metamorphism (M_1) occurred at relatively high pressure and low temperature, and the second (M_2) was related to the formation of a thermal dome. Far from the effect of the thermal dome, phengite (essentially muscovite with some Mg or Fe^{2+} substitution for octahedrally coordinated Al) is of widespread occurrence in the metapelites as the stable white mica phase associated with the M_1 metamorphism. During the second metamorphism, muscovite began to develop, and in rocks close to the thermal dome this mineral is the dominant white mica. At intermediate distances, samples were only partly recrystallized during the second metamorphism, and contain both phengite and muscovite. The pure muscovite yielded a flat release pattern giving an age of ~12 Ma (81-556 muscovite, Fig. 4.18a), interpreted as the time of cooling below the temperature for quantitative retention of radiogenic argon in muscovite after the M_2 metamorphism. Pure phengite showed a monotonically rising age spectrum to a maximum of ~43 Ma (81-580 phengite, Fig. 4.18a), considered to reflect the time since original closure. White mica separates from the intermediate zone revealed convex-upward or hump-backed age spectra (Fig. 4.18b). The key to understanding this behavior comes from the relative degassing characteristics of the pure muscovite and phengite. Figure 4.19 shows the cumulative gas loss with extraction temperature for the two phases. Apparently the younger muscovite loses its gas over a similar but broader temperature interval compared with phengite, thus, dominating the early and late stages of gas release. An artificial mixture of the two pure mineral separates in the proportion observed in some samples from the intermediate zone reveals a similar age spectrum (Fig. 4.18a) as that observed in the natural mixtures, with low initial ages rising to a peak before falling rapidly back toward the muscovite age. In such cases it is clear that the crudely developed

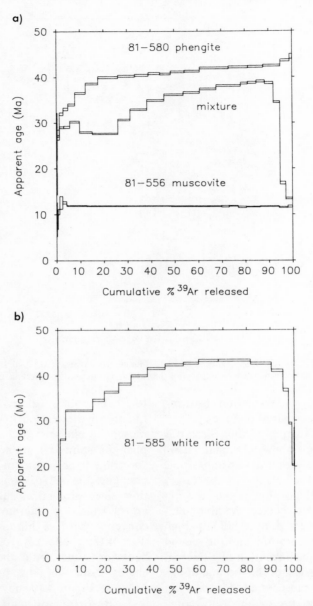

Fig. 4.18. Age spectra measured on white micas from metamorphic rocks, Naxos, Greece (Wijbrans and McDougall, 1986). (a) Age spectra for pure muscovite (81-556) and pure phengite (81-580), and an artificial mixture of 75% phengite (81-580) and 25% muscovite (81-556). (b) Age spectrum for 81-585 white mica, which consists of a natural mixture of phengite and muscovite. Note the similarity between the age spectrum for 81-585 and that for the artificial mixture shown in (a).

plateau of intermediate age has no geological meaning, but is simply the result of the mixing of two phases of different age.

The complexities introduced into age spectra owing to the presence of exsolution features are now well-documented phenomena. In a study of calcic plagioclases from the Broken Hill region of New South Wales, Harrison and McDougall (1981) observed U- or saddle-shaped release patterns (Fig. 4.20) of a kind that had been previously documented in the literature (Dalrymple and Lanphere, 1974; Dalrymple et al., 1975; Lanphere and Dalrymple, 1976; Berger and York, 1981a). In

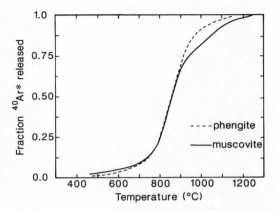

Fig. 4.19. Cumulative gas loss versus temperature diagram for phengite and muscovite from Naxos. Their differences in argon release at high temperature, together with their age contrast, results in age spectra of the kind shown in Fig. 4.18b. After Wijbrans and McDougall (1986).

the Broken Hill examples, ages in the low- and high-temperature portions of the gas release far exceeded the known age of the Earth, indicating unambiguously the presence of excess argon, although the central portions of the age spectra revealed geologically meaningful ages in the samples least contaminated by excess argon. A curiosity of this type of pattern is how the excess argon, probably introduced during a younger thermal event, was the last gas to be released from the plagioclases during the extraction experiment. Calculation of diffusion coefficients from the ^{39}Ar and ^{37}Ar released from potassium and calcium sites, respectively, during the experiments assisted in the identification of three reservoirs of ^{40}Ar (Fig. 4.21). Upon imaging the bytownite using a transmission electron microscope (TEM), Huttenlocher exsolution (Fig. 4.22) was observed. The model for Huttenlocher development is the growth of distinct phases that form alternating lamellae: a homogeneous anorthite-rich ($\sim An_{88}$) phase, and a complex, modulated albite and anorthite phase termed "e" plagioclase from the electron diffraction pattern identification [see Smith (1974) for a review]. This latter material is characterized in bytownite feldspars as alternating pure albite and anorthite lamellae with a wavelength of ~ 50 Å, the anorthite region being about three times the width of the albite zones. Each couple is separated from the next by what is termed an

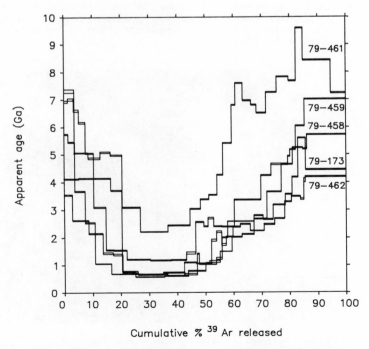

Fig. 4.20. Age spectra of calcic plagioclases from amphibolites, Broken Hill, Australia. After Harrison and McDougall (1981). The saddle-shaped release patterns are characteristic of excess ^{40}Ar and result from the contaminating argon being held in two different lattice sites with contrasting behavior.

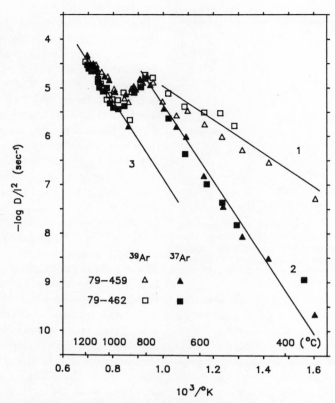

Fig. 4.21. Arrhenius plot of reactor produced isotopes from two of the five samples for which age spectra are given in Fig. 4.20. Three separate diffusion domains are evident, and these can be correlated with features seen in the transmission electron micrograph in Fig. 4.22. Similar slopes for domains 2 and 3, indicating an activation energy of ~35 kcal/mol, relate to anorthite-rich lamellae. The offset of the plots is accounted for by a difference of a factor of ~10 in lamellae width in the two domains. After Harrison and McDougall (1981).

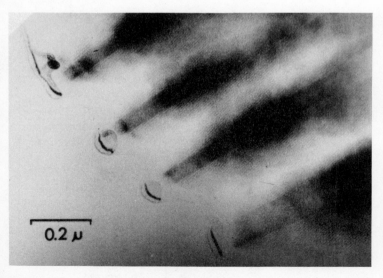

Fig. 4.22. Transmission electron micrograph of 79-459 plagioclase showing both modulated "e" plagioclase, alternating pure albite (domain 1) and anorthite (domain 2), and the coarser transitional anorthite (domain 3), seen as the darker bands about 0.06 μm (600 Å) wide on average. The latter type of domain is typically 10 times the width of the "e" structure—in agreement with the estimate obtained directly from Fig. 4.21. After Harrison and McDougall (1981).

antiphase boundary, seen by Wenk (1979) as a surface of large scale diffusion. The coherence between the three reservoirs observed from the Arrhenius plot, and the exsolved phases imaged in the TEM, was taken as evidence that the excess argon was trapped mainly in two reservoirs, the antiphase boundaries, and the large (~ 600 Å) anorthite lamellae. Harrison and McDougall (1981) reasoned that the low-temperature dependence of anion vacancy transfer would allow excess argon into these regions at low temperatures, but that at the high temperatures of laboratory extraction they would be trapped in the structure relative to the much more mobile atoms in cation sites. Supporting evidence for this model was offered by Claesson and Roddick (1983), who showed that the release of excess argon from calcic plagioclases was accompanied by ^{38}Ar produced by a (n,γ) reaction on ^{37}Cl, and therefore probably originating in an anion site. Zeitler and Fitz Gerald (1986) reported similar saddle-shaped age spectra for some potassium feldspars. They showed that much of the excess ^{40}Ar is concentrated near grain boundaries, with a significant proportion located in more retentive sites than the ^{40}Ar*, under the conditions of a laboratory step heating experiment, as found also for plagioclase. Cogent arguments were advanced for the hypothesis that this excess ^{40}Ar, released at high temperature in the vacuum system, also originates from anion sites.

Harrison and Fitz Gerald (1986) observed a

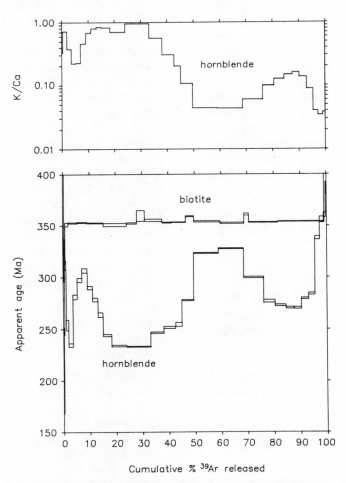

Fig. 4.23. Coexisting hornblende and biotite release patterns from an amphibolite near North Walpole, New Hampshire. After Harrison and Fitz Gerald (1986). Calculated K/Ca ratios for the hornblende vary antipathetically with the oscillation in ages, an indication that the complexity of this spectrum is due to hornblende–cummingtonite exsolution.

dramatic contrast between age spectra of coexisting biotite and hornblende separates from an amphibolite at North Walpole, New Hampshire. The biotite age spectrum was essentially undisturbed, yielding a K–Ar age of 343 Ma, representing cooling from the Acadian Orogeny (Naylor, 1971). The hornblende age spectrum was considerably more complex with three complete cycles from high to low and back to high ages, with the oldest ages concordant with biotite and the youngest ages as low as 200 Ma (Fig. 4.23). Virtually coincident with the cyclic pattern was a sympathetic variation in the apparent K/Ca ratio, with low values associated with the oldest ages in the hornblende. Electron microprobe analysis and TEM imaging provided evidence of cummingtonite exsolution from hornblende on at least three scales. Clearly, the finest lamellae of cummingtonite had lost ^{40}Ar* at temperatures <250°C, whereas the Mg-rich biotite (see Section 5.9.3.2) remained essentially unaffected by this event.

Thus, it will be evident that although the presence of mixed phases may invalidate assumptions of the simplest interpretative model, proper sample characterization in many cases allows identification of separate reservoirs, leading to meaningful geochronological interpretation.

4.7 RESOLUTION WITHIN THE AGE SPECTRUM

The stepwise nature of ^{40}Ar/^{39}Ar heating experiments provides only discrete discontinuous age spectra. Given infinitely small steps, these would resemble model spectra, and the time of thermal events affecting single-phase samples, especially those that have experienced only minor argon loss, could be identified. However, argon is extracted from real samples in relatively few finite steps, so that the measured ^{40}Ar*/^{39}Ar$_K$ ratio is the gas composition averaged over the portion of the near surface concentration profile sampled in that step. In cases where the ^{40}Ar*/^{39}Ar$_K$ ratio is changing rapidly, such as in the first several percent of ^{39}Ar released from samples that have experienced minor ^{40}Ar* loss, the calculated age represents a mixture between the original crystallization age and the age at which ^{40}Ar* loss occurred. The apparent age will always reflect, and change with, the choice of the times and temperatures in the laboratory extraction schedules, and normally can only be regarded as providing a maximum age for the younger thermal event that caused the ^{40}Ar* loss. Increasing the number of steps in the early stages of the experiment allows the apparent age for the gas released initially to be determined with progressively greater fidelity. This is important because it is implicit from diffusion theory that the first infinitesimal fraction of argon released from an ideal sample should yield the age at which the sample was last degassed. For real samples and extraction schedules, it is possible in some cases to extrapolate low-resolution data to determine an outgassing age. A problem to be kept in mind, however, is that because many geological samples appear to contain excess argon in varying amounts near the grain boundaries or in other nonretentive sites, such extrapolations must be done wisely if useful data are to be obtained.

An experimental artifact that may obscure natural age gradients was pointed out by Hanson et al. (1975), who noted that thermal gradients within the extraction crucible might cause premature release of gas originating in the interiors of some grains, whereas in the cooler regions of the furnace, other grains were only beginning to outgas. This would tend to homogenize or blur nonuniform ^{40}Ar* distributions, giving in some cases an impression of a geologically meaningful plateau age.

4.8 EXCESS ARGON

4.8.1 Introduction

Potassium–argon ages much greater than are geologically plausible often are observed, and in some cases they are older than the age of the Earth. This led workers in the mid-1960s to conclude that in certain circumstances argon with isotopic ratios (^{40}Ar/^{36}Ar) much greater than that of atmospheric argon could be incorporated in silicate minerals and melts [see Dalrymple and Lanphere (1969) for summary]. With no independent criterion with which to assess whether or not a sample contains excess ^{40}Ar, that is ^{40}Ar not produced from *in situ* decay of ^{40}K (see Table 2.1), this recognition further weakened the ability to unambiguously interpret K–Ar mineral ages, especially from plutonic and metamorphic environments. The potential of the ^{40}Ar/^{39}Ar age spectrum technique for resolving the internal argon isotope distribution within certain minerals provided hope that samples containing excess ^{40}Ar might be clearly identified. For samples that have not come to equilibrium with the ambient argon pressure by diffusion uptake, the apparent age gradient resolved using this method often allows meaningful chronological data to be obtained from part of the gas release. Note, however, that by an argument similar to that of Huneke (1976), the existence of this contaminating concentration gradient within the crystal would have an effect on the plateau portion of the age spectrum, raising the apparent age, even if only nominally. The variable success of the age spectrum approach in recognizing samples containing excess ^{40}Ar is intimately related to our understanding (or lack of it) of the siting and release kinetics of excess ^{40}Ar, as well as the behavior of minerals in a vacuum heating environment. Early studies of samples known to contain excess ^{40}Ar (Lanphere and Dalrymple, 1971, 1976; Pankhurst et al., 1973; Dalrymple et al., 1975) revealed an age spectrum pattern common to a number of minerals, including some biotites, plagioclases, alkali feldspars, and pyroxenes, that has become known as a "saddle-shaped" age spectrum (high initial ages decreasing to a minimum, and then rising to high apparent ages in the last portion of the gas release), and in some cases a flat release pattern for biotite. An example of the latter produced a well-defined plateau with an apparent age of 5030 Ma, for a biotite separated from the Isua supracrustal rocks of West Greenland (Pankhurst et al., 1973). Lanphere and Dalrymple (1976) were the first to note that the minimum in the saddle of the former type could approach, and presumably attain in certain cases, a geologically meaningful age. Although these initial results suggested that the technique might not fulfill its promise, recent work has further defined characteristics of age spectra, from certain minerals, that can be ascribed unambiguously to the presence of excess ^{40}Ar.

4.8.2 Excess ^{40}Ar uptake by a homogeneous phase

The Turner model of episodic ^{40}Ar loss by diffusion assumes that all grain surfaces are maintained at a zero concentration of ^{40}Ar. The relatively low solubility of argon in silicates assures that in most cases this condition is satisfied, but circumstances clearly arise in which sufficiently high concentrations of gas occur in the intergranular region that allow significant uptake of argon, usually extremely radiogenic, by the crystal. For the case in which the argon partial pressure outside the grain remains constant, the form of the resulting age spectrum is the mirror image of the theoretical age spectra of Turner (1968), for a homogeneous aggregate showing varying degrees of argon loss.

A good example of this behavior was given by Harrison and McDougall (1980b), who examined hornblendes from a Paleozoic gabbro that had experienced a reheating owing to the intrusion of a neighboring Cretaceous granitic body. Although ^{40}Ar/^{39}Ar age spectra from several of the hornblendes revealed loss profiles (see Section 4.3), the majority of the age spectra displayed extremely steep concentration profiles with initial ages as great as 3500 Ma decreasing to ~370 Ma within the first few percent of gas release. These steep profiles were interpreted to be excess ^{40}Ar uptake profiles that had been frozen into the margins of the hornblende crystals following a relatively short interval at high temperature, during which time ^{40}Ar* released from minerals lower in the crust resided in the pore space. Etching experiments on one of these affected hornblendes supported this interpretation. Figure 4.24 shows the age spectrum measured on hornblende sample 78-589. It has the characteristics of ^{40}Ar* uptake superimposed on a small degree of ^{40}Ar* loss, together with a well-defined plateau at 366 Ma (and an isochron age of 366 ± 4 Ma). In this case, ages as old at 2.4 Ga were found for the gas released in the early stages of the experiment from the

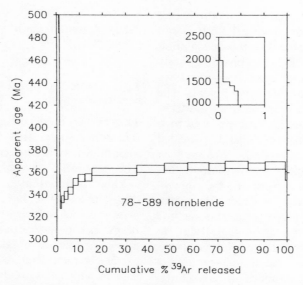

Fig. 4.24. Age spectrum of a homogeneous hornblende that has taken up excess ^{40}Ar. After Harrison and McDougall (1980b). Note the expanded scale on the inset. The excess ^{40}Ar is situated entirely at the grain margins.

sample. The ages subsequently decreased to plausible values within a few percent of ^{39}Ar release. The good correspondence of age spectra from these samples to a simple diffusion model was in part because of the sluggish kinetics of amphibole unmixing, which, for these igneous hornblendes, prevented significant exsolution (see Section 4.6). The dramatic gradient found was revealed by a high resolution analysis, in this case 25 steps. Had only a few steps been performed on this sample, it is likely that recognition of the gradient would have been effectively obscured, especially in view of the overlapping of the excess ^{40}Ar profile by a small diffusion loss profile. It is obvious that fewer extraction steps will tend to mix argon from regions with contrasting age, and perhaps result in a spurious interpretation of the resulting pattern.

4.8.3 Excess ^{40}Ar uptake by a mixture

Pure mineral separates often behave as complex, multimineralic systems. The widespread existence of miscibility gaps in common mineral solid solution series can make it difficult to obtain a single-site (with respect to ^{40}Ar* diffusion), single-dimension mineral separate necessary for the simplest interpretative model to be valid. In slowly cooled rocks, all feldspars, most amphiboles and pyroxenes, and some micas will exsolve, or unmix separate phases, with dimensions small enough to preclude physical separation. An example relevant to this discussion is described in Section 4.6. Briefly, Harrison and McDougall (1981) found saddle-shaped age spectra in calcic plagioclase separated from granulites at Broken Hill, New South Wales (Fig. 4.20). The results were interpreted in terms of three exsolved phases having different diffusion behavior and offering two choices of siting of the excess ^{40}Ar in cation or anion vacancies. The low-temperature dependence of the anion sites allows relatively high argon diffusivity at the inferred temperature of excess ^{40}Ar incorporation of about 300–400°C. Thus, a different siting of excess ^{40}Ar relative to ^{40}Ar*, ^{39}Ar, and ^{37}Ar is indicated, as the last three have been shown to behave congruently (cf. Dalrymple and Lanphere, 1974), being located in or having access to cation sites in undisturbed samples. At high extraction temperatures, the relatively low diffusion jump probability (see Section 5.9) of the argon in anion sites causes the movement of this component to be retarded in the crystal, resulting in extremely high apparent ages late in the gas release.

A similar argument was extended by Harrison and McDougall (1981) to explain the saddle-shaped age spectra of amphiboles from Broken Hill that also contained excess ^{40}Ar, presumably introduced during the same event that resulted in contamination of the coexisting plagioclases. These amphiboles, known to contain hornblende–cummingtonite exsolution (Binns, 1964), yielded high ages for the gas released in the early stages of the experiments, in several cases greater than the age of the Earth. With further heating, ages decreased to a minimum value or plateau, depending upon the resolution of the experiment, thereupon rising to ages far in excess of the known age of the major metamorphism at 1660 Ma. An example of these results is shown in Fig. 4.25, which illustrates the salient features of this kind of age spectrum. In this case, the apparent ages decrease from a maximum of ∼4000 Ma to a plateau over 20% of the ^{39}Ar release of indicated age 1577 Ma, before rising to ages as great as 2300 Ma. The 1577 Ma age, common to a number of samples, is geologically plausible as the time of closure (see Section 5.9.3.3) for argon in the amphibole, because a Pb–Pb isochron on apatites from the region yielded the same age (1565 ± 20 Ma) within uncertainties (Gulson, 1984). An isotope correlation plot of the six steps in the plateau found in the age spectrum for the amphibole produced a well-fitting isochron with an age (1573 ± 5 Ma) and trapped argon component (^{40}Ar/^{36}Ar = 348 ± 58) indistinguishable from the plateau age and atmospheric argon composition, respectively. Although this sample has absorbed a significant quantity of excess ^{40}Ar, probably during a thermal event dated at ∼500 Ma, a portion of the crystal degassed in the step heating method apparently was unaffected and revealed chronologically meaningful data. As anion diffusion (Farver and Giletti, 1985) has a lower temperature dependence in hornblende than argon diffusion (Harrison, 1981), a similar, although more speculative argument could be extended by analogy with the coexisting plagioclase, that is, the high ages found for the gas released in the final stages of the step heating experiment correspond to excess ^{40}Ar trapped in anion sites. Although this model is, as yet, the only credible explanation for this phenomenon, it should be stressed that it is speculative, as no direct evidence exists for the behavior proposed for diffusion from anion and cation sites.

It is appropriate in a discussion of excess ^{40}Ar in mixed phases to include the work of Dalrymple et al. (1975) and Lanphere and Dalrymple (1976) on whole rock diabase (dolerite) samples from Liberia that have in-

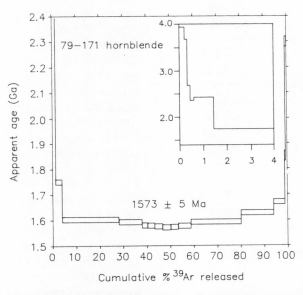

Fig. 4.25. Age spectrum of an exsolved hornblende into which excess ^{40}Ar has diffused. After Harrison and McDougall (1981). Note the expanded scale on the inset. The excess ^{40}Ar is released both in the initial and final stages of gas release. The intermediate region is not contaminated and yields a high precision age.

corporated excess ^{40}Ar. The release patterns are typically characterized by a saddle-shape, but detailed interpretation is difficult owing to the complexity of the assemblages. However, as plagioclase separates from a number of these samples yielded similar U-shaped age spectra, it is likely that much of the excess ^{40}Ar is carried in this phase.

4.8.4 Excess ^{40}Ar and flat release patterns

The ability of biotite to incorporate large amounts of excess ^{40}Ar has been recognized for many years. It is now equally well documented that biotites containing excess ^{40}Ar can exhibit flat release patterns (Pankhurst et al., 1973; Roddick et al., 1980; Foland, 1983) that are not distinguishable in any obvious way from undisturbed patterns. The implication of this result is that either the excess ^{40}Ar is uniformly distributed within the mica structure (i.e., the ^{40}Ar/^{39}Ar age spectrum method works for this mineral), or that experimental artifacts have homogenized and obscured an original gradient, so that the technique in fact does not work in this case. Additional information, treated in Section 4.11, points to the second explanation, that in general, meaningful concentration distributions within biotite are not revealed by step heating *in vacuo*. Pankhurst et al. (1973) suggested that irradiation damage may be responsible for this behavior, although Roddick et al. (1980) argued against this possibility. Structural breakdown of the biotite during heating in the vacuum system, with dehydration and delamination, seems to be the most likely cause for the observed behavior (Hanson et al., 1975; Foland, 1983). Finally, the question of whether isotope correlation diagrams can help elucidate the composition of the trapped argon component has been the subject of diverse opinion (Roddick et al., 1980; Foland, 1983; see Section 4.13). We present an example in Section 4.13 that indicates that even though spatial information is lost from biotite during extraction, there can exist separate, thermally activated reservoirs in biotites that can be revealed by isochron analysis.

4.8.5 Excess ^{40}Ar and variable age spectra

There are many examples of ^{40}Ar/^{39}Ar age spectra from samples, known to contain excess ^{40}Ar, that do not correspond to one of the three types discussed above. Biotite release patterns that are concave upward (Foland, 1983) or saddle-shaped (Lanphere and Dalrymple, 1976) have been described. Pyroxenes showing saddle-shaped spectra or generally increasing ages with progressive ^{39}Ar release are not uncommon (Lanphere and Dalrymple, 1976; Harrison and McDougall, 1981). Clearly, much remains to be understood about the behavior of excess ^{40}Ar in minerals, and many phases so affected may be inappropriate for step heating analysis.

4.9 RECOIL DISTRIBUTION OF ^{39}Ar

An important assumption of interpretative models is that the ^{39}Ar$_K$ produced during irradiation is distributed in a manner similar to the ^{40}K in a sample. However, for the commonly expected neutron energy distribution in a reactor (Fig. 3.4), the ^{39}K(n,p)^{39}Ar reaction results in a spectrum of recoil energies between 0 and 400 keV, with the mean energy in the 100–200 keV range (Turner and Cadogan, 1974). A consequence of this broad span of energies is that the distance ^{39}Ar nuclei recoil varies between essentially 0 and about 0.5 μm. Davies et al. (1963) measured the penetration distance of ^{41}Ar in aluminum for a variety of incident energies. Their results are shown in Fig. 4.26, recalculated assuming a density of 3.0 g/cm^3. Mean recoil distances in the 0–200 keV energy range vary between 0 and 1800 Å (0.18 μm). Turner and Cadogan (1974) integrated these data with the expected recoil energy spectrum for ^{39}Ar to yield an empirical description of the depleted layer of ^{39}Ar produced by recoil adjacent to the surface of a potassium-bearing mineral. The concentration, C, of ^{39}Ar is given by the expression $C = 1 - 0.5 \exp(-x/x_0)$, where x is depth from the surface and x_0 is the mean depth of the depletion layer, 0.082 μm. Subsequent studies (Huneke and Smith, 1976; Huneke, 1976) have in general borne out this prediction. Conclusions from these studies are that ^{39}Ar recoil transfer can be expected to affect age spectra of samples in which potassium is principally located in fine grained phases adjacent to potassium-poor minerals, causing more than one apparent plateau in the same age spectrum

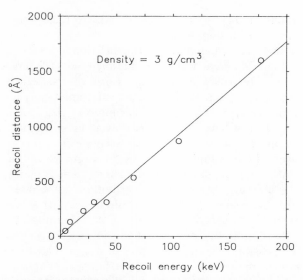

Fig. 4.26. Most probable ^{39}Ar recoil distance as a function of recoil energy. After Davies et al. (1963). The range of recoil energies shown reflects the spectrum expected from the ^{39}K(n,p)^{39}Ar reaction in many nuclear reactors.

in some cases. However, two recent studies remind us not to view the mean depletion or penetration depth of 0.08 μm as a barrier to observing much finer scale structure. Harrison and McDougall (1981) and Harrison and Fitz Gerald (1986) recognized ^{39}Ar and ^{37}Ar degassing patterns in finely exsolved (<0.08 μm) silicates, implying that significant fractions of both these isotopes were not displaced from their original lamellae. This result is expected because the spectrum of recoil energies includes extremely low values, especially if the distribution of emergent proton directions is anisotropic (Turner and Cadogan, 1974).

The recoil energy in the natural decay of ^{40}K to ^{40}Ar of ~28 eV (Brandt and Voronovskiy, 1967) is well above typical activation thresholds for argon in silicate lattices of ~1–3 eV. Therefore it is likely that both ^{40}Ar* and ^{39}Ar$_K$ will occupy similar kinds of positions within the body of a crystal, generally different from the original sites of the parent potassium atoms. This appears to be borne out by ^{40}Ar/^{39}Ar studies on undisturbed geological samples (e.g., Dalrymple and Lanphere, 1974), which often reveal flat age spectra implying identical siting of ^{40}Ar* and ^{39}Ar$_K$.

Typically, analysis of terrestrial samples is undertaken on monomineralic aggregates with particle dimensions much greater than the recoil distance, for reasons discussed in detail in previous sections. Nevertheless, it is of some importance to consider possible recoil effects on age spectra, as during irradiation ^{39}Ar$_K$ may be lost from the near-surface regions of crystals in a fashion that reflects the omnidirectional probability of recoil. Thus, it is expected that there will be a smooth function from no loss at the maximum recoil distance below the surface to half of the ^{39}Ar$_K$ concentration at the surface, given a planar boundary (Turner and Cadogan, 1974). However, in a sample composed of large grains essentially homogeneous in potassium, it seems likely that such loss at the grain edges would be virtually totally compensated for by the flux of incident ^{39}Ar$_K$ atoms from adjacent grains. Provided that few collisions occur in the intergranular region, a good assumption if the samples are irradiated under vacuum, then the distribution in energies of the incoming particles would be identical to the flux outward. The resulting ^{40}Ar*/^{39}Ar$_K$ composition close to the grain surfaces would be essentially unaffected, although it is possible that the buried atoms are differently bound in the crystal structure as a result of lattice damage.

In part, a test of this hypothesis comes from an age spectrum of a hornblende from the McClure Mountain Complex, Colorado, chosen as an interlaboratory standard by Alexander et al. (1978) for its simple, undisturbed geological environment. The age spectrum (Harrison, 1981) is characterized by gas fractions that define, within uncertainty, a plateau age of 515 Ma, except for the fifth step, which makes up only 0.37% of the ^{39}Ar released and gives an apparent age about 7% greater. The first step, which contains only 0.08% of the total ^{39}Ar, yields an age that is statistically indistinguishable from the remainder of the spectrum suggesting no net loss of ^{39}Ar. Subsequent gas fractions making up 0.46, 0.61, and 0.59% of the ^{39}Ar release yield ages similar to the first step. These data support the view (Huneke and Smith, 1976) that ^{39}Ar recoil artifacts in age spectra of coarse grained ($\geq 100\,\mu$m diameter) mineral separates are unimportant, illustrating the desirability of obtaining the coarsest possible aggregate. However, for fine grained materials, recoil effects may be significant, and may make interpretation of age spectra difficult.

Halliday (1978) reported ^{40}Ar/^{39}Ar age spectrum analyses on clay minerals from some ore bodies in Ireland, and convincingly demonstrated that, owing to the extremely fine grain size ($\leq 5\,\mu$m), significant amounts of ^{39}Ar$_K$ and ^{40}Ar* were lost from the samples during irradiation. Additionally, several studies using fine grained whole rock samples such as inclusions in meteorites (Villa et al., 1983) and basalts (Clague et al., 1975; Dalrymple and Clague, 1976; Seidemann, 1978) support the view that recoiling ^{39}Ar$_K$ can be released from the crystal lattice. In contrast, Huneke and Smith (1976) showed that in certain cases, recoiling ^{39}Ar$_K$ from a potassic phase could imbed itself in a more tightly bound site in a potassium-poor mineral, in their case, olivine. As a consequence, high ages are found in the early stages of gas release, falling monotonically to unrealistically low apparent ages as the unsupported ^{39}Ar$_K$ is eventually degassed from olivine at high temperature.

The effect on age spectra of lattice damage as a result of irradiation is more difficult to assess, although the observation of flat release patterns from many undisturbed samples (e.g., Dalrymple and Lanphere, 1974) and age gradients in some disturbed samples (e.g., Harrison and McDougall, 1980b) suggests that at moderate neutron doses an effect is not significant. A useful discussion of this topic is given by Horn et al. (1975).

4.10 GRAIN SIZE AND DISTRIBUTION

The diffusion models of Turner et al. (1966) and Gillespie et al. (1982) demonstrate clearly that the form of theoretical age spectra can be changed dramatically by varying the grain size distribution (see Sections 4.2 and 4.3). However, in all models the low temperature gas release in an age spectrum dates the time of ^{40}Ar* loss. An assumption common to all such models is that, regardless of the grain size distribution, all phases in the sample must remain intact during the process of gas extraction in the laboratory. If this assumption is met then the models predict the evolution of the flux of gas at the original grain surfaces. The advisability of analyzing single-phase samples has already been discussed and in most cases requires separating the mineral of interest from a polymineralic system. Doing this generally means the sample needs to be crushed and sieved, so that we are unlikely to recover in our mineral separate grains that have dimensions identical to those in the original sample. Comminution of grains during crushing may well introduce artifacts into the age spectra. The extent of these artifacts depends on whether the original grain surface behaved as the effective diffusion dimension or whether diffusion was controlled by some smaller domain. In many cases it is clear that subgrain mosaics defined by cleavage, linear dislocation features, incoherent exsolution interfaces, or cracks do control the diffusion dimension. In these instances, provided the recovered grains are not crushed smaller than these domains, age spectra characteristic of diffusive loss of ^{40}Ar* may be resolvable. However, should the recovered grain size be smaller than this critical dimension, both cores and rims of grains may be equally exposed and simultaneously contribute to the composition of the gas at that particular step. The result of this would be to homogenize the ^{40}Ar/^{39}Ar age spectrum to yield an artificial age plateau that corresponds to the conventional K–Ar date for the sample.

Practically, there is often a dilemma: the sample must be crushed fine enough to ensure a minimum of multimineral grains to avoid problems discussed in Section 4.6, but the aggregate must be kept as coarse as possible to avoid this grain size effect. Although some studies have investigated effective diffusion radii in several minerals (Harrison and McDougall, 1980b; Harrison, 1981; Harrison et al., 1985), the most encouraging observation is simply that a number of analyses of samples having average grain diameters of 100 μm have revealed $^{40}Ar^*$ gradients characteristic of diffusive loss, which requires that the natural dimension controlling gas loss is equal to or less than the grain size.

Horn et al. (1975) investigated the effect on $^{40}Ar/^{39}Ar$ age spectra of crushing a lunar plagioclase from >35 to <15 μm diameter. Apart from releasing the $^{40}Ar^*$ at lower temperatures, expected from diffusion considerations, the relative release of argon isotopes was not affected. Irradiation with particularly high fast neutron doses ($\sim 2 \times 10^{19} n/cm^2$), however, caused structural damage to the crystals, resulting in a somewhat modified age spectrum.

4.11 PHASE CHANGES IN VACUUM

Current interpretative models generally assume that gas loss occurs during the extraction experiment from inert particles of a given geometry. The question of how well these geometric solutions describe geological materials is subordinate in importance to the presumed nonreactive nature of minerals in the vacuum extraction system. Unlike lunar and meteoritic samples, which are quite at home in such an environment, many geological samples suitable for K–Ar analysis are thermodynamically unstable during heating in a vacuum. The two principal types of phase change that may cause argon release by a mechanism other than volume diffusion are dehydrogenation (or dehydration) and exsolution homogenization.

Both biotite and hornblende, the two minerals most commonly dated by the K–Ar method, contain structural water. These phases are metastable in a vacuum up to temperatures at which the kinetics of proton evolution allow the first-order dehydration reactions to proceed. However, unlike biotite, which begins to break down at relatively low temperatures, hornblende does not begin to lose significant quantities of $^{40}Ar^*$ until the structure is almost completely dehydrogenated (Gerling et al., 1965; Zimmerman, 1972; Hanson et al., 1975). Clearly the loss of hydrogen and argon is decoupled in hornblende, a result that can be appreciated in terms of crystal chemistry, as the dehydrated hornblende exists as a stable phase. The ability to readjust to the vacuum environment without significantly altering the internal argon isotope distribution is a useful property of this mineral that explains its successful application in monitoring thermal disturbances. In addition to several geological examples, the ability of hornblende to preserve and later reveal diffusion gradients has been demonstrated experimentally by hydrothermally treating an amphibole to induce $^{40}Ar^*$ loss and subsequently analyzing the sample by the age spectrum approach (Harrison, 1981).

However, argon loss in biotite apparently accompanies hydrogen loss (Zimmerman, 1972) at rates that imply delamination in addition to dehydration. In natural samples, biotites have shown a great deal of variability in their ability to register later thermal events (Lanphere and Dalrymple, 1971; Berger, 1975; Hanson et al., 1975). An experimental proof of the kind shown for hornblende yields results whose complexity underscores the ambiguity found in geological samples. Harrison et al., (1985) analyzed a biotite by the age spectrum method that had been hydrothermally treated in an environment buffered for oxygen. The sample was demonstrated to have lost $^{40}Ar^*$ by diffusion by several conventional tests, but yielded an age spectrum that was characterized by exceedingly old ages in the early portion of gas release and then decreased rapidly to a chronologically meaningless plateau (Fig. 4.27).

In a test of biotite release patterns from undisturbed, relatively quickly cooled granitoids, Tetley and McDougall (1978) observed significant unexplained deviations from ideal plateaus in many of the samples. Nonetheless, the literature contains a number of biotite release patterns that appear to yield meaningful information. However, the complex behavior of this system makes biotite a poor candidate for unambiguous age spectrum interpretation.

The second significant phase adjustment

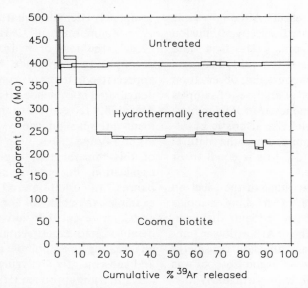

Fig. 4.27. Age spectra for a biotite sample before and after hydrothermal treatment at 700°C and 1 kbar for 9 days. After Harrison et al. (1985). The release pattern does not correspond to theory and suggests artifacts in the extraction experiment have obliterated the diffusion gradient.

during heating is the homogenization of exsolution features. Both alkali feldspar and plagioclase are minerals dated by the K–Ar method that are well known to readily exsolve separate phases during slow cooling. The subsequent distribution of $^{40}Ar^*$ within these phases reflects the partition of potassium between the exsolved phases at the time of unmixing. If in the course of laboratory heating these phases should homogenize, the lamellae boundaries would change or disappear. The release of the argon isotopes from a sample with this new configuration is unlikely to resemble release spectra from the original sample.

Exsolution boundaries in calcic plagioclases are seldom incoherent, but even if antiphase boundaries behave as channels of rapid diffusion (Wenk, 1979), it has been shown experimentally that these features are unlikely to rehomogenize, except at very high temperature (>1200°C), on time scales appropriate to an age spectrum experiment (Tagai and Korekawa, 1981).

However, intermediate alkali feldspars that contain perthite lamellae may remix during heating. Additionally, the microcline lamellae undergo an inversion to an orthoclase structure before melting incongruently with the adjacent albite. Experimental evidence suggests that, as a result of both the rapid diffusion of argon in low-temperature alkali feldspars and the lamellae boundaries defining the effective diffusion dimension, the bulk of argon is released from perthites before any significant phase change occurs (Harrison and McDougall, 1982). However, at temperatures of ~800°C, mixing of perthite lamellae can occur (Goldsmith and Laves, 1954), rendering useless diffusion parameters calculated above temperatures of ~750°C (Harrison et al., 1986). Although this process will tend to reduce concentration gradients of $^{40}Ar^*$, age spectra commonly appear to reflect preservation of such gradients even in results from gas released at temperatures above that of remixing. Thus, it is concluded that the homogenization effects are small on the time scale of a typical $^{40}Ar/^{39}Ar$ step heating experiment.

4.12 BEHAVIOR OF MINERALS AND WHOLE ROCKS

4.12.1 High-temperature alkali feldspar

Alkali feldspars of the sanidine–anorthoclase series require rapid cooling for their preservation in the high-temperature structural state, so that they are found only in rocks at or near the Earth's surface. As a consequence, age spectra from high-temperature alkali feldspars rarely show evidence of overprinting events. The homogeneous nature of these phases results in predictable behavior during a step heating experiment and flat age spectra. McDougall (1981, 1985) has analyzed numerous anorthoclases from pumices in tuffs (related to dating sedimentary sequences containing hominid fossils and other vertebrates) and found them to yield excellent plateaus and to be highly retentive of argon. Indeed, in conventional K–Ar dating, it has been found that temperatures of ~1600°C must be maintained for at least 45 min to extract all the argon from such feldspars (McDougall et al., 1980; McDougall, 1985; McDowell, 1983). Another good example of successful dating of high-temperature alkali feldspar using the $^{40}Ar/^{39}Ar$ age spectrum technique was reported by Hess and Lippolt (1986) on sanidines separated from Carboniferous tuff beds.

4.12.2 Low-temperature alkali feldspar

Low-temperature alkali feldspars are increasingly being used for $^{40}Ar/^{39}Ar$ analysis despite statements in the literature that they do not quantitatively retain $^{40}Ar^*$ at temperatures found near the Earth's surface. The success of $^{40}Ar/^{39}Ar$ age spectrum studies on these minerals relates to their well-mannered behavior during vacuum heating, and the ability of perthitic alkali feldspars to record the low-temperature end of thermal histories. Several studies have demonstrated clearly the sensitivity of low-temperature alkali feldspars to mildly elevated temperatures (Albarède et al., 1978; Maluski, 1978; Berger and York, 1981a; Harrison and McDougall, 1982). However, the French school (Albarède et al., 1978; Maluski, 1978) tends toward the view that mild deformation accompanying heating can cause a reduction in grain size sufficient to permit variable loss of accumulated $^{40}Ar^*$. Harrison and McDougall (1982) concluded from an analysis of slowly cooled microclines that the U-shaped age spectra observed in the earlier studies resulted from an excess ^{40}Ar uptake profile superimposed on a diffusion loss profile; the latter is commonly the product of slow cooling. However, in the study by Albarède et al. (1978), there is other geological and isotopic evidence for a thermal event subsequent to initial crystallization and cooling of the low-temperature alkali feldspars, and the disturbed age spectra probably reflect this second event followed by slow cooling.

Calculations based on ^{39}Ar loss during vacuum heating, as well as geological considerations, led Berger and York (1981a) and Harrison and McDougall (1982) to infer that microcline age spectra often record evidence for cooling between ~250 and 150°C. Zeitler (1985) demonstrated that the plateau portion in U-shaped age spectra is virtually concordant with the coexisting zircon fission track age, thought to represent the time at which a temperature below about 200°C was achieved. Harrison and Bé (1983) utilized this low-temperature sensitivity to study the thermal history of a sedimentary basin by analyzing detrital microclines from deep drill cores (see Section 6.3.3). These grains act essentially as monitors of the degree of heating experienced during burial, as the transport rate of argon is an exponential function of temperature. Additionally, the final stages of gas release can reveal the provenance age. Gillespie et al. (1982, 1983) analyzed microcline concentrates from granitoid xenoliths that had been incorporated in young basalt lava flows. The resulting age spectra are complex, with indications of three separate reservoirs for $^{40}Ar^*$. A likely explanation is that during heating in the basalt, inversion of the microcline to orthoclase begins, as well as homogenization of perthite lamellae, resulting in a complex mixture of structural states and domain dimensions.

The response of perthitic alkali feldspar to low-temperature heating as well as its behavior during vacuum degassing are now well understood and reveal much about the relationship between exsolution structure and argon retention properties. It was suggested previously (Sardarov, 1957; Foland, 1974) that the incoherent boundary between perthite exsolution lamellae acts as a diffusion surface, effectively limiting the maximum diffusion dimension to the width of a lamella. Evidence supporting this view comes from diffusion studies employing the measured ^{39}Ar loss from irradiated samples during vacuum extraction. In homogeneous alkali feldspars, it is commonly observed that a plot of these derived diffusion coefficients (D/l^2) against reciprocal extraction temperature results in well-correlated lines corresponding to one activation energy from ~ 500 to $>1200°C$ (e.g., Harrison and McDougall, 1982). In microperthitic samples, this linear relationship breaks down at about 800°C (e.g., Harrison et al., 1986), the temperature at which microperthite lamellae homogenize by K–Na interdiffusion on laboratory heating time scales (e.g., Goldsmith and Laves, 1954). Above this temperature, the diffusivity appears to decrease—a physically impossible circumstance. The explanation is that merging of many lamellae results in a very large increase in the effective diffusion distance for argon in the evolving mineral structure. Thus, it is the growth of the l^2 term in D/l^2 that results in the decrease in calculated values. Moreover, in cryptoperthites, this transition in diffusion behavior occurs at much lower temperatures ($\sim 600°C$), as would be predicted from the shorter alkali interdiffusion distance (Bruce Idleman, personal communication, 1986). Further evidence for the perthite lamellae model is obtained from the generally superior fit of the diffusion data to a plane sheet model compared with other possible geometric solutions, the observation of oxygen diffusion along perthite boundaries that is three to four orders of magnitude faster than volume diffusion (Giletti and Nagy, 1981), and the adherence of calculated Arrhenius parameters to a diffusion compensation curve (Hart, 1981) when lamellae size is taken into account (Harrison and McDougall, 1982). Geological closure temperatures for anorthoclase, microperthitic microcline, and cryptoperthitic microcline calculated from the straight line portions of these Arrhenius plots of ~ 300, 200, and 120°C, respectively, are found to be internally consistent with a large body of empirical results.

4.12.3 Plagioclase

Dalrymple and Lanphere (1974) recorded two age spectra on plagioclase that revealed broad plateaus. However, many analyses of this complex mineral system tend to yield age spectra characterized by a saddle- or U-shape (Dalrymple et al., 1975; Lanphere and Dalrymple, 1976; Albarède et al., 1978; Berger and York, 1981a; Harrison and McDougall, 1981), indicative of the presence of excess ^{40}Ar. Berger and York (1981a) showed that the shape of the age spectra could be modified by acid leaching of the plagioclase separate prior to irradiation, resulting in reduction of both the early and late release high age portions. Generally, the low potassium content, a tendency for absorption of excess ^{40}Ar, and complex exsolution phenomena restrict use of this mineral to cases in which no other suitable phase is available, for example, in many mafic rocks. It should be noted, however, that plagioclase is the major potassium-bearing phase in many lunar rocks, on which ^{40}Ar/^{39}Ar age spectrum measurements generally have been very successful (see Section 6.2).

4.12.4 Feldspathoids

As noted in Section 2.7.3, leucite appears to be a favorable mineral for ^{40}Ar/^{39}Ar age spectrum analysis from the study undertaken by Radicati di Brozolo et al. (1981). Villa (1986), however, has reported evidence for excess argon in some leucites. No study is known to us in which nepheline has been used for ^{40}Ar/^{39}Ar dating except that of York and Berger (1970), who measured only a single total fusion age.

4.12.5 Biotite

Conventional K–Ar or ^{40}Ar/^{39}Ar total fusion ages on biotites can be very useful in elucidating the time of cooling of igneous or metamorphic rocks below the closure temperature for retention of ^{40}Ar*, especially when measured in conjunction with ages on cogenetic minerals. However, biotite has shown great variability in the form of age spectra for a given geological history. For example, rapidly cooled biotites can exhibit either flat or slightly cusp-shaped release patterns (Tetley and McDougall, 1978; Snee, 1982). Biotites containing excess ^{40}Ar yield flat, convex upward, convex downward, or irregular age spectra (Pankhurst et al., 1973;

Roddick et al., 1980; Foland, 1983). Phlogopites with excess ^{40}Ar give similar irregular ^{40}Ar/^{39}Ar age spectra (Kaneoka and Aoki, 1978). Thermally disturbed biotites that have experienced ^{40}Ar* loss yield flat, convex upward, or irregular release patterns (Berger, 1975; Dallmeyer, 1975a,b; Harrison et al., 1985). These results suggest that the form of age spectra measured on biotites is not particularly diagnostic in allowing selection between different possible interpretations of the kind discussed previously.

4.12.6 Muscovite and phengite

The majority of ^{40}Ar/^{39}Ar age spectra on white micas reported in the literature have been measured on Alpine micas (e.g., Albarède et al., 1978; Frank and Stettler, 1979; Chopin and Maluski, 1980; Hunziker et al., 1986). Generally, the indication is that white micas yield ideal, essentially flat age spectra when undisturbed, and sensible diffusion loss profiles when partially overprinted (Lanphere and Dalrymple, 1971; Hanson et al., 1975; Harrison and McDougall, 1981; Wijbrans and McDougall, 1986). It is not clear why muscovite behaves more systematically than biotite during vacuum extraction of the argon, although Harrison (1983) suggested that the more sluggish dehydration kinetics allowed muscovite to remain metastable in the vacuum to temperatures higher than biotite. Muscovite also is much less likely to incorporate excess argon during a metamorphic episode compared with coexisting biotite (Brewer, 1969; Roddick et al., 1980).

4.12.7 Amphiboles

Hornblende has proved to be a popular mineral for K-Ar dating since Hart (1961, 1964) first demonstrated its usefulness, despite its rather low content of potassium, generally in the 0.1-1.5% range. Hornblende has been used extensively for dating because it is remarkably retentive of ^{40}Ar* and it also has a relatively low content of atmospheric argon (Fig. 2.6; McDougall, 1966; Dalrymple and Lanphere, 1969). With the advent of the ^{40}Ar/^{39}Ar age spectrum technique, measurements were soon done on hornblende to test its suitability (Fitch et al., 1969; Lanphere and Dalrymple, 1971). In particular, two nearly simultaneous studies (Hanson et al., 1975; Berger, 1975) were undertaken to assess the response of hornblende (among other minerals) to an episodic thermal event caused by intrusion of young magmatic bodies into two different metamorphic terranes. Despite the clear relation between measured sample age and distance from the intrusive contact, both studies yielded essentially flat age spectra, although the patterns were not devoid of possibly meaningful age gradients. Hanson et al. (1975) and Berger (1975) both concluded that partially outgassed hornblendes could not be distinguished from undisturbed samples by the ^{40}Ar/^{39}Ar technique, and the former authors suggested that experimental artifacts may have obliterated the natural concentration profiles. In a reassessment of the behavior of hornblende to reheating, Harrison and McDougall (1980b) demonstrated that coarse-grained, homogeneous hornblendes could reveal unambiguous chronological information related to the ages of crystallization and reheating, by means of ^{40}Ar/^{39}Ar analysis. However, as discussed in Section 4.6, metamorphic amphiboles may contain a number of domains of different argon retentivity, reflecting exsolution features. As a result, extremely complex age spectra may be obtained (Harrison and Fitz Gerald, 1986).

4.12.8 Pyroxene

As noted in Section 2.7.6, pyroxenes are not very suitable for K-Ar age measurement, although they have provided some useful and meaningful results. Arguably, if this mineral is to be utilized for K-Ar dating, it is probably best measured by the ^{40}Ar/^{39}Ar technique because of its very low potassium content, and because age spectra can provide evidence as to whether excess ^{40}Ar is present. Nevertheless, pyroxene can be regarded as a mineral to be dated as a last resort.

4.12.9 Whole Rocks

The complications in interpretation of age spectra measured on a mixture of phases (see

Section 4.6) make dating whole rock samples less desirable than dating single mineral phases. However, in cases in which the grain size is too small for mineral separation, or where there is a limited amount of sample, it may be necessary to perform $^{40}Ar/^{39}Ar$ analyses on whole rock specimens. Several of these aspects were discussed in Section 2.7.7, but some reiteration and additional comments are appropriate here. Seidemann (1978) attempted to date deep-sea basalts using the incremental release method, but determined that the approach could not overcome the effects of excess $^{40}Ar^*$ commonly found in these rocks. By performing several irradiations on similar samples in breakseal ampoules, Siedemann (1978) showed that up to 25% of the neutron-induced $^{39}Ar_K$ had been lost from the samples during irradiation owing to recoil from fine grained phases or glass. This resulted in overestimates of the ages of the samples. Likewise, Dalrymple et al. (1980) demonstrated that there was $^{39}Ar_K$ loss during irradiation from some partly altered deep-sea basalts. Although the $^{40}Ar/^{39}Ar$ total fusion ages were greater than the conventional K–Ar ages, they argued that the older ages were likely to be more nearly correct, as evidence was adduced that the same phases that had lost $^{39}Ar_K$ during irradiation also had not retained their $^{40}Ar^*$ through geological time. The age spectra for these rocks commonly were irregular, and Dalrymple et al. (1980) suggested that the total fusion ages were probably the best available estimate of age.

Fleck et al. (1977) found $^{40}Ar/^{39}Ar$ age spectrum analyses of Antarctic Mesozoic tholeiites yielded generally discordant patterns that they took to indicate significant inhomogeneity in the distribution of $^{40}Ar^*$ with respect to $^{39}Ar_K$. They correlated modification of their age spectra with progressive devitrification, to which they attributed $^{40}Ar^*$ redistribution, although they acknowledged that $^{39}Ar_K$ recoil may also play a part in certain cases.

In a recent study of hydrothermally altered lavas and intrusives of the Tertiary Mull igneous complex in Scotland, Mussett (1986) reported age spectra on 21 whole rock samples ranging in composition from basalt to rhyolite. Using the criteria of Lanphere and Dalrymple (1978) to distinguish between those age spectra showing acceptable or unacceptable plateaus, Mussett (1986) suggested that the eight samples with satisfactory age plateaus were yielding reliable estimates of the crystallization age. From these results he concluded that the igneous activity on Mull occurred between 60 ± 0.5 and 57 ± 1 Ma ago. As the total fusion ages were much more scattered, yielding apparent ages ranging from 40 to 69 Ma, the measurement of $^{40}Ar/^{39}Ar$ age spectra, followed by appropriate assessment, appears to have been quite successful in distinguishing between reliable and anomalous ages. This is a most encouraging result particularly as the rocks are polymineralic and extensively altered. More details of the mineralogy of these rocks and information as to where the potassium resides would be helpful in applying the age spectrum approach successfully to other altered whole rock samples.

Where dating of whole rock samples is contemplated, careful examination in thin section should be made of each sample to determine its likely suitability for $^{40}Ar/^{39}Ar$ or K–Ar dating. Experience shows that reliable results may be obtained on well-crystallized, preferably holocrystalline, fresh igneous whole rocks. Samples that show extensive alteration, with development of clays and other low-temperature phases, or contain devitrified glass, often yield age spectra that are difficult to interpret in any meaningful way.

Much greater success has been achieved in the $^{40}Ar/^{39}Ar$ dating of whole rock samples, particularly basalts, brought back from the Moon. Reasons for this probably include the highly degassed and anhydrous nature of lunar material, their great antiquity, and a resolve to understand the results in a quantitative framework. A discussion of this work, given in Section 6.2, emphasizes the complexity of dating whole rocks and provides a general cautionary note when interpreting age spectra from a polymineralic aggregate.

4.12.10 Evolution of illite to muscovite

The behavior of the K–Ar isotopic dating system as the mineral illite develops and evolves toward muscovite during diagenesis and metamorphism has been studied in some detail in the Swiss Alps and its northern foreland. Here unmetamorphosed Mesozoic sedimentary rocks can be traced southward into regions where the rocks have been metamorphosed in the greenschist to amphibolite facies during the Late Cenozoic Alpine Orogeny. Of particular note are the comprehensive mineralogical, chemical, K–Ar, and ^{40}Ar/^{39}Ar age spectrum studies of Frank and Stettler (1979) and Hunziker et al. (1986), based upon samples collected from two traverses about 120 km apart.

Frank and Stettler (1979) separated illite or muscovite from a number of samples that were obtained from a traverse in which the Alpine metamorphic grade increases progressively toward the southeast from lower greenschist to amphibolite facies (staurolite zone). Illites formed diagenetically in the Mesozoic and thermally overprinted at 260–300°C during the Alpine Orogeny gave K–Ar ages ranging from 39 to 20 Ma. Age spectra measured on three of these samples, which have between 45 and 72% of the 2M$_1$ polymorph, showed monotonically rising patterns with increasing gas release, yielding maximum ages as great as 140 Ma. Apparent ages for the gas released in the first step of each experiment ranged from 7.9 to 11.5 Ma. Frank and Stettler (1979) interpreted these patterns to be the result of loss of radiogenic argon by diffusion during the Alpine Orogeny, and that the apparent ages for the initially released gas in each case provided an estimate of the time since the illites began to retain argon quantitatively owing to cooling of the terrane. An age spectrum on the 2–3 μm fraction of an illite that is in the 2M$_1$ polymorphic form and that reached a temperature of ∼380°C during the Alpine Orogeny gave a gently rising pattern for the bulk of the gas release with ages mainly in the 7.6–10.3 Ma range, indicative of nearly complete resetting. At still higher grades, where the temperature attained ∼530°C, a muscovite yielded a nearly flat age spectrum with a plateau at 11.8 ± 1.0 Ma over 95% of the ^{39}Ar released. This was interpreted as the time since the temperature decreased below the closure temperature for retention of argon in the muscovite. Frank and Stettler (1979) concluded that resetting of the K–Ar system is controlled by the progressive transformation of the illite from the 1Md to the 2M$_1$ polymorph in conjunction with diffusional loss of radiogenic argon, both thermally activated mechanisms.

In the complementary study by Hunziker et al. (1986) diagenetic illite of the 1Md polymorphic type from virtually unmetamorphosed Triassic sediments in the Jura region and from boreholes that pass through the Late Oligocene Molasse sequence into the Triassic yielded K–Ar ages on the <2 μm fraction ranging from 189 Ma to as young as 82 Ma. A plot of measured K–Ar age versus the present-day temperature at the level from which the samples were taken in the boreholes reveals a decrease in apparent age with increasing temperature. From this the authors infer that loss of radiogenic argon has been occurring from the illites by volume diffusion, and that by extrapolation a temperature estimate of 260 ± 30°C is derived for essentially complete outgassing of the <2-μm illite grains. The older K–Ar ages are from illites that have remained near the surface, and probably relate to early diagenesis. The younger K–Ar ages are from illites from the boreholes and reflect outgassing of radiogenic argon owing to the thermal effects associated with burial by the younger (Oligocene) sediments.

A ^{40}Ar/^{39}Ar age spectrum on the 2–6 μm size fraction of an illite regarded as nonmetamorphic yielded a typical argon loss profile with an apparent age of 32 Ma for the gas released in the initial step (11% of the ^{39}Ar), rising progressively to about 350 Ma with increasing gas release, although somewhat younger ages were found for the last few percent of gas release. From comparison with the K–Ar age of 143 ± 5 Ma on the same sample and the ^{40}Ar/^{39}Ar total gas age of 164 ± 5 Ma it is inferred that some loss of ^{39}Ar occurred during or subsequent to irradiation, probably by recoil. This illite was dominantly composed of the 1Md polymorph with about 20% in the 2M$_1$ form. It was separated from a sample obtained from a borehole at a depth of 5245 m where the temperature currently is ∼160°C.

The traverse along which Hunziker et al.

(1986) collected is about 120 km northeast of that of Frank and Stettler (1979), but the Triassic sediments show a similar transition to progressively more strongly metamorphosed rocks to the southeast across the Glarus Alps. Within the anchimetamorphic zone, the transition zone between those of diagenesis and metamorphism, illite changes from the dominant 1Md polymorph to the $2M_1$ form as the region affected by greenschist facies metamorphism is approached. Hunziker et al. (1986) showed that the $<2\,\mu m$ fractions of illite from the anchizone rocks yielded K–Ar ages ranging from 56 to 30 Ma, generally decreasing as the proportion of the $2M_1$ polymorph increases. They confirmed the findings of Frank and Stettler (1979) that the resetting of the illite during the Alpine metamorphism occurs by diffusional loss of radiogenic argon concurrently with lattice reorganization. Illite gives way to muscovite in rocks affected by metamorphism in the greenschist facies, and the K–Ar ages on the $<2\mu m$ white mica fraction continue the trend to lower ages, within the 33–15 Ma range. Hunziker et al. (1986) reported $^{40}Ar/^{39}Ar$ age spectra on several coarser fractions of illite and muscovite from the same samples of Triassic sediments affected by metamorphism at greenschist facies grade as were measured by the K–Ar method, with somewhat equivocal results. Thus, a sample of illite with $>80\%$ in the $2M_1$ polymorphic form yielded an essentially flat age spectrum with a plateau segment of age 40.7 ± 1.4 Ma over 85% of the gas release, on the $2-6\,\mu m$ size fraction. The results indicated that there was no significant loss of ^{39}Ar by recoil. However, as the $<2\,\mu m$ fraction yielded concordant K–Ar and Rb–Sr ages of ~ 30 Ma, which Hunziker et al. (1986) interpreted as reflecting the main phase of Alpine metamorphism in the Glarus Alps, they suggested that the high plateau age was caused by the presence of inherited argon, supported by recognition of some detrital mica in the concentrates. The $2-6\,\mu m$ size fraction of a white mica from a slightly higher grade sample, comprising mainly phengitic muscovite, gave a nearly flat age spectrum with a plateau over about 60% of the gas release at 26.6 ± 0.6 Ma, regarded as a geologically reasonable age in relation to the Alpine metamorphism in the region. In contrast the $6-20\,\mu m$ size fraction of white mica from the same rock exhibits a monotonically rising age spectrum with individual ages ranging from 7.5 to 75 Ma in a typical argon loss profile. Again the authors note the presence of some detrital mica in the concentrate, and from the results it is evident that the temperature was high enough for a sufficiently long period to enable complete resetting of the finer fraction, but not of the coarser grain size material, during the Alpine metamorphic phase.

Overall, these studies on illites and muscovites from Triassic sediments in the Swiss Alps and environs clearly show that temperatures greater than $\sim 250°C$ for a significant time interval are required for illites to transform to the $2M_1$ polymorphic form and to lose much of their pre-existing radiogenic argon. Complete resetting of the K–Ar system occurs only subsequent to reconstitution to muscovite, initially in the finer grained fractions. Resetting is controlled both by volume diffusion of argon and structural changes in the white micas, as has also been demonstrated in Naxos by Wijbrans and McDougall (1986). The importance of identifying the various phases present in the rocks is again well illustrated by these studies. Especially in low grade metamorphic terranes different generations of closely related minerals commonly occur together owing to incompleteness of reactions. Because of the presence of such mixed phases, the isotopic age signatures may be complex. Where recrystallization has occurred, ages on white micas commonly reflect the time of cooling below the closure temperature for argon rather than the time of recrystallization.

4.13 ISOTOPE CORRELATION DIAGRAMS

Our remarks on the interpretation of $^{40}Ar/^{39}Ar$ step heating results have been restricted to the presentation of data using the age spectrum format. Implicit in our discussions of this approach for terrestrial samples is the assumption that any trapped argon present is of atmospheric composition (by trapped we mean argon present and remaining in the sample during crystallization or metamorphism). Despite the apparent unlikelihood of plutonic rocks containing atmospheric abundances of argon isotopes, this often appears to be the case. Rocks from deep crustal sources can reveal geologically meaningful plateau ages even though significant corrections for nonradiogenic ^{40}Ar are made for all heating fractions. This may reflect the near-atmospheric isotopic composition from at least a portion of the Earth's mantle influencing a degassed lower crust or the presence of recycled atmospheric argon (Dymond and Hogan, 1973). However, it is not uncommon for minerals to contain trapped argon with ^{40}Ar/^{36}Ar $\gg$ 296. In these cases complex release patterns emerge and simple interpretation via the age spectrum method is not possible.

Merrihue and Turner (1966) recognized that the ^{40}Ar/^{39}Ar step heating approach offered an additional dimension to K–Ar dating—the potential to reveal the composition of trapped argon by use of an isochron, or three isotope ratio plot. Their method was to array the data by plotting the total ^{40}Ar, *in situ* produced daughter plus all other contributions (excluding neutron-induced ^{40}Ar), against ^{39}Ar (a measure of the radioactive parent concentration via J and the constant natural abundance of ^{40}K/K), normalizing both to a primordial isotope of the daughter element, in this case, ^{36}Ar (Fig. 4.28). This approach is analogous to the Rb–Sr isochron method in which a linear correlation of data from a coeval suite of samples reveals both the geological age (proportional to the slope of the line) and the initial isotopic composition of the daughter product element (given by the intercept of the line with the daughter element isotope ratio axis). However, instead of dispersion along the line resulting from samples with variable parent to daughter ratios as in Rb–Sr dating, this effect in the ^{40}Ar/^{39}Ar isochron system may be achieved through results from a single sample because of mixing between two end-member reservoirs, one reservoir comprising the radiogenic ^{40}Ar (relative to the ^{36}Ar) and the other consisting of the trapped argon component. Using this plot, Merrihue and Turner (1966) found well-correlated lines for their meteorite data that yielded ages that appeared to be sensible and meaningful (Fig. 4.29). Lanphere and Dalrymple (1976) tested this method on terrestrial samples known to contain excess ^{40}Ar and found considerable scatter of their data with few apparently systematic trends. They concluded that the method was not well suited for terrestrial material containing excess argon, although Jessberger (1977) pointed out what he believed were some meaningful data alignments. With experience it has become clear that samples that have taken up excess

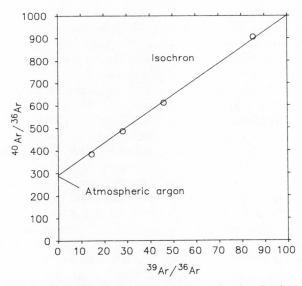

Fig. 4.28. Schematic isochron plot. The sample age is proportional to the slope of the correlation line.

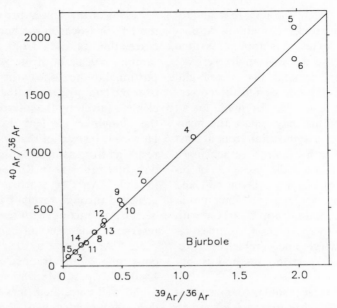

Fig. 4.29. Isochron plot of the Bjurböle chondrite from which an apparent age of ~4.3 Ga is obtained. From Merrihue and Turner (1966).

^{40}Ar in a nonuniform fashion are not good candidates for isochron treatment. These samples have, in effect, a continuum of non-radiogenic argon compositions that result from the mixing of the actual trapped argon with the later, and spatially variable, excess ^{40}Ar that has diffused into the grain boundaries. This serves to emphasize that the presence of more than two components of different isotopic composition within a sample generally results in poor alignment of data on the isochron diagram. In such circumstances, however, the three isotope ratio plot can be useful in indicating that the simple two component model does not hold.

In a timely paper, Roddick (1978) pointed out several potential pitfalls of isochron analysis that could result from experimental artifacts. Of these, he stressed that an additional argon component introduced during gas extraction, the system blank, commonly was forgotten during data reduction. As a result of degassing of the furnace and other vacuum components, all extraction and analysis systems have such a blank, or background, that they impart to the sample gas prior to isotopic analysis. Owing to fractionation or the memory of an exotic argon composition, the background need not be of atmospheric argon isotopic composition. As the published argon blanks for extraction systems applied to terrestrial samples vary by more than three orders of magnitude (e.g., Roddick, 1978), this contribution to the sample argon will range from considerable to trivial. In the nontrivial cases, the blank component should first be subtracted from the measured isotope ratios before being plotted on an isochron diagram.

A drawback of the conventional isochron plot is that, in general, the isotope measured with the poorest precision, ^{36}Ar, is common to both axes. A result is that the errors associated with both axes are highly correlated, and may give rise to misleading linear correlations. Another limitation stems from the open-ended nature of this diagram. That is, samples in which no ^{36}Ar has been detected plot at infinity. A consequence of this is that radiogenically enriched samples will dominate the regression and preclude precise determination of the trapped argon composition. These problems are largely circumvented by an alternate form of isochron analysis in which ^{36}Ar/^{40}Ar is plotted against ^{39}Ar/^{40}Ar (Turner, 1971b; Roddick et al., 1980), thus using ^{40}Ar as the reference isotope (Fig. 4.30). Because ^{40}Ar usu-

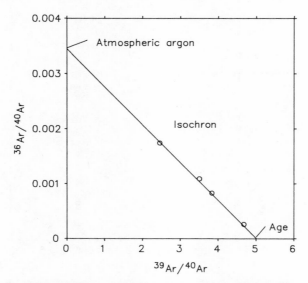

Fig. 4.30. Schematic illustration of the inverse isochron plot. The two intercepts correspond to the pure trapped and pure radiogenic components in the sample.

ally is the most abundant isotope and therefore can be measured very precisely, the correlation between errors in both ratios is small and often negligible. Sometimes called the inverse isochron, this approach results in a closed array of data with a negative slope in which the age is given by the x-intercept (the inverse $^{40}Ar^*/^{39}Ar_K$ ratio, and comprising the potassium-derived radiogenic argon and neutron-induced ^{39}Ar) and the trapped composition corresponds to the y-intercept (the inverse $^{40}Ar/^{36}Ar$ ratio), again providing that the composition of the argon is dominated by these two reservoirs. If other isotopic components are present, such as excess argon, or significant amounts of cosmogenic argon (in the case of extraterrestrial samples in particular), then simple linear arrays will be compromised. A more detailed account of the properties of correlation plots of this kind is given by Turner (1971b), and Kelley et al. (1986) discuss correlation in multicomponent argon systems.

A simple example of the use of this isotope correlation diagram is shown in Fig. 4.31 from the work of McDougall (1985). The Pliocene to Pleistocene sedimentary sequence exposed adjacent to the eastern shores of Lake Turkana, Kenya, contains abundant tuff horizons, in some of which anorthoclase-bearing pumice clasts occur. Alkali feldspars separated from the clasts have been extensively investigated using the K–Ar and $^{40}Ar/^{39}Ar$ methods to establish numerical age constraints for the sediments from which numerous hominid and other vertebrate fossils have been recovered. Figure 4.31 shows the high degree of correlation present among all 16 heating steps obtained from an anorthoclase from the Ninikaa Tuff. The age of 3.01 ± 0.02 Ma fits sensibly into the regional stratigraphy. The initial $^{40}Ar/^{36}Ar$ intercept of 296.4 ± 4.2 is indistinguishable from atmospheric argon. In a simple case such as this, the isochron analysis provides confidence that the sample has not been thermally disturbed and contains trapped argon only as a result of atmospheric contamination. However, interpretation is not always so simple.

An illustration of the complexity that can be encountered with terrestrial samples is the work of Heizler and Harrison (1988), who found disturbed release spectra in biotite (Fig. 4.32) from the Buckskin Peak Quartz Diorite, Oregon. The biotite spectrum gives a distinct peak in the mid-portion of gas release at ages older than the maximum possible emplacement age of 160 Ma. The explanation for both the old age and complexity in release pattern is apparent when the individual gas fractions are

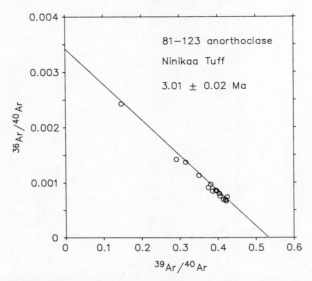

Fig. 4.31. Inverse isochron of anorthoclase from Koobi Fora, Kenya, showing simple mixing between atmospheric argon and a purely radiogenic component. After McDougall (1985).

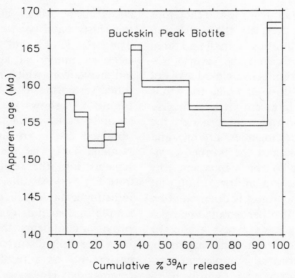

Fig. 4.32. Disturbed release spectra for a biotite from a small granitoid with a maximum emplacement age of 160 Ma, suggesting the presence of excess argon. After Heizler and Harrison (1988).

plotted on an isochron diagram (Fig. 4.33). The data show two sequential linear arrays indicating two distinctly different trapped argon compositions with essentially no mixing between them. The apparent ages of about 150 Ma are consistent with other mineral dates from this terrane. These results underscore the fact that age spectrum plots are simply model ages and that complex patterns may reflect an inappropriate choice of trapped argon composition. Using the ^{36}Ar/^{40}Ar ratios indicated by extrapolation of the linear trends in Fig. 4.33 to correct for nonradiogenic argon, we have replotted the age spectrum for the biotite in Fig. 4.34. This well-defined plateau illustrates the sensitivity of some ages to the choice of trapped argon isotope composition and serves to caution the reader not to prejudge the geochron-

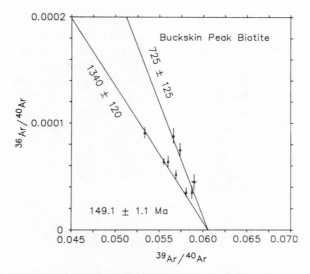

Fig. 4.33. The inverse isochron for the Buckskin Peak Quartz Monzonite reveals the presence of two distinct trapped argon components in biotite separated in thermally distinct reservoirs. The numbers adjacent to each correlation line indicate the trapped $^{40}Ar/^{36}Ar$ ratio. After Heizler and Harrison (1988).

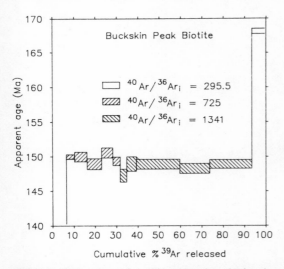

Fig. 4.34. Age spectrum from Fig. 4.32 replotted using the trapped argon components revealed by isochron analysis (Fig. 4.33). The flat release pattern at ~150 Ma over 85% of gas release remind us that age spectra calculated assuming trapped argon of atmospheric composition are model ages. After Heizler and Harrison (1988).

ological merit of a $^{40}Ar/^{39}Ar$ analysis based solely on the age spectrum. This phenomenon is not restricted to biotite; Heizler and Harrison (1988) documented several thermally distinct trapped argon components in hornblende, muscovite, and K-feldspar, and Hanes et al. (1985) found two trapped argon components in an altered pyroxenite.

Phillips and Onstott (1986) plotted $^{40}Ar/^{39}Ar$ step heating data on a $^{36}Ar/^{40}Ar$ versus $^{39}Ar/^{40}Ar$ correlation diagram and inferred an excess argon component in a phlogopitic mica from a South African kimberlite. However, their data contain only one linear array of three successive increments which comprise less than 1% of the total gas and blank contribution to these steps is as high as 85% of the measured ^{36}Ar.

There are several choices of algorithm available to calculate the slope and intercepts of the arrayed data. The simplest is a "least-squares" regression in which all variation is assumed to occur on only one axis—a geochronologically unlikely circumstance. The best line is obtained by minimizing the squares of the deviations of the data from this line in either the x or y axis. The preferred approaches take into account analytical errors on both axes (McIntyre et al., 1966; York, 1966), and permit the consideration of correlation between variables common to both axes (York, 1969). Goodness of fit parameters such as MSWD (McIntyre et al., 1966) or SUMS (York, 1969), where MSWD = SUMS/$(n-2)$, where n is the number of data points, indicate how well data fit the least-squares calculated straight line. Where the data fit the line to within experimental error a MSWD value of one is the expectation. A value much less than one suggests experimental errors have been overestimated. These comments apply in cases in which there are many data points available and with realistic assigned errors. For small numbers of data, the value of the index can significantly exceed unity before the presence of excess scatter in the results, related to some form of geological error, can be identified. Thus Brooks et al. (1972) suggested that for Rb–Sr data a value of MSWD > 2.5 normally would indicate scatter of results about the calculated line greater than that to be expected from experimental error alone. For $^{40}Ar/^{39}Ar$ step heating data, analyzed according to the isochron approach, a similar cutoff value seems appropriate (Dalrymple and Lanphere, 1974; Fleck et al., 1977; Roddick, 1978; Roddick et al., 1980). Thus a MSWD > 2.5 may indicate that the simple isochron model is not appropriate for a particular data set, requiring further consideration as to why there is excess scatter about the line.

5. DIFFUSION THEORY, EXPERIMENTS, AND THERMOCHRONOLOGY

5.1 INTRODUCTION

Although diffusion is probably the simplest of all natural transport mechanisms in principle, our understanding of the details of this phenomenon in most cases is only approximate, and in the instances in which experiment and theory have resolved the atomistic behavior, the descriptions are bewilderingly complex. In spite of the unlikelihood of fully understanding mechanisms of diffusion in silicate structures of interest to geologists, a simple phenomenological appreciation of this process provides us with a powerful tool to solve a variety of petrological and geochemical problems.

To obtain data on conditions experienced by rocks deep in the crust, geologists have traditionally utilized environmental monitors that have their basis in equilibrium exchange. That is, it is assumed that the system attains its lowest energy state for the mixture of chemical components present. Two limitations of this approach are that the kinetics, or reaction pathways, may tend to inhibit the attainment of this balance from being reached, and that even given no impediment to equilibrium, the information obtained is of a static nature. For example, the results from equilibrium thermometers and barometers tell us only about peak conditions obtained, but relate nothing about the rates of descent or ascent of the particular segment of crust. It is this latter dynamic information that sheds light on the time scales of geological events that in turn can help constrain the underlying physical processes involved.

Recently, a new generation of metamorphic petrologists has adopted a different approach. By focusing on the reaction pathways, the *degree* to which a system has attempted to attain equilibrium can be calibrated to reveal absolute cooling rate data (e.g., Lasaga et al., 1977). However, long before this conversion, geochronologists recognized that most mineral ages from deep crustal rocks were, in effect, kinetic thermometers; that is, the apparent age recorded by the system corresponds to the temperature at which the daughter product ceased to be lost from the crystal by diffusion. In recent years, advances have been made that allow us to quantify this long held view, and extract both thermal and temporal information from geochronometric systems. That the $^{40}Ar/^{39}Ar$ dating method has been in the forefront of this movement is not surprising because of the virtually unique dimension that the step heating, or age spectrum provides.

In this chapter we shall outline the theoretical basis of diffusion and heat flow, present useful mathematical solutions for geologists, review the relevant diffusion literature, and illustrate the power of combining these analogous processes with a geological example. Because the theories of heat conduction and molecular diffusion are parallel, many of our early derivations will employ the conduction description because we can rely on the reader's everyday experience and intuition for assistance.

5.2 THE PROCESS OF DIFFUSION

The basis of conductive heat flow or diffusion theory is that random molecular motions tend to equilibrate heat or concentration in a finite volume. As a result of the random nature of these movements, the diffusing substance will tend to move from regions of high concentration to regions of low concentration. This is because the same fraction of diffusant initially in the high- and low-concentration areas moves in the direction of the other region, resulting in a net loss of atoms or heat from the initially enriched portion and vice versa. Consider a many-storied hotel in plan view consisting of a checkerboard of interconnected rooms, each with doors on all four walls. Initially, the rooms are all occupied, but as the noon checkout time approaches, each occupant must randomly choose a door through which to leave. For those adjacent to the outer wall of the hotel, there exists a one-in-four chance (two-in-four for those on the corner) of falling to the pavement below. Although entirely without direction, soon after mid-day a gradient of room occupation can be perceived with available rooms appearing near the boundary and spreading inward. Eventually, the entire floor will be empty of patrons. In effect, the driving force of *net* diffusion is a concentration gradient, although the diffusion process proceeds at a rate often independent of the magnitude of the chemical potential.

5.3 PHENOMENOLOGICAL BASIS OF DIFFUSION THEORY

Heat transfer in solids occurs by conduction, or the diffusion of thermal energy by random molecular motions, exactly analogous to the random walk description of molecular diffusion. This parallel was recognized by Adolf Fick (1855) who provided diffusion with a theoretical basis by adapting Fourier's Law of Heat Conduction (Fourier, 1822) to mass flux. The equation is based on the hypothesis that the rate of transfer of heat per unit area is proportional to the thermal gradient:

$$(5.1) \qquad q/A_x = F = -K \frac{\partial T}{\partial x}$$

where q is the heat transfer rate, A_x is cross-sectional area, F is the flux of heat across this unit surface, K is the proportionality constant called thermal conductivity, T is the temperature, and x is the space coordinate. Fick's First Law is a restatement of Eq. (5.1) replacing temperature with concentration, C, and thermal conductivity with the diffusion coefficient, D. The minus sign indicates that heat or mass flow is in the direction of decreasing temperature or concentration, that is, down the gradient. Figure 5.1 illustrates this steady-state

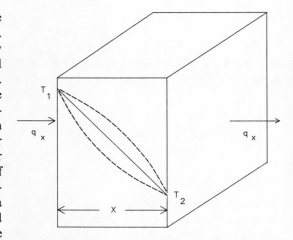

Fig. 5.1. Schematic representation of Fourier's Law showing that for one-dimensional steady-state heat flow, the heat flux at any point along x is proportional to the temperature gradient. The dashed curves represent the thermal profiles within the solid for the cases of heat sources (convex) and sinks (concave).

relationship in which a temperature drop from T_1 to T_2 occurs over a slab of thickness x yielding a heat flux, F. However, the same expression is valid for the nonuniform temperature distributions shown as dashed curves in Fig. 5.1. In these cases, $\partial T/\partial x$ is the tangent to the convex or concave curve at the point of interest. Curvature in the temperature profile for a constant K implies the occurrence of sources or sinks of heat. Equation (5.1) is the fundamental equation from which we can calculate the heat flux across the surface of the earth, or rearranging, the relationship with which to experimentally obtain values of the thermal conductivity of rocks.

The general problem of unsteady-state conductive heat transfer within a solid involves the prediction of the temperature within a solid region of interest as a function of the space coordinates and time, (x,y,z,t). To derive an equation that can be solved for the temperature distribution, $T(x,y,z,t)$, we apply the law of conservation of energy to a differential control volume:

$$(5.2) \qquad E_{\text{IN}} + E_{\text{GENERATED}} = E_{\text{OUT}} + E_{\text{STORED}}$$

where E is thermal energy that either enters (IN), leaves (OUT), is produced (GENERATED), or stored (STORED) in the control volume $dxdydz$. Figure 5.2 shows this

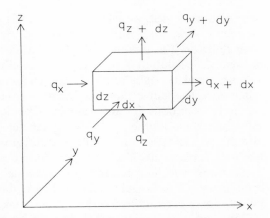

Fig. 5.2. Representation of energy conservation in a volume element.

three-dimensional inventory of energy, where, for example, q_x is the heat entering the volume along x, and q_{x+dx} is the heat leaving the infinitesimal region along x. Therefore

(5.3) $\quad E_{IN} = q_x + q_y + q_z$

(5.4) $\quad E_{OUT} = (q_{x+dx}) + (q_{y+dy}) + (q_{z+dz})$

However, q_{x+dx} can be put in terms of q_x if we view q_x as a function of the conductive heat transfer rate along x, and expanding this function in a Taylor series about x to $x + dx$. Thus,

(5.5) $\quad q_{x+dx} = q_x + \dfrac{\partial q_x\, dx}{\partial x} + \dfrac{\partial^2 q_x (dx)^2}{\partial x^2\, 2!} + \dfrac{\partial^3 q_x (dx)^3}{\partial x^3\, 3!} + \cdots \dfrac{\partial^n q_x (dx)^n}{\partial x^n\, n!}$

In fact, there are partial derivatives of q_x with respect to y and z, as well as mixed partial derivatives. However, since q_{x+dx} has an invariable y and z value, these terms go to zero. Since the dx is infinitesimal, terms containing $(dx)^2$ and $(dx)^3$ may be neglected when compared to the second term on the right-hand side of the equation. Thus,

(5.6) $\quad q_{x+dx} = q_x + \dfrac{\partial q_x}{\partial x} dx$

(5.7) $\quad q_{y+dy} = q_y + \dfrac{\partial q_y}{\partial y} dy$

(5.8) $\quad q_{z+dz} = q_z + \dfrac{\partial q_z}{\partial z} dz$

Therefore,

$E_{OUT} = q_x + \dfrac{\partial q_x}{\partial x} dx + q_y + \dfrac{\partial q_y}{\partial y} dy + q_z + \dfrac{\partial q_z}{\partial z} dz$

(5.9)

The amount of energy produced within the volume $dxdydz$ is

(5.10) $\quad E_{GENERATED} = Q\, dxdydz$

where Q is the heat production per unit volume (e.g., radioactive decay, latent heat, strain heating), and the rate of change of stored energy is given by

(5.11) $\quad E_{STORED} = \rho(\partial U/\partial t)\, dxdydz$

where ρ is density and U is internal energy. Substitution into the original conservation Eq. (5.2) gives

(5.12)
$$q_x + q_y + q_z + Q\, dxdydz$$
$$= q_x + \dfrac{\partial q_x}{\partial x} dx + q_y + \dfrac{\partial q_y}{\partial y} dy + q_z$$
$$+ \dfrac{\partial q_z}{\partial z} dz + \rho \dfrac{\partial U}{\partial t} dxdydz$$

which reduces to

(5.13)
$$Q\, dxdydz = \dfrac{\partial q_x}{\partial x} dx + \dfrac{\partial q_y}{\partial y} dy + \dfrac{\partial q_z}{\partial z} dz$$
$$+ \rho \dfrac{\partial U}{\partial t} dxdydz$$

From thermodynamics, the equation relating internal energy to specific heat (c_p) as a function of temperature (T) can be written

(5.14) $\quad \dfrac{\partial U}{\partial t} = c_p \dfrac{\partial T}{\partial t}$

This allows us to recast the heat flux partial derivatives in terms of T using Fourier's Law of Heat Conduction (5.1). Thus,

(5.15) $\quad q_x = -K(\partial T/\partial x)\, dydz$

(5.16) $\quad q_y = -K(\partial T/\partial y)\, dxdz$

(5.17) $\quad q_z = -K(\partial T/\partial z)\, dxdy$

Partial differentiation of Eqs. (5.15), (5.16), and (5.17) with respect to x, y, and z, respectively, yields

$$\frac{\partial q_x}{\partial x} = -K\frac{\partial^2 T}{\partial x^2}dydz \tag{5.18}$$

$$\frac{\partial q_y}{\partial y} = -K\frac{\partial^2 T}{\partial y^2}dxdz \tag{5.19}$$

$$\frac{\partial q_z}{\partial z} = -K\frac{\partial^2 T}{\partial z^2}dxdy \tag{5.20}$$

Substituting Eqs. (5.14), (5.18), (5.19), and (5.20) into Eq. (5.13) and dividing through by $dxdydz$ gives

$$K\left(\frac{\partial^2 T}{\partial x^2} + \frac{\partial^2 T}{\partial y^2} + \frac{\partial^2 T}{\partial z^2}\right) + Q = \rho c_p \frac{\partial T}{\partial t} \tag{5.21}$$

but $K/(c_p\rho) = \kappa$, the thermal diffusivity with units of (dimension)^2/time. Thus,

$$\frac{\partial^2 T}{\partial x^2} + \frac{\partial^2 T}{\partial y^2} + \frac{\partial^2 T}{\partial z^2} + \frac{Q}{K} = \frac{1}{\kappa}\frac{\partial T}{\partial t} \tag{5.22}$$

Equation (5.22) is the second-order partial differential equation (i.e., it contains two or more independent variables) whose solution, given specific boundary and initial conditions, yields the temperature as a function of the three space coordinates and time for an isotropic body. Of course, this representation can be modified to other coordinate systems, for example, polar or cylindrical. The diffusion analogue, or Fick's Second Law (for a constant molecular diffusivity and no production term), is simply

$$\frac{\partial C}{\partial t} = D\left(\frac{\partial^2 C}{\partial x^2} + \frac{\partial^2 C}{\partial y^2} + \frac{\partial^2 C}{\partial z^2}\right) \tag{5.23}$$

Clearly, the diffusion coefficient, D, combines the separate elements of the thermal diffusivity, that is, the conductivity is the molecular permeability and the specific heat corresponds to the solubility of the diffusant.

5.4 METHODS OF SOLUTION FOR CONSTANT D (OR κ)

Provided that the diffusion coefficient is constant, solutions of the diffusion Eq. (5.23) can be obtained for a number of initial and boundary conditions covering most of the concentration distributions and geometries of interest to geologists. We shall see that these solutions have two distinct types: either an infinite trigonometric series or a solution involving the error function. For diffusion in cylindrical coordinates, the solution is of the form of Bessel's equation, and the tabulation of this function replaces the trigonometric series. In this section we derive solutions for diffusion in a plane sheet and semi-infinite media using the separation of variables and Laplace transform methods, respectively. These solutions form the basis from which solutions for spherical, cylindrical, and rectangular geometries are obtained. These derivations follow that of Crank (1975); however, many geologists are not versed in the methods of solution of higher order partial differential equations and to make this section more accessible, Appendixes A.5.1, A.5.2, and A.5.3 may assist readers with no more than first year calculus.

5.4.1 Plane sheet geometry

In one dimension, Fick's Second Law has the form

$$\frac{\partial C}{\partial t} = D\frac{\partial^2 C}{\partial x^2} \tag{5.24}$$

Many important geological problems can be adequately modelled by considering heat or mass transport in only one direction, for example, heat flow across a thinning lithosphere (McKenzie, 1978) or the diffusion of argon from perthite lamellae (Harrison and McDougall, 1982). This geometry of linear flow in the solid bounded by two parallel planes is variously referred to as a plane sheet or infinite slab. Solution of Eq. (5.24) for the case of diffusion of a uniformly distributed substance within a slab of thickness l and boundaries maintained at zero concentration (C_0) is given in Appendix A.5.1 using the separation of variables method. The resulting concentration distribution after time, t, is described by

$$C = \frac{4C_0}{\pi}\sum_{n=0}^{\infty}\frac{1}{(2n+1)} \\ \times \exp[-D(2n+1)^2\pi^2 t/l^2]\sin\frac{(2n+1)\pi x}{l} \tag{5.25}$$

which for the case of a slab of thickness $2l$ can be written

$$(5.26) \quad C = \frac{4C_0}{\pi} \sum_{n=0}^{\infty} \frac{(-1)^n}{(2n+1)} \times \exp[-D(2n+1)^2\pi^2 t/4l^2] \cos\frac{(2n+1)\pi x}{2l}$$

5.4.2 Semi-infinite medium

It is often useful, both as a good description of certain diffusion experiments (e.g., Harrison and Watson, 1983) and as a useful approximation, to consider the flow of mass or heat across the boundary of two semi-infinite media. A venerable example of the latter is Kelvin's original treatment of the age of the Earth, which he described as a semi-infinite solid bounded at the surface by a plane maintained at 0°C. A solution of Eq. (5.24), Fick's Second Law in one dimension, for the case of flow across a plane separating two half-spaces can be obtained by considering the half-space to consist of an infinite number of line sources of individual width $\partial\chi$ with a mass $C_0 \partial\chi$ (Crank, 1975). The concentration at some distance after time t resulting from this source is

$$(5.27) \quad C = \frac{C_0 \partial\chi}{\sqrt{4\pi Dt}} \exp(-\chi^2/4Dt)$$

and the sum over the entire population of elements is given by

$$(5.28) \quad C = (C_0/\sqrt{\pi}) \int_{x/\sqrt{4Dt}}^{\infty} \exp(-\eta^2)\, \partial\eta$$

where $\eta = \chi/\sqrt{4Dt}$. As this integral has no solution, a mathematical function called the error function, written erf where

$$(5.29) \quad \text{erf } z = (2/\sqrt{\pi}) \int_0^z \exp(-\eta^2)\, \partial\eta$$

is tabulated to provide a solution (e.g., Crank, 1975). This function has the properties

$$\text{erf}(-z) = -\text{erf } z, \quad \text{erf}(0) = 0,$$
$$\text{erf } \infty = 1, \quad 1 - \text{erf } z = \text{erfc } z$$

where erfc is the complementary error function.

The solution of Eq. (5.28) is given by

$$(5.30) \quad C = (1/2)C_0 \,\text{erfc}(x/\sqrt{4Dt})$$

An equivalent second solution using the Laplace transform method is given in Appendix A.5.2. This technique has some remarkable properties, permitting solution of a number of complex problems in diffusion and heat flow.

5.4.3 Spherical and cylindrical geometries

Using coordinate transformations, Eq. (5.23) can be easily modified to describe radial flow in a sphere and in an infinite circular cylinder. The details of the conversion to spherical coordinates are given in Appendix A.5.3 as an illustration of the kind of mathematical substitution applied in these cases. The resulting concentration distribution for a sphere of radius a with an initially uniform concentration C_1 held in an infinite reservoir of zero concentration, C_0 is given by

$$(5.31) \quad C = \frac{C_1 2a}{\pi r} \sum_{n=1}^{\infty} \frac{(-1)^n}{n} \sin\frac{n\pi r}{a} \times \exp(-Dn^2\pi^2 t/a^2)$$

and the expression for the equivalent case in cylindrical coordinates is

$$(5.32) \quad C = \frac{C_1 2}{a} \sum_{n=0}^{\infty} \exp\frac{(-D\alpha_n^2 t)J_0(r\alpha_n)}{\alpha_n J_1(a\alpha_n)}$$

where $J_0(x)$ is the Bessel function of the first kind of order zero, $J_1(x)$ is the Bessel function of the first order, and α is a root of $J_0(a\alpha_n) = 0$.

5.5 NUMERICAL APPROACHES FOR VARIABLE D

The solutions presented thus far are for the case where D or κ do not vary with time. Whereas thermal diffusivity is well approximated as constant in the temperature interval 200–1000°C (see Section 5.8.5), the diffusion constant is a sensitive function of temperature, restricting rigorous application of our solutions to the geologically uninteresting (or implausible) case of constant temperature. However, provided that there is no thermal gradient across the solid, we can accommodate a time-dependent diffusion coefficient by replacing Dt with $\int_0^t D(t')\,dt'$ in Eqs. (5.26), (5.30), (5.31), and (5.32).

Zeitler (1988) discussed a numerical scheme (described in Section 4.5) that accommodates complex initial distributions and temperature histories.

5.6 CALCULATION OF EPISODIC ^{40}Ar LOSS

We have derived expressions for the concentration distributions of heat or a diffusing substance in a plane sheet and sphere [Eqs. (5.26) and (5.31)]. The solution for an infinite cylinder is given in Eq. (5.32). However, in many cases, usually because we cannot directly image the distribution of the diffusant in the solid of interest, we require a solution for the fractional amount of heat or substance that has left (or entered) the solid in a given time. This fraction represents the approach from zero loss ($f = 0$) at $t = 0$ to total equilibration at $t = \infty$.

To obtain expressions for the fractional loss (or uptake) in these three geometries, we assess the starting concentration, subtract the amount remaining at time t (described by integrating the concentration distribution between $r = 0$ to $r = a$ at t), and dividing by the original amount present, C_1.

For example, the mass (M_0) of uniformly distributed ^{40}Ar initially in a sphere is the integral over the radius of differential concentrations $\delta M = 4\pi r^2\,dr\,C(r)$, or

$$(5.33) \quad M_0 = \int_0^a \delta M = (4/3)\pi a^3 C_1$$

From the integration of Eq. (5.31) from $r = 0$ to $r = a$, the mass remaining in the sphere after time t is

$$
\begin{aligned}
(5.34)\quad M_t &= \int_0^a 4\pi r^2 C_1 \frac{2a}{\pi r}\sum_{n=1}^{\infty}\frac{(-1)^n}{n} \\
&\quad \times \sin\frac{\pi n r}{a}\exp(-Dn^2\pi^2 t/a^2)\,dr \\
&= \frac{8a^3}{\pi}C_1 \sum_{n=1}^{\infty}\frac{1}{n^2}\exp(-Dn^2\pi^2 t/a^2)
\end{aligned}
$$

Thus fractional loss, f, is

$$(5.35)\quad f = \frac{M_0 - M_t}{M_0} = 1 - \frac{6}{\pi^2}\sum_{n=1}^{\infty}\frac{1}{n^2}\exp(-Dn^2\pi^2 t/a^2)$$

Expressions for fractional loss from all three geometric solutions as well as convenient approximations circumventing the need for the infinite series are presented in Table 5.1. Also see Fechtig and Kalbitzer (1966) for a useful discussion of the applications of these ex-

TABLE 5.1. *Exact and approximate solutions of the diffusion equation for sphere, infinite cylinder, plane sheet, and cubic geometries*[a]

Geometry	Equation	Validity
Sphere (radius a)	$f = 1 - (6/\pi^2)\sum_{1}^{\infty}(1/n^2)\exp(-n^2\pi^2 Dt/a^2)$	All f
	$f \simeq 1 - (6/\pi^2)\exp(-\pi^2 Dt/a^2)$	$0.85 \leq f \leq 1$
	$f \simeq (6/\pi^{3/2})(\pi^2 Dt/a^2)^{1/2} - (3/\pi^2)(\pi^2 Dt/a^2)$	$0 \leq f \leq 0.85$
Infinite cylinder (radius a)	$f = 1 - 4\sum_{1}^{\infty}(1/\alpha_n^2)\exp(-\alpha_n^2 Dt/a^2)$	All f
	$f \simeq 1 - 9/13\exp(-5.78 Dt/a^2)$	$0.60 \leq f \leq 1$
	$f \simeq (4/\sqrt{\pi})(Dt/a^2)^{1/2} - (Dt/a^2)$	$0 \leq f \leq 0.60$
Plane sheet (half-width l)	$f = 1 - (8/\pi^2)\sum_{0}^{\infty}[1/(2n+1)^2]\exp[-(2n+1)^2\pi^2 Dt/4l^2]$	All f
	$f \simeq 1 - (8/\pi^2)\exp(-\pi^2 Dt/4l^2)$	$0.45 \leq f \leq 1$
	$f \simeq (2/\sqrt{\pi})(Dt/l^2)^{1/2}$	$0 \leq f \leq 0.60$
Cube (edge a)	$f = 12(Dt/a^2\pi)^{1/2} - 48(Dt/a^2\pi) + 64(Dt/a^2\pi)^{3/2}$	$0 \leq f \leq 0.97$

[a]*Sources*: Crank (1975), Jain (1958), Jost (1960), and Reichenberg (1953).

pressions to argon loss calculations. Graphical results of these equations are shown in Fig. 4.6.

5.7 SLOW COOLING

The mathematical expressions presented so far have been limited to the description of episodic loss during an isothermal heating event, although a time-dependent D can be numerically accommodated within the expressions, lending some geological realism. However, dynamic thermal histories have a more complex form requiring a somewhat greater degree of sophistication in analysis. The problem of isotopic closure during slow cooling is extremely relevant as most mineral chronometers begin life not as if instantaneously resetting a stop watch, but as a shifting balance between accumulation and loss of daughter product in a cooling terrane. In a remarkable paper in 1973, M. H. Dodson of the University of Leeds presented an analytical solution of the daughter product distribution within a solid following a cooling history linear in $1/T$. The form of this thermal history is commonly expected in crustal rocks and is discussed later, giving this analysis tremendous potential in extending our interpretations of mineral ages.

5.7.1 Dodson's model

Ideally, the age of a mineral or rock calculated from the accumulated products of radioactive decay corresponds to a point in time at which a completely mobile daughter product becomes fixed. This case is well approximated by volcanic rocks that erupt at temperatures sufficiently high that isotopic equilibration usually is obtained, and cool to relatively low temperatures, at which the daughter product is immobile, in geologically brief spans of time. A more complicated transition occurs in deeper crustal rocks that cool slowly from high temperatures at which the daughter product (e.g., argon) escapes as fast as it is formed. At low temperatures, diffusion is so negligibly slow that accumulation of radiogenic isotopes can be thought of as complete in well-crystallized materials. Between these two states there is a continuous transition over which accumulation eventually balances loss, then exceeds it. Figure 5.3 shows how a calculated age in this situation relates to the transition interval—the apparent age is the extrapolation of the total accumulation part of the curve to the time axis, which implicitly corresponds to a temperature via the assumption of linear increase in $1/T$.

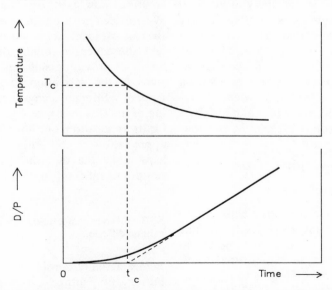

Fig. 5.3. Diagrammatic depiction of closure model, where T_c is closure temperature and t_c is mineral age. During slow cooling from high temperature, the daughter to parent ratio (D/P) passes from a completely open state (i.e., $D/P = 0$) through a zone of partial accumulation, shown by the curved portion of the curve in the lower diagram, until the temperature is sufficiently low that diffusion loss ceases and the D/P ratio grows at a constant rate with time. After Dodson (1973), with permission.

Thus, the effective closure temperature (T_c) is defined as the temperature of the system at the time represented by its apparent age (Dodson, 1973). In spite of the fact that closure does not occur at this time, the prior accumulation and subsequent loss balance to give this definition an actual thermal significance. What follows is an annotated derivation of Dodson's formulas. Emphasis on this model underscores our feeling that this is a unique and powerful contribution to isotope geochemistry and should be widely accessible to earth scientists.

5.7.2 The mathematics of slow cooling

Most diffusion processes are thermally activated and obey the Arrhenius relationship

$$(5.36) \qquad D = D_0 \exp(-E/RT)$$

where D_0 is the frequency factor (D at infinitely high temperature), R is the gas constant, T is the absolute temperature, and E is the activation energy of the diffusion process. Within the range of E for geological materials, the diffusion coefficient varies strongly with even subtle changes in T, due to the exponential relationship. For example, for $E = 65$ kcal/mol, a change in temperature from 500 to 490°C causes a drop in diffusivity of nearly a factor of two. Assuming a cooling history proceeds linearly in $1/T$, or at least can be approximated as such over the closure interval, the decrease in the diffusion coefficient has the form of a simple exponential decay. Dodson introduced a time constant, τ, which corresponds to the time taken for D to diminish by a factor e^{-1}, or to drop to 37% of its previous value. Thus we can write

$$D = D_0 \exp(-E/RT_0 - t/\tau)$$
$$(5.37) \qquad = D(0) \exp(-t/\tau)$$

where T_0 and $D(0)$ are the initial temperature and diffusion coefficient, respectively, at time $t = 0$. For example, when t corresponds to twice the time constant, the exponential coefficient reduces the initial diffusivity to 14% of its original value. Note from Eq. (5.37) that

$$(5.38) \qquad \tau = R \bigg/ \left(E \frac{d\,1/T}{dt} \right) = -RT^2 \bigg/ \left(E \frac{dT}{dt} \right)$$

This relationship holds provided that the cooling interval is short with respect to the half-lives of the decay systems used—a good approximation in many geological situations.

Summarizing, the exponential decay of the loss coefficient with time, a consequence of the form of the Arrhenius law, allows for a closed mathematical solution to this problem. Dodson's original 1973 paper derives the solution of the accumulation–diffusion–cooling equation by using appropriate substitutions and variable boundary conditions that reduce the problem to an equation identical to Fick's Second Law. This can then be solved from the general infinite series expressions given by Carslaw and Jaeger (1959) for specified geometries. However, this derivation is given in an appendix, the majority of the paper being devoted to a conceptually simpler analogy and a heuristic derivation that allow an intuitive appreciation of the closure concept. Our experience is that even with this assistance, we were slow to completely understand Dodson's mathematics. It is worth noting that this contribution is the only acknowledged solution in the section "Time dependent diffusion coefficients" in the classic treatise, *The Mathematics of Diffusion*, by John Crank (although other solutions now exist, e.g., Lasaga et al., 1977). We proceed with an annotated derivation of the case of closure for a first-order loss reaction and subsequently volume diffusion, with the intent of providing sufficient linkage between steps in the derivation to accommodate all readers. Although such a detailed understanding of the closure concept is not a prerequisite for its application, our hope is that a complete appreciation of this important work will at least make the user more aware of the assumptions inherent in the mathematics.

5.7.3 Closure temperature for first-order loss and volume diffusion

Before actually considering the spatial distribution of isotopes during slow cooling, Dodson (1973) first derived the general form the closure equation would take if loss proceeds by a first-order reaction, that is, a reaction rate involving only the first power of daughter concentration. Although this derivation is covered in detail in Appendix A.5.4, the salient points can be revealed quite simply. The time varying concentration of some species, x, that is spontaneously created from parent C_p (with a production constant λ_x), but reacts away with the temperature- (and thus time-) dependent coefficient $k = k_0 \exp(-E/RT)$, is expressed as the difference of the production and loss terms. That is,

$$\left(\frac{dx}{dt}\right)_{net} = \left(\frac{dx}{dt}\right)_{prod} - \left(\frac{dx}{dt}\right)_{loss}$$

(5.39) $\qquad = \lambda_x C_p - k(0)\exp(-t/\tau)x$

assuming a linear increase in $1/T$. Provided that cooling begins from a temperature high enough to ensure an initial high value of the reaction coefficient to equilibrate the system to zero concentration of x, this expression can be solved using useful approximations to yield an equation for the closure temperature of x (see Appendix A.5.4).

The closure temperature, T_c, for first-order loss is given by

(5.40) $\qquad T_c = \dfrac{E/R}{\ln(\tau \gamma k_0)}$

where $\gamma = \exp(c)$ where c is Euler's constant (0.5772).

The importance of this result is largely in the form it takes, giving us some feeling for the type of expression that results from slow cooling, mindful of the assumptions required to simplify the exponential integral. Note that because τ contains the term T_c (see Section 5.7.5), this equation must be solved iteratively.

The derivation (Appendix A.5.4) for closure temperature in a first-order loss system gives us an expectation of the general form the closure equation will take in slow cooling systems. Although not mathematically rigorous, Dodson (1973) presented a simple and intuitive way of looking at the closure process in terms of volume diffusion. His method was to view the lumped loss parameter in the infinite series solution (e.g., $Dn^2\pi^2 t/a^2$ for the sphere) as equivalent to the loss coefficient k in the first-order loss closure Eq. (5.40).

With a constant diffusion coefficient and zero surface concentration, the fraction F (or $1 - f$) of diffusant initially of uniform distribution that *remains* in a system after time t is given by

(5.41) $\qquad F = \sum\limits_{n=1}^{\infty} (B/\alpha_n^2)\exp(-\alpha_n^2 Dt/a^2)$

where $B = 6$, $\alpha_n = n\pi$ for a sphere of radius a; $B = 4$, α_n is the nth root of $J_0(x)$ [see Eq. (5.32)] and is approximately $(n - 1/4)\pi$ for a cylinder of radius a; and $B = 2$, $\alpha_n = (n - 1/2)\pi$ for a plane sheet of thickness $2a$. In Eq. (5.41), if the term $\alpha_n^2 D/a^2$ in each summation step represents, in effect, a loss coefficient with "weight" B/α_n^2, then the term corresponds to k of the previous section and the "weight" is the portion of the total decay product that is represented. If we subdivide the entire system into n subsystems, the closure temperature, T_{cn}, for the nth subsystem is given by

(5.42) $\qquad \dfrac{E}{RT_{cn}} = \ln(\gamma\tau\alpha_n^2 D_0/a^2)$

The bulk closure temperature of the system will simply be the average of the infinite set of T_{cn}. Because E/RT varies linearly with time, the amount of decay product remaining in the nth subsystem after a long period of time will be linearly related to E/RT_{cn}. Thus, for the system as a whole it is reasonable to determine the weighted arithmetic mean of E/RT_{cn} using the weight factor B/α_n^2. From Eq. (5.42) we obtain

$$\frac{E}{RT_c} = \sum_{n=1}^{\infty} \frac{B}{\alpha_n^2} \ln(\gamma\tau\alpha_n^2 D_0/a^2)$$

$$= \sum_{n=1}^{\infty} \frac{B}{\alpha_n^2} [\ln(\gamma\tau D_0/a^2) + 2\ln\alpha_n]$$

$$= \ln(\gamma\tau D_0/a^2) \sum_{n=1}^{\infty} \frac{B}{\alpha_n^2}$$

(5.43) $\qquad + 2B \sum\limits_{n=1}^{\infty} \dfrac{\ln\alpha_n}{\alpha_n^2}$

However, B and α_n are related by $\sum_{n=1}^{\infty}(B/\alpha_n^2) = 1$ making this equation

(5.44) $\qquad \dfrac{E}{RT_c} = \ln(\gamma\tau D_0/a^2) + 2B \sum\limits_{n=1}^{\infty} \dfrac{\ln\alpha_n}{\alpha_n^2}$

If we call

$$g = \exp\left[2B \sum_{n=1}^{\infty} \frac{\ln\alpha_n}{\alpha_n^2}\right]$$

we can rewrite Eq. (5.44) as

(5.45) $\quad E/RT_c = \ln(\gamma\tau D_0/a^2) + \ln g$

or

(5.46) $\quad E/RT_c = \ln(\gamma g\tau D_0/a^2)$

Remembering that $\gamma = 1.78$, we can combine $\gamma g = A$ making our final equation

(5.47) $\quad E/RT_c = \ln(A\tau D_0/a^2)$

To obtain numerical values for A, we substitute the appropriate values into

$$A = \exp\left(c + 2B \sum_{n=1}^{\infty} \frac{\ln \alpha_n}{\alpha_n^2}\right)$$

yielding $A = 55$ for the sphere, 27 for the cylinder, and 8.7 for the plane sheet. Substituting in the constituents of τ [Eq. (5.38)] we get

(5.48) $\quad \dfrac{E}{RT_c} = \ln\left(\dfrac{ART_c^2 D_0/a^2}{E\, dT/dt}\right)$

5.7.4 Dodson's method of solution of the accumulation–diffusion–cooling equation in terms of heat conduction with variable boundary conditions

Given a temperature history cooling linearly in $1/T$, the general diffusion–cooling–accumulation equation (from 5.39) in dimensionless form is given by Dodson (1973) as

$$\frac{\partial}{\partial \theta}\left(\frac{C}{C_0}\right) = \frac{\tau D(0)}{a^2} \exp{-\theta} \nabla^2\left(\frac{C}{C_0}\right)$$

(5.49) $\quad\quad\quad + \lambda\tau \exp{-\theta\tau\lambda}$

where $\theta = $ dimensionless time t/τ, $C = $ concentration of daughter product, $C_0 = $ initial concentration of parent, and $D(0) = D_0 \exp(E/RT_0)$ is the initial value of the diffusion constant. If we define a new variable,

(5.50) $\quad q = 1 - (\exp{-\lambda\tau\theta}) - (C/C_0)$

then substituting we can rewrite Eq. (5.49), where

(5.51) $\quad\quad\quad M = \tau D(0)/a^2$

in the following way

$$\frac{\partial}{\partial\theta}(1) - \frac{\partial}{\partial\theta}(\exp{-\lambda\tau\theta}) - \frac{\partial}{\partial\theta}(q)$$

$$= \frac{\tau D(0)}{a^2}(\exp{-\theta})[\nabla^2(1) - \nabla^2(\exp{-\lambda\tau\theta})$$

(5.52) $\quad - \nabla^2(q)] + \lambda\tau \exp{-\lambda\tau\theta}$

which reduces to

$$\lambda\tau(\exp{-\lambda\tau\theta}) + \frac{\partial}{\partial\theta}(q) = \frac{\tau D(0)}{a^2}(\exp{-\theta})(\nabla^2 q)$$

(5.53) $\quad\quad\quad + \lambda\tau \exp{-\lambda\tau\theta}$

which further reduces to

(5.54) $\quad \dfrac{\partial}{\partial\theta}(q) = \dfrac{\tau D(0)}{a^2}(\exp{-\theta})(\nabla^2 q)$

or

(5.55) $\quad \dfrac{\partial q}{\partial \theta} = M(\exp{-\theta})(\nabla^2 q)$

where M is defined by Eq. (5.51). If $u = M(1 - \exp{-\theta})$, then

(5.56) $\quad \dfrac{\partial q}{\partial u} = \nabla^2 q$

which has the form of Fick's Second Law in one dimension with a time-dependent boundary condition

(5.57) $\quad q_s = 1 - (1 - u/M)^{\lambda\tau}$

and the initial condition $q = 0$ at $u = 0$. Solutions to these equations for the various diffusion geometries follow from the general infinite series expressions given by Carslaw and Jaeger (1959) and yield the concentration distribution within solids following slow cooling.

5.7.5 Closure profiles in minerals

Dodson (1973) presented, but did not evaluate, expressions for the limiting concentration distributions within minerals following slow cooling. The implication of these equations is that although a mineral may have an overall closure temperature following slow cooling, in fact, each radial position within the crystal has a unique closure temperature, the volume average of which is the parameter we have been discussing thus far. With isotopic techniques available with which to measure age gradients within crystals (e.g., ion probe, $^{40}Ar/^{39}Ar$ age spectrum) it seems profitable to have access to the complete description of closure profiles within minerals.

For example, the concentration expression for the plane sheet was given by Dodson (1973) as

$$q_\infty = 2 \sum_{n=1}^{\infty} (-1)^{n+1} \frac{\cos[(n-1/2)\pi x]}{(n-1/2)\pi}$$

(5.58) $$\times \left\{1 - \frac{\Gamma(\lambda\tau+1)}{[(n-1/2)^2\pi^2 M]^{\lambda\tau}}\right\}$$

where $q_\infty = \lambda\tau\theta_\infty - (C_\infty/C_0)$. Since $\lambda\tau \ll 1$, the term inside the braces is approximately $1 - (1/1)$. This requires special attention. Dodson (1986) has pointed out a suitable approximation upon noting that from Weierstrasse's definition of the gamma function,

(5.59) $\quad \Gamma(1 + x) = \exp(-cx) \quad$ for $x \ll 1$

Therefore, an expression such as $\Gamma[(1 + x)]/y^x$ becomes

$$\frac{\exp(-cx)}{\exp(x \ln y)} = \exp[-x(c + \ln y)]$$

$$= \exp(-x \ln(\gamma y))$$

(5.60) $$\simeq 1 - x \ln(\gamma y)$$

where $\gamma = \exp(c)$.

The term within the braces in Eq. (5.58) now becomes

$$1 - \frac{\Gamma(\lambda\tau+1)}{[(n-1/2)^2\pi^2 M]^{\lambda\tau}}$$

(5.61) $$\simeq \lambda\tau \ln\left[\gamma(n-1/2)^2\pi^2\tau \frac{D(0)}{a^2}\right]$$

making Eq. (5.58)

$$\lambda\tau\theta_\infty - \frac{C_\infty}{C_0} = 2\lambda\tau \sum_{n=1}^{\infty} (-1)^{n+1} \frac{\cos[(n-1/2)\pi x]}{(n-1/2)\pi}$$

(5.62) $$\times \ln\left[\gamma(n-1/2)^2\pi^2\tau \frac{D(0)}{a^2}\right]$$

The infinite summation (S) can be split as follows (Dodson, 1986)

(5.63) $\quad S = S_1(x) \ln\left[\gamma\tau \frac{D(0)}{a^2}\right] + 2S_2(x)$

where

$$S_1(x) = \sum_{n=1}^{\infty} (-1)^{n+1} \frac{\cos[(n-1/2)\pi x]}{(n-1/2)\pi}$$

(5.64) $\quad \equiv 1/2 \quad$ for all x

and

$$S_2(x) = \sum_{n=1}^{\infty} (-1)^{n+1} \frac{\cos[(n-1/2)\pi x]}{(n-1/2)\pi}$$

$$\times \ln[(n-1/2)\pi]$$

The left-hand side of Eq. (5.62) can be written

$$\lambda\tau\theta_\infty - C_\infty/C_0 = \lambda\tau\theta_\infty - \lambda(t_\infty - t_c)$$

$$= \lambda\tau\theta_\infty - \lambda\tau(\theta_\infty - \theta_c)$$

(5.65) $$= \frac{E}{RT_c}\lambda\tau - \frac{E}{RT(0)}\lambda\tau$$

Making the final expression for closure temperature as a function of position in a crystal

(5.66) $\quad \dfrac{E}{RT_c} = \ln(\gamma\tau D_0/a^2) + 4S_2(x)$

The summations of $4S_2(x)$ (Dodson, 1986) for a variety of positions within plane sheet, spherical, and cylindrical solids are given in Table 5.2, which allow closure temperatures to be calculated at intervals within crystals following slow cooling. For example, consider the case of a microcline that cooled at about 5°C/Ma through the closure interval for argon, and for

TABLE 5.2. *Closure function* $4S_2(x)$

x	Plane sheet	Cylinder	Sphere
0.00	0.41194	1.02439	1.38629
0.05	0.41653	1.03831	1.38980
0.10	0.43036	1.03990	1.40039
0.15	0.45367	1.05944	1.41826
0.20	0.48685	1.08728	1.44371
0.25	0.53047	1.12393	1.47723
0.30	0.58535	1.17011	1.51946
0.35	0.65254	1.22676	1.57131
0.40	0.73347	1.29515	1.63392
0.45	0.82999	1.37695	1.70887
0.50	0.94458	1.47439	1.79824
0.55	1.08057	1.59051	1.90484
0.60	1.24254	1.72949	2.03262
0.65	1.43699	1.89732	2.18721
0.70	1.67352	2.10291	2.37704
0.75	1.96710	2.36026	2.61543
0.80	2.34291	2.69317	2.92511
0.85	2.84826	3.14674	3.34950
0.90	3.58951	3.82350	3.98796
0.95	4.90636	5.05471	5.16455
0.96	5.33878	5.46577	5.56120
0.97	5.90027	6.00366	6.08275
0.98	6.69734	6.77401	6.83402
0.99	8.06977	8.11483	8.16128
0.995	9.44913	9.47507	9.49661
Volume average	1.58611	2.71862	3.43012

which an E of 35 kcal/mol and a D_0/a^2 of 500/s have been determined (see Section 5.9.3.5). The bulk closure temperature, assuming a plane sheet geometry, is

$$T_c = \frac{E/R}{\ln\left[\dfrac{ART_c^2 D_0/a^2}{E\,dT/dt}\right]} = \frac{(35{,}000\,\text{cal/mol})/(1.987\,\text{cal/mol}\cdot{}^\circ\text{K})}{\ln\left[\dfrac{8.7\cdot 1.987\,\text{cal/mol}\cdot{}^\circ\text{K}\cdot(438{}^\circ\text{K})^2\cdot 500/\text{s}}{35{,}000\,\text{cal/mol}\cdot 1.58\times 10^{-13}{}^\circ\text{K/s}}\right]} = 165{}^\circ\text{C}$$

As mentioned earlier, this equation is iterative in T_c. That is, a trial value inserted into the argument of the logarithm will return a second-order estimate of T_c. Because the logarithm dampens sensitivity to variations in the iterative process, this loop converges rapidly, usually in two iterations. For example, had we estimated T_c within the argument to be 500°C rather than 165°C, our first iteration yields a T_c of 153°C. On reinserting this value we obtain $T_c = 165$°C.

Returning to the calculation of closure profiles, note that the average of the age gradient recorded in the first few percent of gas release during an $^{40}\text{Ar}/^{39}\text{Ar}$ age spectrum experiment using this sample corresponds to grain margin closure equivalent to some 10 Ma later at a temperature of

$$T_c = \frac{E/R}{\ln(\gamma\tau D_0/a^2) + 4S_2(x)}$$

$$= \frac{(35{,}000\,\text{cal/mol})/(1.987\,\text{cal/mol}\cdot{}^\circ\text{K})}{\ln(1.78\cdot 5.55\times 10^{13}\,\text{s}\cdot 500/\text{s}) + 7}$$

$$= 115{}^\circ\text{C}$$

where τ reflects the local closure temperature, also obtained by iterative convergence.

The implication of Eq. (5.66) is that a single mineral sample can provide finite segments of cooling curves rather than only a single temperature–time point.

5.7.6 Summary of Dodson's closure temperature model

The closure temperature model applies only in those circumstances in which simplifying assumptions are met. The requirement of slow cooling means that the time constant (over which the diffusivity drops by a factor e^{-1}) must be much greater than the initial value of the characteristic diffusion time, $a^2/D(0)$. In other words, diffusion must be initially so rapid that argon is not retained on a timescale equivalent to τ. Although the assumption of linear cooling in $1/T$ is required to make the mathematics tractable, the calculated closure temperature is somewhat insensitive to the choice of this parameter and Dodson (1973) has argued that the tangent to the cooling curve at the point of closure is probably adequate. It should be pointed out that where thermal contrasts occur in the crust, as in the intrusion of igneous bodies into cool country rocks, the form of cooling is essentially linear in $1/T$. Dodson (1973) points out that no significant errors will be introduced by neglecting the steady-state daughter concentration that will likely exist at the commencement of cooling as this component corresponds to the approximation made in Eq. (A.5.4.16). Other obvious violations of model interpretations include the presence of inherited argon, subsequent episodic $^{40}\text{Ar}^*$ loss, and mineral recrystallization. To expand on this latter point, should the mineral of interest form below the closure temperature, for example, during low grade metamorphism, then the closure model is no longer a valid description of the daughter retention history. Similarly, if cooling is accompanied by differential stress resulting in recrystallization, then the likelihood of diffusion being the rate-limiting daughter product transport mechanism during such structural reconstitution is remote.

In our discussions thus far, we have taken for granted the existence of good quality diffusion data that can be used in conjunction with the episodic loss or slow cooling models. Indeed, this entire approach to the interpretation of geochronology pivots on the availability of meaningful kinetic data. In Section 5.9, we discuss the criteria for performing these measurements and review the literature.

5.8 SOLUTIONS OF THE HEAT FLOW EQUATION

5.8.1 Introduction

Earlier in this chapter, we derived the general equation of unsteady-state heat flow in a solid from energy balance considerations. Having developed the diffusion analogue, we now return to the problem of heat conduction to obtain solutions describing heat flow in likely crustal configurations. Given both thermochronological data and a meaningful thermal model, a complete description of the thermal evolution of certain geological phenomena, such as orogeny and igneous intrusion, may be possible.

The simplest approach is to consider analytical solutions that describe end-member processes such as simple uplift or subsidence at a constant rate, and then to acquaint ourselves with more realistic, compound numerical models that can combine features, for example, igneous intrusion, with concurrent crustal uplift. We will find that our earlier derived diffusion models for various geometries are of little use in the consideration of heat flow, as a uniform boundary condition in these solutions was of a constant zero concentration—an unlikely scenario for heat flow in the crust. Rather, we will view geometric features in the crust as having different initial conditions (e.g., high temperature), but essentially identical diffusive properties as their host.

5.8.2 Simple uplift

Carslaw and Jaeger (1959, p. 388) present an expression for the effect of uplift on the vertical temperature distribution $T(z, t)$ in a homogeneous half space, where z is depth below the Earth's surface and t is the time since the onset of uplift at constant rate w. At $t = 0$ the temperature distribution is assumed to be

(5.67) $\quad T(z, 0) = T_s + (Gz/K)$

where T_s is temperature at the Earth's surface, G is the geothermal flux, and K is the thermal conductivity. This assumed initial temperature distribution corresponds to steady-state conditions if there is no contribution from radioactive heat generation. If uplift is matched by a rate of erosion w at the Earth's surface, then $T(0, t) = T_s$ for all time and the temperature distribution at an arbitrary depth and time is

$$T(z, t) = T_s + (Gz/K) + (Gwt/K)$$
$$+ (G/2K)(z - wt)\exp(-wz/\kappa)$$
$$\times \mathrm{erfc}[(z - wt)/2(\kappa t)^{1/2}]$$
(5.68) $\quad -(z + wt)\,\mathrm{erfc}[(z + wt)/2(\kappa t)^{1/2}]$

For a rock sample that at time $t = 0$ was located at a depth z_0 the cooling curve $T'(t)$ for that sample is obtained by simply substituting $z = z_0 - wt$ into Eq. (5.68); thus, $T'(t) = T(z_0 - wt, t)$.

Cooling by uplift and erosion yields a cooling curve that is quite different in character from that for cooling by thermal conduction *in situ*. For generality we shall express Eq. (5.68) in terms of the dimensionless variables

$T^* = (T - T_s)/T_s$ (dimensionless temperature)
$z^* = z/h$ (dimensionless depth)
$t^* = \kappa t/h^2$ (dimensionless time)
$\beta = wh/K$ (uplift parameter)
$\phi = Gh/KT_s$ (geothermal flux parameter)

where h is the depth of the sample at time $t = 0$. For most geological uplift problems $0 < \beta < 10$ and $0 < \phi < 10$ (for T_s in Kelvin). As an example, if the sample is originally at depth $h = 10^4$ m and $w = 0.01$ m year^{-1} = 3.16×10^{-10} m/s and $\kappa = 10^{-6}$ m^2/s, then $\beta = 3.16$; if $G = 0.05$ W/m, $K = 2.0$ W/m°K, and $T_s = 280°$K, then $\phi = 0.89$.

In dimensionless variables [Eq. (5.68)] becomes

$$T^*(z^*, t^*)/\phi = z^* + \beta t^* + 1/2[(z^* - \beta t^*)$$
$$\times \exp(-\beta z^*)$$
$$\times \mathrm{erfc}\{(z^* - \beta t^*)/(2t^{*1/2})\}$$
$$- (z^* + \beta t^*)$$
(5.69) $\quad \times \mathrm{erfc}\{(z^* + \beta t^*)/(2t^{*1/2})\}]$

and the cooling curve is obtained by substituting $z^* = 1 - \beta t^*$ into Eq. (5.69). The results are shown in Fig. 5.4 for a range of values of the uplift parameter β. Cooling by uplift and erosion yields a cooling curve that is characterized by slow cooling at first and rapid cooling only as the sample approaches the Earth's surface.

A different approach to this problem was described by Royden and Hodges (1984) who suggested that the solution to the conduction equation in a moving medium with a production term, given some initial temperature distribution and fixed uplift rate, could yield the temperature–depth path for an uplifting lithosphere. The selection of input parameters, particularly uplift rate, could be constrained for a given terrane by using the closure temperatures and mineral age data from a variety of phases as a test of the calculations.

Alternatively, Eq. (5.69) can describe the temperature distribution within the crust that results from subsidence, a condition that can lead to daughter product redistribution in potassium-bearing phases (e.g., Harrison and Bé, 1983). In this case the "uplift" rate is negative with rapid subsidence causing strongly perturbed temperature distributions from the initial linear, equilibrium thermal gradient.

Results from Fig. 5.4 are equally valid for uplift and subsidence. A somewhat more realistic numerical model of conductive heat flow during uplift was presented by Parrish (1982). This model incorporated an exponential distribution of heat-producing elements within the crust as well as an allowance for a variable uplift history and reduced heat flow.

5.8.3 Finite tabular pluton without uplift

Another analytic solution of geological interest is that for the cooling of a finite vertical dike of width $2d$ and length $2L$ buried at a depth h_1 below the surface and with its lower surface at depth h_2. The dike is aligned so that its strike direction is parallel to the y axis and $y = 0$ is a plane of symmetry. If the dike, initially at temperature T_1, is injected into material at temperature T_0 the temperature distribution at an arbitrary point (x, y, z) at time t is

$$T(x,y,z,t) = T_0 + 1/8(T_1 - T_0)$$
$$\times \{\text{erf}[(x + d)/2(\kappa t)^{1/2}]$$
$$- \text{erf}[(x - d)/2(\kappa t)^{1/2}]\}$$
$$\times \{\text{erf}[(y + L)/2(\kappa t)^{1/2}]$$

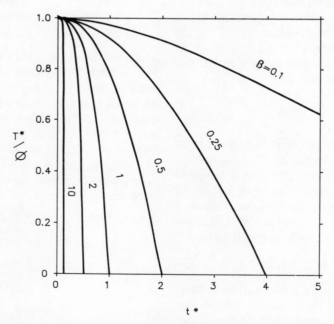

Fig. 5.4. Cooling curve for an uplifted sample. Erosion and uplift rates are assumed equal. The vertical axis is dimensionless temperature normalized by dimensionless geothermal flux; the horizontal axis is dimensionless time. Different curves correspond to different values of dimensionless uplift parameter, β. After Harrison and Clarke (1979).

$$- \text{erf}[(y-L)/2(\kappa t)^{1/2}]\}$$
$$\times \{\text{erf}[(z-h_1)/2\kappa t)^{1/2}]$$
$$- \text{erf}[(z-h_2)/2(\kappa t)^{1/2}]$$
$$+ \text{erf}[(z+h_1)/2(\kappa t)^{1/2}]$$
(5.70) $$- \text{erf}[(z+h_2)/2(\kappa t)^{1/2}]\}$$

This result can be appreciated in terms of the derivation of Eq. (5.30) (or A.5.2.17), the expression for the transport of heat or mass from a semi-infinite medium. In effect, Eq. (5.70) is the product of the superposition of two extended initial distributions in each of the three axes that define by overlap the finite volume $-d \leqslant x \leqslant d$, $-L \leqslant y \leqslant L$, $h_2 \leqslant z \leqslant h_1$. In addition, an extra term defines a mirror source reflected across $z = 0$ to simulate the effect of the Earth's surface. Although the pluton is intruded into an isothermal region, the effect of a geothermal gradient can be accounted for by simply adding this component, calculated from Eq. (5.67), to the result generated for (x, y, z, t) using Eq. (5.70). Cooling histories generated from this model (Fig. 5.5) are characterized by a strong concave upward appearance resulting from the strong thermal contrast between pluton and country rock. This model is useful in considering the cooling behavior of igneous bodies intruded high into the crust and allowed to cool in the absence of simultaneous crustal uplift.

A variety of solutions for cylindrical and spherical intrusions into isothermal media are given in Carslaw and Jaeger (1959) including solutions involving composite thermal properties (e.g., a contact aureole). Other references that provide solutions for specific geometric or boundary conditions are Jaeger (1957, 1959, 1961, 1964), Lovering (1935, 1936), and Crank (1975).

5.8.4 Dike with uplift and heat generation

Because granitoid intrusion is often contemporaneous with orogeny-related regional uplift, a solution of the unsteady-state heat flow equation that accounts for both dike intrusion and uplift is warranted. In addition, the previous model did not allow consideration of latent heat of crystallization (a moving boundary problem) other than simply raising

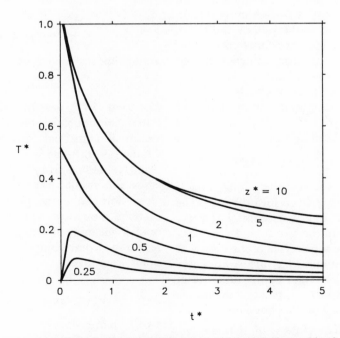

Fig. 5.5. Cooling curves for the central axis of an infinitely long and deep buried dike resulting from cooling by thermal conduction. Results are expressed in terms of dimensionless temperature, $T^* = (T - T_0)/(T_1 - T_0)$, and dimensionless time, $t^* = \kappa t/d^2$, for a variety of dimensionless depths, $z^* = z/h_1$. Depth and dike half width are equal. After Harrison and Clarke (1979).

the initial temperature to account for the additional thermal energy. A numerical model that allows these three features to be considered simultaneously was described by Harrison and Clarke (1979).

Because of its complexity the results cannot be expressed in closed mathematical form and the solution must be obtained by numerical methods. The basic assumptions that lead to this model are the following:

1. Before injection of the pluton there is a uniform distribution of radioactive materials within the region ($0 \leqslant z \leqslant Z$) and heat is released at a constant rate Q per unit volume. We can neglect the fact that Q decreases slowly with time provided the time interval of interest is substantially less than the half lives of radioactive isotopes present.

2. A steady heat flux G is assumed to flow into the base of the slab at $z = Z$ so that for all time

$$\partial T(x,Z,t)/\partial z = G/K$$

3. The surface of the slab is at $z = 0$ and this surface is maintained at a constant temperature T_s. This assumption implies that the rate of erosion at the Earth's surface must remain approximately equal to the rate of uplift $w(t)$, where $w(t)$ is some specified function of time. (Climate change would also influence T_s, but this effect would not be large.)

4. Before injection of the pluton at $t = 0$, the slab $0 \leqslant z \leqslant Z$ is in thermal equilibrium and there is no uplift prior to $t = 0$; thus, for $t > 0$ the temperature distribution is

(5.71) $T(x,z,t) = T_s + Gz/K = Qz(Z - z/2)$

5. At time $t = 0$ an infinitely long tabular dike of width $2d$ is injected into the slab. The dike is rectangular in cross section, has strike direction parallel to the y axis, and dips vertically. At $t = 0$ the top surface of the dike is at $z = h_1$ and the bottom surface at $z = h_2$, where $h_2 < Z$. The vertical faces of the dike are at $x = \pm d$ so that $x = 0$ is a plane of symmetry and $\partial T(0,z,t)/\partial x = 0$ on this boundary.

6. The initial temperature of the injected dike is T_1 and the dike material has the same thermal properties as the surrounding host rock except that the rate of radioactive heat production per unit volume is taken to be Q_1, a constant that may differ from Q, and the thermal contribution of latent heat is approximated by increasing Q_1 for a period of time appropriate to the crystallization interval.

The basic equation for two-dimensional heat transfer with uplift and heat generation is

(5.72) $\dfrac{\partial^2 T}{\partial x^2} + \dfrac{\partial^2 T}{\partial z^2} - \dfrac{w}{\kappa}\dfrac{\partial T}{\partial z} - \dfrac{1}{\kappa}\dfrac{\partial T}{\partial t} = \dfrac{-Q}{KJ^*}$

where J^* is the mechanical equivalent of heat. At $t = 0$, the boundaries of the dike are

$$x = \pm d \qquad h_1 \leqslant z \leqslant h_2$$

and

$$z = h_1 \qquad -d \leqslant x \leqslant +d$$

and

$$z = h_2 \qquad -d \leqslant x \leqslant +d$$

The initial conditions are

$T(x,z,t) = T_s + Gz/K + Qz(Z - z/2)$
outside dike
$= T_1$
inside dike

The boundary conditions are

$$T(x,0,t) = T_s$$
$$\partial T(x,Z,t)/\partial z = G/K$$

and owing to the symmetry plane at $x = 0$

$$\partial T(0,z,t)/\partial x = 0$$

Equation (5.72), together with boundary and initial conditions, is then solved using conventional numerical methods. Harrison and Clarke (1979) found good agreement between results obtained from this model in limiting cases and results from both of the previous closed form solutions, giving some confidence that the numerical scheme is stable and can predict meaningful temperature distributions resulting from complex intrusion–cooling–uplift scenarios.

5.8.5 Model parameters

5.8.5.1 Introduction

Before using any of the preceding thermal models to describe a particular geological environment, an important limitation must be assessed—whether or not heat transfer was dominantly diffusive. The existence of fluids in porous rock provides the potential for a much more rapid, and much less well constrained, mechanism of heat flow. Heating a fluid from below or the side in a porous medium results in a density instability that causes the fluid to rise, transporting heat with an efficiency that can be many orders of magnitude greater than conduction. Unfortunately, heat transfer of this nature is far less predictable than simple diffusion as subtle changes in parameters such as rock permeability and scale of convection result in dramatic changes in heat flow (e.g., Ribando et al., 1978; Walker and Homsy, 1978). However, in cases in which insufficient pore space or fluid is present to allow enhanced heat transfer, the simple, geometric nature of heat flow described by Eq. (5.22) allows a framework within which reconstruction of geological temperature histories is possible.

5.8.5.2 Thermal Properties and Dimensions

The component parts of the thermal diffusivity, rock density, heat capacity, and conductivity, are all amenable to laboratory measurement. Although these parameters change somewhat with temperature, it is often possible to predict these variations, and compensatory changes in the numerator and denominator of the definition of κ buffer against radical variations in the thermal diffusivity with changing temperature. Using available thermodynamic and conductivity data, Wells (1980) produced expressions for the specific heat (c) and thermal conductivity (K) as a function of temperature, T.

$$c = 753 + 0.46T - 1.45 \times 10^7/T^2 \quad \text{(in J/kg)}$$
(5.73)

$$K = (0.311 + 1.72 \times 10^{-4}T)^{-1}$$
$$+ 1.05 \times 10^{-3}(|T\text{-}800|$$
$$+ T\text{-}800) \quad \text{(in W/m-°K)}$$
(5.74)

Thus, the thermal diffusivity at 600 and 1000°C varies only between 7 and 8×10^{-7} m²/s, respectively, although the specific heat, for example, increases by about 40%.

Heat generation and latent heat terms are also measurable quantities and can be estimated sufficiently precisely for the purposes of most thermal models. However, it may be necessary to first assess the actual need for the latter of these two parameters. For example, in wishing to link thermochronological data from a cooling diapir with a realistic description of heat flow from this body, the initial condition of the pluton must first be assessed. A recent treatment of the intrusion mechanics of diapirs by Mahon et al. (1988) suggests that granitoid bodies become emplaced at mid-crustal levels at uniform temperatures when about 40% crystallized. The relatively rapid rate of emplacement allows the pluton to be viewed as being instantaneously intruded at this temperature—an assumption inherent in both of the preceding dike models. This being the case, only a small allowance for latent heat is required. The dimensions of the system may be largely available from a plan view of the terrane being modelled. Less clear are questions of how deep a sample was prior to the onset of uplift, what the aspect ratio of a pluton is, or whether the shape of the feature requiring description is amenable to a geometric solution of the conduction equation.

Initial temperatures for parts of the system may be obtainable via external sources such as experimental investigations, constraints from metamorphic petrology, and direct geothermometry.

Many input parameters may not be well constrained. It is obviously the responsibility of the user to assure that the simulation is either insensitive to large variations in these parameters or to view the experiment as a forward model and attempt to achieve concordance with the thermochronological data by reasonable excursions in these input variables.

5.9 DIFFUSION STUDIES AND RESULTS

5.9.1 Background

Up to this point, the description of diffusion has been largely mathematical with little appreciation of the atomistic mechanisms of diffusion. In part, this reflects how little we know about details of the latter in complex silicates, but we can at least speculate on possible mechanisms. Because of its inert nature, the diffusion of argon is a special case—its intralattice migration is effectively controlled by the work required to distort the neighboring atoms to allow passage of the argon atom into a new site, and not by any local binding energy (Sardarov, 1961). There are four possible schemes by which transfer of atoms can occur in a lattice. These mechanisms are illustrated for the case of a simple lattice in Fig. 5.6. The first mechanism illustrated involves simple exchange of adjacent atoms in a perfect lattice. A variant of this is ring exchange in which a circular group of atoms moves simultaneously in the same clockwise sense moving into a site vacated by the advancing neighbor. The remaining three types of diffusion jumps all involve the existence of point defects, or imperfections in the crystal lattice. The simplest of these is the vacancy transfer in which an atom moves into an adjacent Schottky defect. The interstitial or Frenkel mechanism describes the case of an atom sited in between normal lattice sites moving to a new interstitial location by squeezing past atoms in regular sites. Finally, the interstitialcy mechanism operates when an interstitial atom displaces a normally sited atom into an interstitial location. As a result, the site centered on the original interstitial moves, in effect, twice as far as either atom.

Atoms oscillate within their equilibrium lattice positions about 10^{12} times every second. Above absolute zero (0°K) there is a finite probability of an atom having sufficient local thermal energy to jump from its current position to an adjacent site by one of the means just discussed. As temperature is raised, the probability of some atoms in the Boltzmann distribution obtaining the energy threshold required to overcome the potential barrier increases exponentially.

Empirically, the manner in which diffusivity varies with temperature is described by the Arrhenius relationship [Eq. 5.36],

$$D = D_0 \exp(-E/RT)$$

An Arrhenius-type plot is shown in Fig. 5.7 for illustration as to how the coefficients of activation energy (E) and frequency factor (D_0) can be extracted from a linear array of diffusion data on such a diagram. In this example, the natural logarithm of the diffusion coefficient D is plotted against the reciprocal absolute temperature. By taking the natural logarithm of

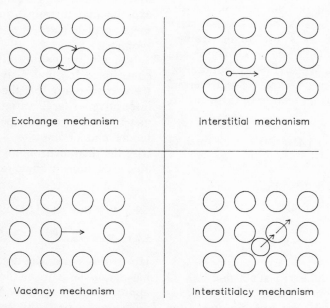

Fig. 5.6. Schematic illustration of the four possible mechanisms of diffusion transport: Exchange, vacancy, interstitial, and interstitialcy.

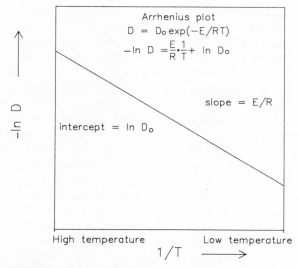

Fig. 5.7. Schematic Arrhenius plot illustrating the relationship of parameters in the equation of a straight line.

both sides of Eq. (5.36) we obtain

$$\ln D = \ln D_0 - \frac{E}{R}\frac{1}{T} \tag{5.75}$$

or an equation of the form $y = mx + b$ where y is the $\ln D$ coordinate, E/R is the slope of the line in Fig. 5.7, $1/T$ is the x axis and $\ln D_0$ is the y axis intercept of the line. In other words, the slope of the line generated by the data in Fig. 5.7 is proportional to the activation energy and the y axis intercept is the natural logarithm of the frequency factor.

An example of the calculation of a diffusion coefficient from argon loss data is given in Appendix A.5.5 using the results for argon diffusion in orthoclase of Foland (1974).

The theoretical basis for the Arrhenius equation can be appreciated in terms of the increase of Schottky (or Frenkel) defects, or vacancies, with rising temperature, which grow as $n \propto \exp - (\Delta H/RT)$, where n is the number of vacancies and ΔH is the enthalpy of formation of 1 mol of defects. By taking an atom from within the crystal and removing it to the surface the potential energy of the crystal is increased as the atom placed on the outside is subject to only half the attractive forces as within. The source of this energy is thermal and the quantity required to affect this transfer is ΔH. Thus, the temperature of the crystal is raised, and atoms diffuse faster due to both an overall shift in the Boltzmann function and an increase in the number of potential vacancies.

5.9.2 Experimental criteria

Perhaps the most problematic aspect of $^{40}Ar/^{39}Ar$ thermochronology is the requirement for knowledge of the natural diffusion behavior of argon in silicates. The reasons for this are manifold and will be enumerated in this introduction. Experimental concerns aside, we in earth sciences share the unique challenge of wishing to understand material properties on time scales well beyond those available from direct laboratory experiment. Fortunately for us, the reactions of common interest often exist over a broad stability field and have a systematic temperature dependence that permits us to perform experiments above the temperatures of direct interest and to extrapolate the results downslope; for example, argon diffusion in biotite (Section 5.9.3.2). However, these experiments do not free us of the concern that our extension of the data may have no meaning if, for example, a diffusion mechanism with a lower temperature dependence is encountered over the extrapolated portion. Although we can design experiments that utilize geological events as a check on our results, we must proceed with considerable caution in our interpretation of the diffusion literature.

Because of its essential incompatability, $^{40}Ar^*$ produced *in situ* in natural samples has been emphasized in virtually all studies of the diffusion of argon in silicates. Although recent advances in microanalysis will allow single grain experiments (Sutter and Hartung, 1984; York et al., 1981, 1984), all published investigations have used sized aggregates as the starting material. Several authors (e.g., Giletti, 1974a; Harrison, 1981) have outlined the criteria required for a successful diffusion experiment. These are (1) the phase must remain stable throughout the duration of the experiment; (2) the effective diffusion length scale must be known or be derivable from the experiment; (3) the shape of the particles in the aggregate must conform to a solution of the diffusion equation; (4) the aggregate must contain effectively one grain size; (5) only one mineral phase can be present; (6) the initial distribution of argon must be uniform; and (7) heating must be isothermal. Certainly, violations of several of these criteria are not devastating if additional information is available. For example, if the grain size (item 4) or initial argon distribution (item 6) are nonuniform but known, the mathematical model can be modified to allow for these variations. However, it is difficult to see how violation of the stability criterion (item 1) could be overcome without prejudice to the experiment. As the majority of samples dated by the $^{40}Ar/^{39}Ar$ method are hydrous (e.g., micas and amphiboles) and may contain iron (e.g., biotite and hornblende), this criterion demands hydrothermal treatment with the capability of buffering the environment with respect to oxygen.

A typical experiment of the kind pioneered by Bruno Giletti involves isothermal, hydrothermal treatment of a closely sized, encapsulated, monomineralic aggregate for periods as long as many months. This heating induces the $^{40}Ar^*$ initially present within the sample to diffuse to the grain boundary where it is effectively lost to an infinite reservoir at zero concentration. With respect to the latter point, in reality, the $^{40}Ar^*$ evolved from the sample mixes with argon of atmospheric composition in the capsule and is prevented from reentering the solid by the highly unfavorable partition coefficient. The amount of $^{40}Ar^*$ remaining in the sample after the experimental heating is determined and compared to the concentration in the starting material. Thus, the fraction of $^{40}Ar^*$ lost is determined allowing calculation of a model diffusion coefficient using a geometrically appropriate solution of Fick's Second Law. This type of experimental design does not violate any of the specified criteria but has not been widely used perhaps in part because of the limited availability of experimental facilities for long periods of time.

However, not all diffusion experiments have utilized this approach. Early experiments meant to reveal transport rates of argon in minerals used a myriad of methods and produced a substantial literature of largely conflicting results. Indeed, by 1969, Alan Mussett, reviewing the then current argon diffusion literature of over 50 papers, concluded that there were no criteria of retentivity that could be generally applied and emphasized the ambiguity of results produced by inappropriate experimental methods. The central problems were lack of sample characterization and failure to keep the specimen stable throughout the duration of the experiment. As an example of the former, diffusion coefficients for feldspar compiled by Mussett (1969) disagree by over seven orders of magnitude at any one temperature (Fig. 5.8). The reason for this apparent discrepancy does not lie in dramatic diffusivity contrasts between structurally similar minerals, but that the diffusion radius chosen to extract D from the fractional loss expression was the measured particle size, heedless of the virtually omnipresent exsolution structures in plagioclase and low-temperature alkali feldspars. As an example of the failure to keep the sample stable throughout the experiment, one need go no further than Mussett's feldspar compilation. The microcline data in Fig. 5.8 show a snakelike variation in activation energy resulting from homogenization of perthite lamellae at $\sim 800°C$ (see Harrison et al., 1986) and the eventual inversion to orthoclase structure. Failure to keep the microcline structure intact results in a complex and uninterpretable pattern. A more common example is the use of vacuum step heating for hydrous phases such as biotite or hornblende (Berger and York,

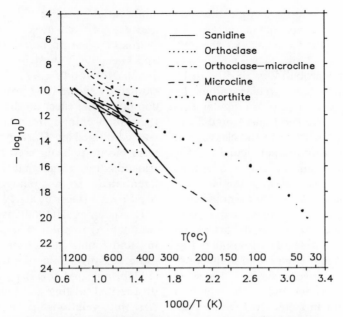

Fig. 5.8. Arrhenius plot showing results of diffusion studies using feldspars as of 1969. The divergent behavior results from several artifacts, including improper choice of diffusion radius and homogenization of exsolution structures. From Mussett (1969), with permission.

1981a). This method is practically irresistable as the diffusion experiment can be performed on-line as a byproduct of the age spectrum analysis. Unfortunately, heating minerals containing water in a vacuum results in structural decomposition through dehydration, and, thus, results obtained are not likely to represent transport rates in the original structure. Arrhenius parameters calculated from this approach suffer from both this shortcoming and a degree of arbitrary data selection due to the general scatter of calculated diffusion coefficients, especially for biotite.

No attempt is made here to reference all argon diffusion studies as the compilation of Mussett (1969) provides a source for much of this early work. Rather, we shall concentrate on the studies that have met the criteria of a successful diffusion experiment, enumerated earlier in this chapter.

5.9.3 Laboratory studies of argon diffusion in natural silicates

5.9.3.1 Introduction

Before beginning our discussion of experimental results, it is worth recalling Fig. 4.6 from Chapter 4. This illustration shows the relationship between fractional loss and the dimensionless parameter, Dt/a^2, for the conventional solutions of the diffusion equation—spherical, infinite cylinder, plane sheet, and cubic geometries. Clearly, for a thermal event to proceed with an intensity Dt/a^2, the sphere loses the most argon and the plane sheet the least. This is simply because, with the highest surface to volume ratio, the bulk of the argon in a sphere is much closer to the grain surface than in other geometries. However, for the plane sheet, the initial mass distribution is identical to the concentration distribution. Thus, the result of a fine experiment but inappropriate choice of diffusion model could be a calculated value of D that is wrong by as much as an order of magnitude. Examples of appropriate choices for the diffusion model might be as follows. A hornblende that cleaves into an aggregate with an essentially rectangular cross section and an aspect ratio of five could be modelled using an infinite cylinder model provided that the particle radius defines what we shall call the effective diffusion radius—the grain or subgrain dimension that controls argon loss from the volume. Because of the elongate nature of the sample, diffusion out the ends can be approximated as zero leaving radial diffusion from the axis the only important diffusion path. The noncircular cross section could probably be dealt with by obtaining an equivalent circular radius by averaging the approximately rectangular surface. If it can be determined that argon diffuses much more rapidly parallel to the cleavage of biotite compared with perpendicular to the cleavage, then an infinite cylinder model may be an appropriate model despite the fact that the mica resembles a sheet. A homogeneous feldspar that cleaves into roughly cubic particles can be modelled using the approximation for cubic geometry (Table 5.1). Also note in Fig. 4.6 that a cube of half edge a and a sphere of radius a have relatively similar degassing behavior (Jain, 1958) and use of the equation for a sphere would in some cases be an adequate approximation.

Finally, planar structures such as perthite lamellae in alkali feldspar can be satisfactorily described as a plane sheet, as the incoherent lamellar boundaries appear to define the effective diffusion dimension.

5.9.3.2 Biotite–Phlogopite

Biotite, being the mineral most commonly dated by the K–Ar method, has been the focus of a great many argon diffusion studies using laboratory heating (e.g., Brandt et al., 1967; Kotlovskaya and Tovarenko, 1974; Zimmerman, 1972; Gerling et al., 1966; Kotlovskaya, 1975; Aprub et al., 1973; Brandt and Voronovskiy, 1967; Gerling and Morozova, 1957; Melenevskiy et al., 1978; Hanson, 1971; Norwood, 1974; Berger et al., 1979; Harrison et al., 1985). Reported activation energies for the diffusion of ^{40}Ar in biotite vary over 90 kcal/mol corresponding to a large range of closure temperatures. However, with the exception of Melenevskiy et al. (1978), Brandt et al. (1967), Norwood (1974), and Harrison et al. (1985), these laboratory measurements have violated the important criterion of keeping the phase stable throughout the duration of the experiment and are likely invalid measures of the transport rate of ^{40}Ar in biotite. Studies using natural settings were unable to define sufficiently precise diffusion parameters because of the large uncertainties in the thermal models proposed (Hart, 1964; Hanson and Gast, 1967; Westcott, 1966; Hurley et al., 1962). Norwood's (1974) hydrothermal diffusion study showed the probable existence of a compositional effect on the rate of diffusion in the annite–phlogopite solid solution series that was confirmed by Harrison et al. (1985). The term biotite is being used to denote iron-rich trioctahedral micas that are arbitrarily differentiated from phlogopite in having Mg:Fe < 2:1 (Deer et al., 1963).

Evernden et al. (1960) investigated the diffusion of ^{40}Ar in phlogopite using vacuum heating and found this phase to be significantly more retentive than biotite. However, Giletti (1974b) repeated this experiment utilizing hydrothermal heating and found a very different Arrhenius relationship. In the discussion that follows, we shall focus on the studies of Giletti (1974b) and Harrison et al. (1985) as they reflect most of the advantages and limitations of this approach to defining diffusion behavior in micas.

Giletti (1974b) carefully sized flakes of phlogopite ($Ann_4 = 4\%$ annite component) that had been cut from a single mica book to obtain three size fractions with a range of a^2 of ~ 20. After repeated determinations of the ^{40}Ar* content of the starting material by isotope dilution methods, aliquants were hydrothermally heated for between 20 h and 72 days. The fluid to solid weight ratios in the charges were about one. Following isothermal heating at temperatures between 600 and 900°C, well within the stability field (Wones, 1967; Wones and Eugster, 1965), the runs were quenched, and the mica was recovered and analyzed by isotope dilution for the ^{40}Ar* content. The results of 19 such experiments produced a well-defined line with a slope proportional to an activation energy of 57.9 ± 2.6 kcal/mol and a frequency factor of $0.75^{+1.7}_{-0.52}$ cm^2/s (Fig. 5.9). Giletti (1974b) then compared the results with the earlier study of Evernden et al. (1960) and found very different behavior in the Arrhenius relations (Fig. 5.10). Where the vacuum step heating data showed a wavy distribution, the hydrothermal data essentially defined a straight line, typically about two orders of

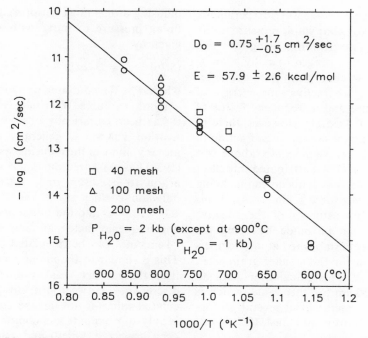

Fig. 5.9. Diffusion data for argon in phlogopite. From Giletti (1974b), with permission.

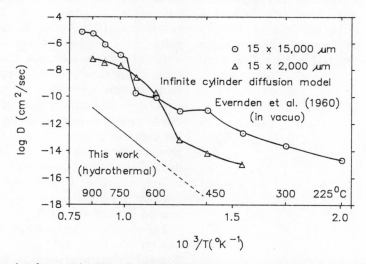

Fig. 5.10. Diffusion data for argon in phlogopite showing comparison of vacuum derived data (indicated as Evernden et al., 1960) and hydrothermal data [where this work refers to Giletti (1974b)]. Reproduced with permission.

magnitude lower in diffusivity than the data obtained in vacuum. Giletti (1974b) concluded that his experiments had revealed the natural diffusion behavior of ^{40}Ar in phlogopite whereas the vacuum-derived results reflected the structural breakdown of the phase during heating.

Harrison et al. (1985) performed experiments similar in most respects to Giletti's study on a biotite (Ann_{56}). An additional consideration for this iron-bearing system is that the environment must have an oxygen fugacity regulated at an appropriate value to maintain stability of the mica (Wones and Eugster, 1965). This was accomplished by use of various oxygen buffers that provided an f_{O_2} equivalent to the quartz–fayalite–magnetite buffer. At 2 kbar total pressure, an Ann_{56} biotite would be stable in this environment up to ~850°C. Experiments were performed at 1 kbar water pressure in the 600–750°C temperature range using four different grain sizes that ranged in a^2 by a factor of 12. The three smallest grain sizes defined a single line with an activation energy of 47.0 ± 2.1 kcal/mol and a frequency factor of $0.077^{+0.21}_{-0.06}$ cm^2/s (Fig. 5.11). However, three of the four runs using the coarsest size fraction of 202 μm radius plotted significantly above this line. Harrison et al. (1985) attributed this effect to the intrinsic effective diffusion radius being smaller than the measured grain radius. If this is correct, then an estimate of the effective diffusion radius can be made by using the diffusion coefficient established at any temperature by the data for the smaller grain sizes, and solving for a. This calculation gives an estimate of ~150 μm for this value, which is similar to that extracted from several geological studies (e.g., Hanson and Gast, 1967; Westcott, 1966).

Two runs at high pressures (14 kbar) plotted well below the other data and indicate a pressure effect on the diffusion of ^{40}Ar in biotite. A modified form of the Arrhenius equation that accounts for the extra work diffusing atoms must perform against the confining pressure to move within the lattice is given by

(5.76) $\quad D = D_0 \exp[-(E + PV)/RT]$

where P is the confining pressure and V is the activation volume. Weertman (1970) noted that it has been empirically observed that the activation volume is generally similar to the molar volume of the diffusing species. Henshaw (1958) has measured the molar volume of solid argon to be 22 cm^3/mol. Using an average diffusion coefficient for the 14 kbar runs of 1.1×10^{-13}/s and the 1 kbar activation energy and frequency factor, an activation volume of 14 cm^3/mol can be calculated from Eq. (5.76). This is similar to the molar volume calculated from the van der Waals radius of argon. The discrepancy between the observed and calculated values is not unexpected as the theory is only approximate and compressibilities have been neglected. Giletti and Tullis (1977) concluded that the effect of pressure on ^{40}Ar diffusion in phlogopite up to 15 kbar may not be significant. However, of their two data points, only one has a lower pressure datum with which to compare. For that sample there exists an apparent pressure effect equivalent to an activation volume of ~10 cm^3/mol. The calculated activation volume of ~14 cm^3/mol can be used in estimating the pressure effect on the diffusion of argon in biotites at pressures greater than 1 kbar. However, for most crustal thermal gradients, biotites will not be expected to retain argon at pressures greater than ~4 kbar, and the correction that need be applied is relatively small.

The systematic temperature dependence we observe in these laboratory results allows down-slope extrapolations to geologically important temperatures and, thus, access to extremely long time scales. However, it is not without reservations that this is done as it is possible that a diffusion mechanism with a lower temperature dependence could be intersected, rendering these calculations meaningless. For example, Takeda and Morosin (1975) observed that the rate of increase of the a and b axis lengths in phlogopite decreased at about 400°C, and the flattening angle of the K octahedra increased at about the same temperature. Such changes in the physical properties

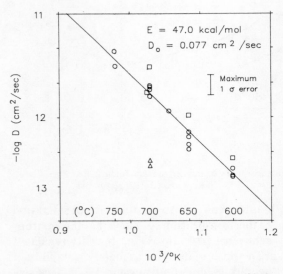

Fig. 5.11. Diffusion data for argon in a biotite (Fe/Fe + Mg = 56) treated hydrothermally at 1 kbar under controlled f_{O_2} conditions. After Harrison et al. (1985). Circles represent grain radii of less than 100 μm and squares indicate a grain radius of 200 μm. Triangles represent 100 μm radius grains heated under a pressure of 14 kbar.

of mica may influence the transport rate of argon. Although it would be useful to obtain diffusion estimates from geological settings, results from studies utilizing the thermal effects of intrusions to provide diffusivities of ^{40}Ar in biotite (Hart, 1964; Westcott, 1966; Hanson and Gast, 1967) over geological time are subject to large uncertainties in the heat flow calculations and are not useful in constraining the Arrhenius plot at lower temperatures.

Proceeding with this caution stated, we can use these diffusion parameters to calculate closure temperatures for micas compositionally similar to the one used by Harrison et al. (1985). For example, using the equation derived by Dodson (1973) for the case of cooling linear in $1/T$, we can calculate closure temperatures for an Ann_{56} biotite (Harrison et al., 1985) cooling at 100°C/Ma, 10°C/Ma, and 1°C/Ma to be 340, 310, and 280°C, assuming an effective diffusion radius of 150 μm.

The more general application of these data to calculating closure temperatures for other biotite solid solutions is limited by the existence of a compositional effect on argon diffusion. In addition to the two studies just described, Norwood (1974) reported ^{40}Ar diffusion coefficients obtained using hydrothermal methods on Ann_{37}, Ann_{63}, and Ann_{68}, although the last two compositions were run at only two temperatures. Results of all five samples are shown in Fig. 5.12, a plot of activation energy versus annite content of the mica. The clear correlation between these parameters indicates that the diffusivity of argon is an extremely sensitive function of composition. Clearly, the mica composition must be known prior to estimation of a closure temperature.

5.9.3.3 Hornblende

Of all the phases commonly dated by the K–Ar method, hornblende has been shown to be the most retentive of ^{40}Ar*. As a result, amphiboles, and hornblende in particular, have been the focus of a number of diffusion studies utilizing vacuum step heating, dynamic tempering (differential linear heating), and isothermal vacuum heating (Amirkhanov et al., 1960; Zimmerman, 1972; Berger and York, 1979, 1981a,b; Berger et al., 1979; Kotlovskaya, 1964; Kotlovskaya and Tovarenko, 1964; Hart, 1960, 1964; Gerling et al., 1965, 1966; Batyrmurzayev et al., 1971; Batyrmurzayev and Voronovskiy, 1976).

Again, results of these studies likely have no validity as the dehydration of the phase during heating very probably affects the transport rate of ^{40}Ar. For example, Brandt et al. (1967) showed that considerable loss of ^{40}Ar* from riebeckite occurred during heating in air at 750°C, but, under 4 kbar water pressure, no ^{40}Ar* was lost. The only comprehensive hydrothermal argon diffusion study of hornblende was reported by Harrison (1981). Sized fractions of two hornblende separates that yielded plateau-type age spectra were hydrothermally treated for durations as long as 12 weeks in the 750–900°C temperature range. The two hornblende samples, which had contrasting Fe/Mg ratios, defined a line (Fig. 5.13) with some scatter (possibly related to composition) with an activation energy of 66.1 kcal/mol and a frequency factor of 0.061 cm^2/s. The selection of a diffusion model was problematic as the particles were rectangular, but, for simplicity, a spherical model was chosen. Use of an infinite cylinder model would be as appropriate, but would not alter the activation energy and would change the frequency factor only by a factor of about two.

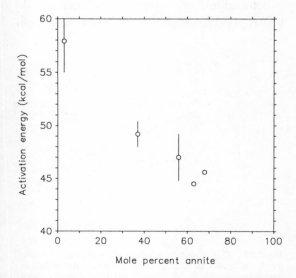

Fig. 5.12. Activation energy for argon in annite–phlogopite micas varies strongly with composition in the solid solution series. After Harrison et al. (1985). An implication of these data is that mica composition must be known before a closure temperature can be assessed.

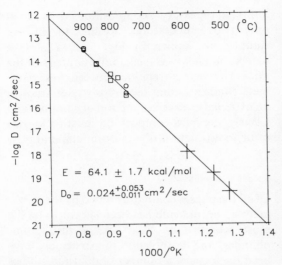

Fig. 5.13. Arrhenius plot of experimentally derived diffusion data in a magnesian hornblende (squares) and ferro hornblende (circles). After Harrison (1981). Crosses represent geologically derived results.

to laboratory experiments. In fact, an experiment using the smallest grain size of this study would need to be maintained at 500°C for about 100 years before meaningful ^{40}Ar* loss would occur.

Linear regression of both laboratory and geological data sets yielded an activation energy of 64.1 ± 1.7 kcal/mol and a frequency factor of $0.024^{+0.052}_{-0.011}$ cm^2/s. Using these parameters, the effect of both grain radius and cooling rate on closure temperature (T_c) is demonstrated in Fig. 5.14. The range of cooling rates shown from 1 to 10^3°C/Ma covers most plutonic and metamorphic conditions. Closure temperature is plotted against cooling rate for two different grain radii, 40 and 80 μm, which correspond to the minimum effective diffusion radius estimated by Harrison (1981) and the value used by Harrison and McDougall (1980b) in their diffusion calculations, respectively. From Fig. 5.14, it appears that a T_c of between 500 and 550°C would be characteristic for such a hornblende in a batholith cooling at about 100°C/Ma.

Experiments of Gerling et al. (1965), using the dynamic tempering method, indicated a strong compositional effect on the retentiveness of amphiboles. They concluded that with increasing Mg content, the activation energy increased dramatically up to a value of ~200 kcal/mol for pargasite. O'Nions et al. (1969) gave a geological example that appeared

In most circumstances, the geological relevance of experimental diffusion data must be argued with some caution in that extrapolations of the data must be made over many orders of magnitude without a priori knowledge that the diffusion mechanism dominant in the laboratory experiments will be significant in the low-temperature regime. However, Harrison and McDougall (1980b) derived a geologically based diffusion law for argon in hornblende by quantitatively assessing the thermal effect experienced by hornblende gabbro samples at varying distances away from an intrusive granite heat source. Data generated in this way yield an E of about 60 kcal/mol and a D_0 of 10^{-3} cm^2/s, placing them, within uncertainty, directly on the extrapolation of the laboratory data to lower D values (Fig. 5.13). The coincidence of these two data sets indicates that the same diffusion mechanism controls argon transport in hornblende over the 500–900°C temperature interval.

Despite the larger uncertainties, the importance of the geological determination of diffusion coefficients is not subordinate to the laboratory results, as most calculated closure temperatures (T_c) are in the 500–550°C range, well below the range of temperatures accessible

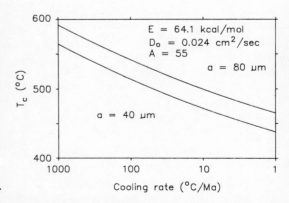

Fig. 5.14. Plot of closure temperature versus cooling rate for hornblendes of two grain radii using the diffusion parameters shown. Note the insensitivity of cooling rate on closure temperature, because of the generally large logarithmic argument in Eq. (5.48).

to show a correlation between argon retentiveness and Mg content of the amphibole. The experimental results of Harrison (1981) failed to confirm this observation, but more recent studies of geological settings continue to suggest a significant compositional effect (Berry and McDougall, 1986; Onstott and Peacock, 1987).

5.9.3.4 Muscovite

As with the two previous minerals discussed, muscovite has been the subject of vacuum heating diffusion studies, many of which are summarized in Mussett (1969). These investigations found a relationship between argon and water loss that suggested that the release of argon was related to structural breakdown. In two early studies, Hart (1960) and Brandt et al. (1967) heated muscovite under water pressure and found much less argon loss than under vacuum conditions, underscoring this point. Probably the most comprehensive hydrothermal diffusion study undertaken on muscovite is reported in an unpublished Brown University thesis by Gary Robbins (Robbins, 1972). Unfortunately, at crustal pressures, the dehydration breakdown of muscovite in the presence of water pressure occurs by about 700°C (e.g., Evans, 1965) limiting the maximum experimental temperature to approximately this value. The lower limit of about 600°C is dictated by the slow diffusivity of argon at and below this temperature, which requires inordinately long heating durations to effect measurable $^{40}Ar^*$ loss. Within this temperature range, Robbins (1972) made 17 runs using three different grain size fractions, and plotted Arrhenius curves for the data calculated using both the infinite cylinder and plane sheet models together with the average grain widths and thicknesses, respectively. He found a good fit of all data to the plane sheet model with an activation energy of 40 kcal/mol and a frequency factor of $6 \times 10^{-7} \, cm^2/s$, although an equal argument could be made for the infinite cylinder calculation with a similar activation energy but a frequency factor some 500 times larger. Although his coarsest grain size plotted above the line defined by the two finer grain size fractions, concordance could be achieved using an effective diffusion radius of 150 μm—the value argued earlier for biotite (Section 5.9.3.2).

5.9.3.5 Feldspars

Although diffusion studies involving feldspars were common among the earliest measurements (e.g., Evernden et al., 1960), interest tapered off when it became apparent that plutonic alkali feldspars were not sufficiently retentive of $^{40}Ar^*$ to be useful in defining ages of orogenic events. Although high temperature alkali feldspar appeared to be more retentive, their near surface environment of formation precluded extracting any meaningful thermal insights by use of diffusion data. Foland (1974) renewed interest in this phase with the publication of a hydrothermal and vacuum heating study of a homogeneous orthoclase. This study revealed that $^{40}Ar^*$ loss from alkali feldspar proceeds at the same rate whether heated in vacuum or under water pressure and reinforced the view that high-temperature feldspars are rather retentive of $^{40}Ar^*$. Foland (1974) explained the known unretentive nature of microcline as resulting from fine scale perthite lamellae acting as the diffusion boundary, reducing the average migration distance for argon atoms to a fraction of the grain size (see Section 4.12.2). The activation energy of 43.8 kcal/mol and frequency factor of $0.0098 \, cm^2/s$ were similar to many earlier estimates (see Mussett, 1969) and predicted closure temperatures on the order of ~150°C for micron scale perthitization. Berger and York (1981a) used the measured ^{39}Ar loss from feldspar during vacuum extraction to calculate diffusion coefficients. When plotted against the reciprocal absolute temperature of the heating step, the Arrhenius relation could be obtained. Using this approach, Harrison and McDougall (1982) noted that microcline differs from orthoclase in both activation energy and diffusion radius with a typical activation energy of about 30 kcal/mol, although values of E obtained using better experimental facilities are typically in the 35–45 kcal/mol range (Harrison et al., 1986).

No systematic diffusion study of plagioclase taking into account the complex exsolution structure has been undertaken. This is a resul of the low potassium content of this phase making it less interesting as a chronometer and the extreme variability in microstructure (with both composition and cooling history).

5.10 COUPLING OF ARGON DIFFUSION AND HEAT FLOW

5.10.1 Introduction

Our discussions thus far have emphasized the similarities between heat conduction and atomic diffusion. We have suggested that the predictability and widespread relevance of heat transfer by conduction could provide thermal and temporal constraints needed to effect diffusion calculations as well as to give insight as to the form of the temperature perturbation. In this section we give an example of how the amalgam of the two methods can reveal information needed to assess the validity of our laboratory diffusion experiments.

5.10.2 $^{40}Ar^*$ loss from microcline at Inyo Domes, California

A rhyolite conduit at Inyo Domes, Long Valley Caldera, emplaced ~600 years ago into rocks of the Sierra Nevada Batholith caused a thermal perturbation in the adjacent country rocks. This site has been extensively cored as part of the Continental Scientific Deep Drilling Program (Eichelberger et al., 1984) and temperature logging has indicated a small trace of the original heat still resident in the conduit. The abundant microcline in the thermal aureole was sufficiently heated by the magma for long enough durations to cause significant $^{40}Ar^*$ loss. The magnitude of this loss, revealed by the $^{40}Ar/^{39}Ar$ age spectrum technique, is a function of the diffusion properties of the mineral and the thermal intensity of the disturbance. If the dimensions, initial conditions, and thermal properties of the dike are known or can be estimated, then the effective D/a^2 value could be extracted from the Dt/a^2 calculated from the $^{40}Ar^*$ loss data and then plotted with

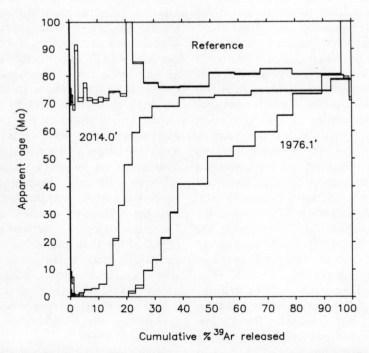

Fig. 5.15. Age spectra of alkali feldspar obtained from 3 m (1976.1 ft) and 7 m (2014.0 ft) from a rhyolite conduit emplaced 600 years ago. The age spectrum identified as Reference is taken far from the thermal effect of the magma and represents the general appearance of the two thermally affected samples prior to intrusion.

the maximum temperature to yield a geologically derived diffusion datum. Our earlier contention (Section 5.9.3.5) that microcline can yield the intrinsic Arrhenius parameters at temperatures below ~800°C can be tested in this way.

^{40}Ar/^{39}Ar age spectrum analyses of microcline from 3 and 7 m from the contact with the 50-m-wide dike reveal about 20 and 50% ^{40}Ar* loss, respectively, compared with that found for a sample obtained from a location well beyond the thermal effect of the intrusion (Fig. 5.15; Ryerson et al., 1985). The release patterns correspond well to our simple model of one activation energy and diffusion radius, although the initially zero ages perhaps indicate a distribution of diffusion radii. However, this effect is not likely to seriously affect our subsequent calculations which assume a single value.

5.10.3 The model

Jaeger (1964) developed the case of ^{40}Ar* loss from a sheetlike mineral adjacent to an infinite magma sheet of thickness $2d$ [Eq. (5.70)]. We have previously described how perthitic alkali feldspars can be viewed as sheetlike lamellae, and the dimensions of the conduit allow a good approximation, the assumption of an infinite dike in two dimensions. Because of the time-varying diffusion coefficient at any distance from the intrusion, a modified form of the Arrhenius equation,

(5.77) $\quad D = D_0 \exp[-E/RT(t)]$

must be incorporated. For the case of a relatively brief episodic loss, no production term need be considered, so we only require solution of Fick's Second Law,

(5.78) $\quad \dfrac{\partial^2 C}{\partial x^2} - \dfrac{1}{D}\dfrac{\partial C}{\partial t} = 0$

together with,

(5.79) $\quad t' = \displaystyle\int_0^t D(t)\,dt$

yielding,

$$\dfrac{\partial^2 C}{\partial x^2} - \dfrac{\partial C}{dt'} = 0$$

For a given initially uniform concentration, C_0, at $t = 0$, the usual boundary and initial conditions are

(5.80) $\quad \begin{aligned} C &= C_0, & -a \leqslant x \leqslant a, & \quad t = 0 \\ C &= 0, & x = \pm a, & \quad t > 0 \end{aligned}$

the fractional ^{40}Ar* loss up to time t is

(5.81) $\quad \begin{aligned} f &= -\dfrac{1}{aC_0}\int_0^t D\dfrac{\partial C}{\partial x}\bigg|_{x=a} dt \\ &= -\dfrac{1}{aC_0}\int_0^{t'} \dfrac{\partial C}{\partial x}\bigg|_{x=a} dt' \end{aligned}$

Carslaw and Jaeger (1959) used the Laplace transform method to evaluate the *average* temperature within a sheet after time t that is exactly analogous to the fraction of argon *remaining* in the mineral. The result

(5.82) $\quad \begin{aligned} f &= \left(\dfrac{4t'}{a^2\pi}\right)^{1/2} + \left(\dfrac{16t'}{a^2}\right)^{1/2} \\ &\quad \times \sum_{n=1}^{\infty}(-1)^n \,\text{ierfc}\,\dfrac{na}{\sqrt{t'}} \end{aligned}$

where ierfc is the integral of the complementary error function. Jaeger (1964) pointed out that up to $f = 0.6$, only the first term of Eq. (5.82) need be used so that

(5.83) $\quad f = \left(\dfrac{4t'}{a^2\pi}\right)^{1/2}$

For the case of a plane sheet intrusion, the value of t' is, from Eqs. (5.70), (5.77), and (5.79),

(5.84) $\quad t' = \left(\dfrac{D_0 d^2}{\kappa}\right)\displaystyle\int_0^{\infty} \exp\left\{-\dfrac{E}{RT_i}\left[1 + \dfrac{T_0}{2T_i}\left\{\text{erf}\dfrac{\zeta+1}{\sqrt{4\tau}} - \text{erf}\dfrac{\zeta-1}{\sqrt{4\tau}}\right\}\right]^{-1}\right\} d\tau$

where $\zeta = x/d$ is the distance from the axis of the intrusion expressed as a ratio to the dike half width (i.e., the contact is at $\zeta = 1.0$), $\tau = \kappa t/d^2$ (i.e., dimensionless time), T_0 is the initial absolute temperature, and T_i is the absolute ambient temperature. This result can be expressed in terms of an equivalent time, t^*, at maximum temperature, T_m, in which

(5.85) $\quad t' = D_0 t^* \exp[-E/R(T_i + T_m)]$

To obtain T_m, Eq. (5.70) in one dimension can be used together with the expression for the time at which maximum temperature is reached in the contact zone (Jaeger, 1964),

(5.86) $\quad \tau_m = \zeta/\ln[(\zeta + 1)/(\zeta - 1)]$

In practice, Eq. (5.85) can be written in the form

$$\ln(D_0 d^2/a^2 \kappa) = \ln(t'/a^2) - \ln(\kappa t^*/d^2)$$
(5.87) $\quad\quad\quad\quad + E/R(T_i + T_m)$

Equation (5.87) can be appreciated graphically by plotting the lumped parameter $\kappa t/d^2$ against ζ for specified values of T_0/T_i and E/RT_i. For our considerations, $T_0/T_i = 4$ and $E/RT_i = 65$ for an activation energy typical for argon diffusion in microcline (Fig. 5.16). For our preferred values, the estimated duration at a maximum temperature of 440°C is 24 years corresponding to a D/a^2 value calculated from

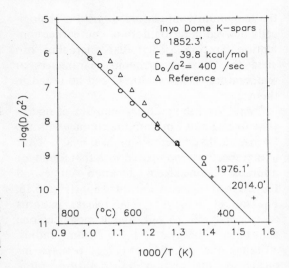

Fig. 5.17. Arrhenius plot derived from measured ^{39}Ar loss from Inyo Domes alkali feldspars and laboratory extraction temperatures.

the measured ^{39}Ar loss during laboratory extraction. These latter data define an $E = 39.8$ kcal/mol and $D_0/a^2 = 400$/s and project close to the geologically derived estimates, confirming the validity of this approach in extracting diffusion parameters (Fig. 5.17).

Appendix A.5.1 SEPARATION OF VARIABLES SOLUTION FOR PLANE SHEET

One method for solving the partial differential equation for one-dimensional diffusion, that is,

(A.5.1.1) $\quad \dfrac{\partial C}{\partial t} = D \dfrac{\partial^2 C}{\partial x^2}$

is to assume that the two concentration variables, time, t, and space, x, are separable. That is, a solution for Eq. (A.5.1.1) can be written as

(A.5.1.2) $\quad\quad C = X(x)T(t)$

where X and T are the separate functions of x and t, respectively. Substitution of Eq. (A.5.1.2) into (A.5.1.1) and rearrangement gives

(A.5.1.3) $\quad \dfrac{1}{T}\dfrac{dT}{dt} = \dfrac{D}{X}\dfrac{d^2 X}{dx^2}$

such that the left-hand side of the equation depends only on t and the right-hand side depends solely on x. We equate both sides to some constant, which we take to be $-\lambda^2 D$. The

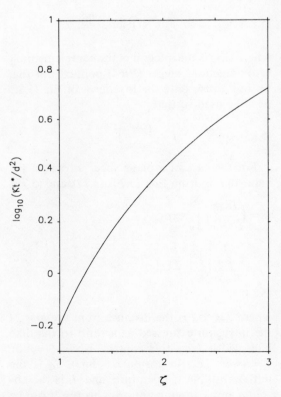

Fig. 5.16. Plot of dimensionless time, $\kappa t^*/d^2$, versus distance x from dike of half width, d (i.e., $\zeta = x/d$), where κ is thermal diffusivity and t^* is equivalent square pulse duration of maximum temperature (see text for details). After Jaeger (1964).

form and sign of this constant reflect the structure of the partial differential equation. For a slightly different choice of constant, the subsequent algebra would be complicated, but eventual analysis would lead us to this convenient expression. The negative sign, for example, implies that the solution of the right-hand side of Eq. (A.5.1.3) will involve a trigonometric result, rather than one involving an exponential series, appropriate for a positive constant. From Eq. (A.5.1.3) and our equivalent constant we have two ordinary differential equations:

$$\text{(A.5.1.4)} \quad \frac{1}{T}\frac{dT}{dt} = -\lambda^2 D$$

$$\text{(A.5.1.5)} \quad \frac{d^2 X}{dx^2} = -\lambda^2 X$$

which have the following solutions.

Equation (A.5.1.4) is the familiar first-order equation we solved in Section 2.5 to yield the age equation. Rewritten as

$$\text{(A.5.1.6)} \quad dT/T = -\lambda^2 D\, dt$$

and integrating, we can easily see the solution is

$$\text{(A.5.1.7)} \quad T = \exp(-\lambda^2 Dt)$$

Equation (A.5.1.5), a second-order ordinary differential equation, yields the solution

$$\text{(A.5.1.8)} \quad X = A \sin \lambda x + B \cos \lambda x$$

which may not be as obvious as the previous solution. In spite of inexperience in solving higher order differential equations, we can appreciate this result by approaching it from the opposite direction. We shall digress momentarily on the nature of differential equations before turning to this solution.

Differential equations arise from a relationship between the variables that involve n essential arbitrary constants, or coefficients. This relationship will give rise to a differential equation of order n, free of constants, by differentiating n times with respect to the independent variable. For our solution [Eq. (A.5.1.8)], A and B are arbitrary constants and λ is a fixed constant. Differentiating we get

$$\text{(A.5.1.9)} \quad dX/dx = A\lambda \cos \lambda x - B\lambda \sin \lambda x$$

and

$$\text{(A.5.1.10)} \quad \begin{aligned} d^2X/dx^2 &= -\lambda^2(A \sin \lambda x + B \cos \lambda x) \\ &= -\lambda^2 X \end{aligned}$$

Note that Eq. (A.5.1.10) is free of arbitrary constants, of the appropriate order, identical to Eq. (A.5.1.5), and that the constants could not have been eliminated by manipulation of the first two equations.

The separate solutions Eqs. (A.5.1.7) and (A.5.1.8) together with Eq. (A.5.1.2) give an expression for concentration

$$\text{(A.5.1.11)} \quad C = (A \sin \lambda x + B \cos \lambda x)\exp(-\lambda^2 Dt)$$

A general solution can be obtained by summing solutions of this type yielding

$$C = \sum_{m=1}^{\infty} (A_m \sin \lambda_m x + B_m \cos \lambda_m x) \exp(-\lambda_m^2 Dt)$$

(A.5.1.12)

where the constants A_m, B_m, and λ_m are determined for particular boundary and initial conditions. For diffusion out of a plane sheet of width, l, of an initially uniformly distributed substance while the boundaries are maintained at zero concentration, these conditions are:

initial

$$\text{(A.5.1.13)} \quad C = C_0, \quad 0 < x < l, \quad t = 0$$

and boundary

$$\text{(A.5.1.14)} \quad C = 0, \quad x = 0, \quad x = l, \quad t > 0$$

These sets of conditions define what is happening to the concentration distribution as a result of changes in the two independent variables. The boundary condition requires that $B_m = 0$ and $\lambda_m = m\pi/l$ as the exponential portion of Eq. (A.5.1.12) will always have some positive value requiring the two trigonometric terms to go to zero at $x = 0$ and $x = l$. Since the cosine of zero is one, B_m must be equal to zero, and λ_m must equal $m\pi/l$ in order that the sine term be zero at $x = l$. The initial condition becomes

$$C_0 = \sum_{m=1}^{\infty} A_m \sin(m\pi x/l) \quad 0 < x < l$$

(A.5.1.15)

Multiplying both sides of Eq. (A.5.1.15) by $\sin(p\pi x/l)$ and integrating from 0 to l using the relationship

$$\int_0^l \sin\frac{p\pi x}{l} \sin\frac{m\pi x}{l} dx = \begin{cases} 0, & m \neq p \\ l/2 & m = p \end{cases}$$

(A.5.1.16)

yields an expression from which we can extract the remaining constant. This is the orthogonality relationship used for finding coefficients of a Fourier sine series (e.g., Boyce and DiPrima, 1965). These two terms are orthogonal and will cancel each other unless $m = p$, in which case the solution is $l/2$. Hence,

$$C_0 \int_0^l \sin\frac{p\pi x}{l} dx = \sum_{m=1}^{\infty} A_m \int_0^l \sin(m\pi x/l)$$
$$\times (\sin p\pi x/l) \, dx$$

$$C_0 \left[-\frac{l}{p\pi} \cos\left(\frac{p\pi x}{l}\right) \right]_0^l = \sum_{m=1}^{\infty} A_m \frac{l}{2}$$

(A.5.1.17) $\quad C_0 \dfrac{2l}{p\pi} = \sum_{m=1}^{\infty} A_m \dfrac{l}{2}$

and the terms for which m is even vanish as $m \neq p$, therefore,

(A.5.1.18) $\quad A_m = \dfrac{4C_0}{m\pi} \quad$ for $m = 1, 3, 5 \cdots$

Thus, the final solution is

$$C = \sum_{m=1}^{\infty} (A_m \sin \lambda_m x) \exp(-\lambda^2 Dt)$$

$$= \frac{4C_0}{\pi} \sum_{n=0}^{\infty} \frac{1}{(2n+1)} \exp[-D(2n+1)^2 \pi^2 t/l^2]$$

(A.5.1.19) $\quad \times \sin\dfrac{(2n+1)\pi x}{l}$

Appendix A.5.2 LAPLACE TRANSFORM SOLUTION FOR SEMI-INFINITE MEDIUM

Appendix A.5.1 described the technique of variable substitution to transform complex differential equations into simpler functions we can deal with using rudimentary calculus. However, for many complex problems, or for solutions for short periods of time, solutions can be obtained only by use of an operator technique. The Laplace transformation is a device that when applied to the diffusion equation, removes the time variable. The resulting ordinary differential equation can then be solved to give the transformed concentration as a function of the space variable, and then returned to an expression involving both space and time and satisfying the initial and boundary conditions. This latter process, for us, simply involves using a table of transforms.

Let $f(t)$ be any known function such that integrations encountered can be performed on it. The Laplace transform of $f(t)$, denoted by $\bar{f}(p)$ is defined by

(A.5.2.1) $\quad \bar{f}(p) = \displaystyle\int_0^\infty \exp(-pt) f(t) \, dt$

where p is a parameter sufficiently large to allow this integral to converge. For example, the transformed form of $f(t) = 1$ is

$$\bar{f}(p) = \int_0^\infty \exp(-pt) \, dt = \left.\frac{\exp(-pt)}{p}\right|_0^\infty = \frac{1}{p}$$

(A.5.2.2)

and the transformed state of $f(t) = \exp(kt)$ is

$$\bar{f}(p) = \int_0^\infty \exp(-pt) \exp(kt) \, dt$$

$$= \int_0^\infty \exp(-(p-k)t) \, dt$$

(A.5.2.3) $\quad = -\left.\dfrac{\exp(-(p-k)t)}{p-k}\right|_0^\infty = \dfrac{1}{p-k}$

Note that in both the transformed results, there is no dependence on t and also that if $k = 0$, the result in Eq. (A.5.2.3) reduces to that of Eq. (A.5.2.2). Reversion of the transformed expressions can be made with reference to a table of transforms, versions of which can be found in Carslaw and Jaeger (1959) and Crank (1975).

We now consider diffusion in a semi-infinite medium, $x > 0$, where the boundary concentration is maintained at C_0, and the initial concentration in the region $x > 0$ is zero. We must solve

(A.5.2.4) $\quad \dfrac{\partial C}{\partial t} = D \dfrac{\partial^2 C}{\partial x^2}$

together with the boundary and initial conditions,

(A.5.2.5) $\quad C = C_0, \quad x = 0, \quad t > 0$

and

(A.5.2.6) $\quad C = 0, \quad x > 0, \quad t = 0$

Multiplying both sides of Eq. (A.5.2.4) by $\exp(-pt)$ and integrating with respect to t from 0 to ∞ yields

$$\int_0^\infty \exp(-pt)\frac{\partial^2 C}{\partial x^2}dt$$

(A.5.2.7) $\quad -\frac{1}{D}\int_0^\infty \exp(-pt)\frac{\partial C}{\partial t}dt = 0$

In this case, the order of differentiation and integration can be reversed giving

$$\int_0^\infty \exp(-pt)\frac{\partial^2 C}{\partial x^2}dt$$

(A.5.2.8) $\quad = \frac{\partial^2}{\partial x^2}\int_0^\infty C\exp(-pt)\,dt = \frac{\partial^2 \bar{C}}{\partial x^2}$

$$\left[\text{remember } \bar{C} = \int_0^\infty C\exp(-pt)\,dt\right]$$

Integration by parts yields

$$\int_0^\infty \exp(-pt)\frac{\partial C}{\partial t}dt = C\exp(-pt)\Big|_0^\infty$$

(A.5.2.9) $\quad + p\int_0^\infty C\exp(-pt)\,dt = p\bar{C}$

$\Big[$Note: $\quad \int \exp(-pt)(\partial C/\partial t)\,dt = udv - \int duv$

where $u = \exp(-pt)$, $v = C$,

$$du = -p\exp(-pt)\,dt \text{ and } dv = \frac{\partial C}{\partial t}dt\Big]$$

But $C\exp(-pt)|_0^\infty$ goes to zero at $t = 0$ due to the initial condition [Eq. (A.5.2.6)] and at $t = \infty$ via the exponential factor, $\exp(-\infty) = 0$. The one-dimensional diffusion equation reduces to

(A.5.2.10) $\quad D\frac{\partial^2 \bar{C}}{\partial x^2} = p\bar{C}$

Transforming the boundary equation in a similar fashion yields

$$\bar{C} = \int_0^\infty C_0 \exp(-pt)\,dt = \frac{C_0}{p}, \quad x = 0$$

(A.5.2.11)

If we rewrite our transformed equation

(A.5.2.12) $\quad \frac{1}{\bar{C}}\frac{d^2\bar{C}}{dx^2} = \frac{p}{D}$

and state that $q^2 = p/D$, then

(A.5.2.13) $\quad \frac{1}{\bar{C}}\frac{d^2\bar{C}}{dx^2} = q^2$

We recognize this as the second-order differential equation solved in Appendix A.5.1 [Eq. (A.5.1.5)] using a trigonometric series. But, the difference is that the constant is positive and, in this case, an exponential solution is most appropriate. The solution is

(A.5.2.14) $\quad \bar{C} = \frac{C_0}{p}\exp(-qx)$

Again, we can appreciate this result by differentiating twice to obtain

(A.5.2.15) $\quad \frac{d^2\bar{C}}{dx^2} = \frac{q^2 C_0}{p}\exp(-qx)$

It can be seen by substituting Eq. (A.5.2.15) into Eq. (A.5.2.13) that Eq. (A.5.2.14) is a solution of the differential equation. From a table of Laplace transforms the returned expression is

(A.5.2.16) $\quad \frac{\exp(-qx)}{p} \to \text{erfc}\,\frac{x}{\sqrt{4Dt}}$

Making the final solution

(A.5.2.17) $\quad C = C_0\,\text{erfc}\,\frac{x}{\sqrt{4Dt}}$

Note that this equation differs from the extended distribution solution [Eq. (5.30)] by a factor of 1/2. This reflects the constant concentration, kept in this case at all times at the boundary, in contrast to the expression [Eq. (5.30)] in which the fixed initial distribution relaxes into the semi-infinite void.

Appendix A.5.3 TRANSLATION TO SPHERICAL COORDINATES

This appendix is intended as an illustration of how coordinate transformation can convert the diffusion or heat flow equation to radially symmetric geometries. In this example, the diffusion equation in three dimensions is translated to spherical coordinates and, using a simple substitution, converted into an equation reminiscent of linear flow in one dimension. The resulting solution for this equation (developed in Appendix A.5.1) can be used directly and, after resubstitution, yields a solution for flow in a sphere with the boundary and initial conditions stated for that solution [Eqs. (A.5.1.13) and (A.5.1.14)].

The diffusion equation in three dimensions is

$$\text{(A.5.3.1)} \qquad \frac{\partial C}{\partial t} = D\left(\frac{\partial^2 C}{\partial x^2} + \frac{\partial^2 C}{\partial y^2} + \frac{\partial^2 C}{\partial z^2}\right)$$

Spherical coordinates are designed to fit situations with central symmetry. The spherical coordinates (r, ϕ, θ) of a point x are its distance, r from the origin, its elevation angle ϕ, and its azimuth angle θ. Relations between the rectangular coordinates (x, y, z) and its spherical coordinates are

$$\text{(A.5.3.2)} \qquad x = r \sin \phi \cos \theta$$

$$\text{(A.5.3.3)} \qquad y = r \sin \phi \sin \theta$$

$$\text{(A.5.3.4)} \qquad z = r \cos \phi$$

But diffusion is only radial, that is, there is no dependence on ϕ or θ, therefore, after substitution Eq. (A.5.3.1) can be written

$$\text{(A.5.3.5)} \qquad \frac{\partial C}{\partial t} = \frac{1}{r^2}\frac{\partial}{\partial r}\left(Dr^2\frac{\partial C}{\partial r}\right)$$

We then make the substitution $u = Cr$, and by differentiating obtain

$$\text{(A.5.3.6)} \qquad \frac{\partial C}{\partial r} = \left(r\frac{\partial u}{\partial r} - u\right)\bigg/r^2$$

$$\text{(A.5.3.7)} \qquad \frac{\partial^2 C}{\partial r^2} = \left(r^2\frac{\partial^2 u}{\partial r^2} - 2r\frac{\partial u}{\partial r} + 2u\right)\bigg/r^3$$

It follows that

$$\text{(A.5.3.8)} \qquad \frac{2}{r}\frac{\partial C}{\partial r} = \left(2r\frac{\partial u}{\partial r} - 2u\right)\bigg/r^3$$

Substituting Eqs. (A.5.3.7) and (A.5.3.8) into Eq. (A.5.3.5) yields

$$\text{(A.5.3.9)} \qquad \frac{\partial C}{\partial t} = \frac{D}{r}\frac{\partial^2 u}{\partial r^2}$$

but once again $C = u/r$,

$$\text{(A.5.3.10)} \qquad \frac{\partial u}{\partial t} = D\frac{\partial^2 u}{\partial r^2}$$

This is recognizable as Fick's Second Law in one dimension and can be solved by separation of variables in a fashion identical to that illustrated in Appendix A.5.1. For the boundary and initial conditions

$$\text{(A.5.3.11)} \qquad u = 0, \qquad r = 0, \qquad t > 0$$

$$\text{(A.5.3.12)} \qquad u = aC_0, \qquad r = a, \qquad t > 0$$

$$\text{(A.5.3.13)} \qquad u = rC_1, \qquad 0 < r < a, \qquad t = 0$$

where C_0 and C_1 are the zero surface and uniform initial concentrations, respectively, the solution of Eq. (5.26) is

$$\text{(A.5.3.14)} \qquad \begin{aligned} C = C_1 \frac{2a}{\pi r} \sum_{n=1}^{\infty} \frac{(-1)^n}{n} \sin\frac{n\pi r}{a} \\ \times \exp(-Dn^2\pi^2 t/a^2) \end{aligned}$$

Appendix A.5.4 CLOSURE TEMPERATURE FOR FIRST-ORDER LOSS

The net rate of increase of the concentration of a radiogenic decay product, x, is given by the rate of production minus the rate of loss, or

$$\text{(A.5.4.1)} \qquad \left(\frac{dx}{dt}\right)_{\text{net}} = \left(\frac{dx}{dt}\right)_{\text{prod}} - \left(\frac{dx}{dt}\right)_{\text{loss}}$$

where $(dx/dt)_{\text{prod}} = \lambda_x C_p$ (λ_x is the decay constant and C_p is the parent concentration). The loss term in Eq. (A.5.4.1) can be equated to the product of a loss coefficient, $k(t)$, and the daughter concentration, x. Because a first-order reaction takes the form of an exponential decay (with a time constant, τ),

$$(dx/dt)_{\text{loss}} = k(t)x = k(0)\exp(-t/\tau)x$$

$$\text{(A.5.4.2)} \qquad [\text{assuming } k = k_0 \exp(-E/RT)]$$

where k_0 is the reaction coefficient at infinitely high temperature and $k(0)$ is the reaction coefficient prior to the onset of cooling. Thus,

$$\text{(A.5.4.3)} \qquad (dx/dt)_{\text{net}} = \lambda_x C_p - k(0)\exp(-t/\tau)x$$

By substituting the dimensionless variable $v(t) = \tau k(0)\exp(-t/\tau)$ and rearranging we get

$$(dx/dt)_{\text{net}} = \frac{dx}{dv}\frac{dv}{dt}$$

where $dv/dt = -k(0)\exp(-t/\tau) = -v/\tau$. Rearranging yields

$$\text{(A.5.4.4)} \qquad \frac{dx}{dv} - x = -\frac{\lambda_x C_p \tau}{v}$$

Multiplying this through by $\exp(-v)$ gives

$$\exp(-v)\frac{dx}{dv} - x\exp(-v)$$

(A.5.4.5)
$$= -\lambda_x C_p \tau \frac{\exp(-v)}{v}$$

Then,

$$\exp(-v)\,dx - x\exp(-v)\,dv$$

(A.5.4.6)
$$= -\lambda_x C_p \tau \frac{\exp(-v)\,dv}{v}$$

or

$$\int \exp(-v)\,dx - \int x\exp(-v)\,dv$$

(A.5.4.7)
$$= -\lambda_x C_p \tau \int \frac{\exp(-v)\,dv}{v}$$

Integrating the first term of the left-hand side by parts

(A.5.4.8)
$$\int u\,dz = uz - \int z\,du$$

(where $u = \exp(-v)$, $du = -\exp(-v)dv$, $dz = dx$, and $z = x$) then

$$\int \exp(-v)\,dx = x\exp(-v) + \int x\exp(-v)\,dv$$

(A.5.4.9)

Substituting Eq. (5.47) into Eq. (5.45) yields

$$x\exp(-v) + \int x\exp(-v)\,dv$$
$$- \int x\exp(-v)\,dv$$

(A.5.4.10)
$$= -\lambda_x C_p \tau \int \frac{\exp(-v)\,dv}{v}$$

or

$$x = -\lambda_x C_p \tau \exp(v) \int \frac{\exp(-v)\,dv}{v} + \text{constant}$$

(A.5.4.11)

where the constant is associated with the integration. An integral of the form $\int_a^\infty [\exp(-u)/u]\,du$ is called an exponential integral and, in definite form, results in an infinite series. The exponential integral is often designated by $E_i(a)$. Results of exponential integrals can be expressed in several ways, for example

$$E_i(a) = -c - \ln a + \frac{a}{1\cdot 1!} - \frac{a^2}{2\cdot 2!} + \frac{a^3}{3\cdot 3!} \cdots$$

(A.5.4.12)

$$E_i(a) = \frac{\exp(-a)}{a}\left(1 - \frac{1!}{a} + \frac{2!}{a^2} - \frac{3!}{a^3} \cdots \right)$$

(A.5.4.13)

where c is Euler's constant (0.5772). We can write Eq. (A.5.4.11) as

$$-x = \lambda_x C_p \tau \exp(v)\, E_i(v) \Big|_{v(0)}^{v(t)}$$

(A.5.4.14)
$$= \lambda_x C_p \tau \exp(v)\, E_i[v(t)] - E_i[v(0)]$$

The integration constant disappears because the integral has been defined between two specific values. In Eq. (A.5.4.14), $v(0)$ represents the value of $v[\tau k(0)]$ at $t = 0$ and $v(t)$ is v at any time, t. To simplify Eq. (A.5.4.14), approximations can be made based on the properties of the exponential integral. For very large v, $E_i(-v)$ tends towards zero. This can be appreciated by evaluation of Eq. (A.5.4.13). From Eq. (A.5.4.12) it is apparent that as v tends towards zero, $E_i(v)$ approaches $(-c - \ln v)$. Defining $\gamma = \exp(c) = 1.78$, then $-\ln v - c = -\ln v - \ln \gamma = -\ln(v\gamma)$. Restating, as v goes to zero, $E_i(v)$ tends toward $-\ln(v\gamma)$. Recalling that we defined $v(t) = \tau k(0)\exp(-t/\tau)$ at $t = 0$, $v(0) = \tau k(0)$, where $k(0) = k_0 \exp(-E/RT_0)$. Note that $v(0)$ will be a very large number when τ and T_0 have large values (slow cooling and high initial temperatures, respectively). Thus, if we constrain the solution to the case in which cooling begins at sufficiently high temperature at a sufficiently slow rate, $v(0)$ tends to be very large, hence, $E_i[v(0)]$ tends towards zero. What constitutes sufficiently high temperature or sufficiently slow cooling is dealt with in Section 5.7.6. In other words, provided we satisfy these two constraints, Eq. (A.5.4.14) can be rewritten

(A.5.4.15) $-x = \lambda_x C_p \tau \exp(v)\, E_i[v(t)] \Big|_{\text{Slow } dT/dt}^{\text{High } T_0}$

At very large times, $v(t)$ tends to zero [via $\exp(-t/\tau)$] implying that $\exp(v)$ will approach unity simplifying the above equation to

(A.5.4.16) $\quad x = \lambda_x C_p \tau \ln[\gamma v(t)] \Big|_{\text{Slow } dT/dt}^{\text{High } T_0}$

Substituting in for $v(t) = \tau k(0) \exp(-t/\tau)$ we get

$$x = \lambda_x C_p \tau \ln[\gamma \tau k(0) \exp(-t/\tau)]$$
$$= \lambda_x C_p \tau \ln[\gamma \tau k_0 \exp(-E/RT_0 - t/\tau)]$$
$$= \lambda_x C_p \tau \ln(\gamma \tau k_0) - \frac{E}{RT_0} - \frac{t}{\tau}$$

Rewriting yields

$$-x = \lambda_x C_p t - \tau[\ln(\gamma \tau k_0) - (E/RT_0)]$$

(A.5.4.17)

and rearranging gives

$$\frac{t}{\tau} = \frac{-x}{\lambda_x C_p \tau} + \ln(\gamma \tau k_0) - \frac{E}{RT_0}$$

The definition of closure temperature (Fig. 5.3) implies that in spite of the continuous transition from open to closed system behavior, we can view the system as having zero concentration of the decay products prior to t_c, and total retention subsequent to t_c. Therefore

(A.5.4.18) $\quad \dfrac{t_c}{\tau} = \ln(\gamma \tau k_0) - \dfrac{E}{RT_0}$

From the definition of τ, we know that at the closure temperature,

(A.5.4.19) $\quad \dfrac{E}{RT_c} = \dfrac{E}{RT_0} + \dfrac{t_c}{\tau}$

Substituting this expression in Eq. (A.5.4.18) and rearranging yields

(A.5.4.20) $\quad T_c = \dfrac{E/R}{\ln(\tau \gamma k_0)}$

Appendix A.5.5 SAMPLE DIFFUSION CALCULATION

As an illustration of how an argon loss datum can be translated into an estimate of the diffusion coefficient, we consider the experimental results of Foland (1974) in which aggregates of a homogeneous K-feldspar were hydrothermally treated to induce argon loss. For example, Run 55 consists of feldspar sized to $63.5 \pm 3\,\mu\text{m}$ radius (a), heated for 5 days, 15 h, and 35 min ($t = 4.881 \times 10^5$ s) at a temperature (T) of 800°C. Upon analysis, this sample contained 4.467×10^{-9} mol $^{40}\text{Ar}^*$/g compared to 2.393×10^{-8} mol $^{40}\text{Ar}^*$/g in the sample prior to treatment. This $^{40}\text{Ar}^*$ deficit corresponds to a fractional loss of 0.813.

For a variety of reasons, Foland (1974) chose a spherical diffusion model to calculate the diffusion coefficient, D [Eq. (5.32); Table 5.1]. By examining Eq. (5.31), it is apparent that the infinite summation of exponential terms must equal 0.3076 (simply $[(1 - 0.813)\pi^2]/6$). For a high degree of equilibration as in our case ($>80\%$), this series converges rapidly allowing a convenient example. Substitution of a trial value for the grouped parameter $D\pi^2 t/a^2$ of 1.186 into the exponential argument returns a functional value of 0.3054. The second term of the series, $\exp(-4 \cdot 1.186)/4$ adds only 0.0022 to this summation yielding a total of 0.3076, the third term is on the order of 10^{-6}, the fourth about 10^{-10}, and so on giving a value of this infinite summation of 0.3076—our required total. Thus, we have assessed the grouped parameter $D\pi^2 t/a^2$ to have a value of 1.186. By dividing by the heating duration and π^2 and multiplying by the square of the grain radius, we obtain a value for D of $9.92 \times 10^{-12}\,\text{cm}^2/\text{s}$ [This value is slightly different from that reported by Foland (1974).]

6. APPLICATIONS AND CASE HISTORIES

6.1 OVERVIEW

In the preceding chapters we have considered many aspects of the ^{40}Ar/^{39}Ar dating method, including the theoretical basis for interpretation of some of the simpler age spectra. By way of illustration a number of age spectra have been discussed, usually quite briefly. In this concluding chapter we wish to present, in some detail, examples of applications of the ^{40}Ar/^{39}Ar method to terrestrial and extraterrestrial rocks in order to provide a better appreciation of the power of the technique as well as its limitations.

6.2 LUNAR GEOCHRONOLOGY

6.2.1 General comment

One of the first applications of the ^{40}Ar/^{39}Ar dating method was to the meteorite Bruderheim, a hypersthene chondrite. As discussed in Chapter 4 (Sections 4.2, 4.3), Merrihue and Turner (1966) and Turner et al. (1966) showed that the age spectrum could be interpreted in terms of a markedly disturbed system with a major outgassing event occurring about 500 Ma ago, when approximately 90% of the radiogenic argon in the sample was lost. This example clearly demonstrated that the ^{40}Ar/^{39}Ar step heating approach could reveal gradients of radiogenic argon in geological samples. Subsequently, the technique has been applied extensively to extraterrestrial samples with marked success, mainly because many of these samples are well crystallized and consist of anhydrous minerals formed at high temperature, and, thus, remain stable during much of the laboratory step heating experiment, facilitating the recovery of gradients of radiogenic argon (see Section 4.11). It is particularly appropriate, therefore, to present here examples of the application of the ^{40}Ar/^{39}Ar dating method to the elucidation of the history of these important and unique objects. We have chosen to discuss results on lunar basalts from Mare Tranquillitatis, sampled during the *Apollo 11* mission, in some detail, but data from several other samples also will be discussed.

6.2.2 Mare Tranquillitatis geochronology

6.2.2.1 Introduction

On July 20, 1969, the *Apollo 11* lunar module landed near the southwestern margin of Mare Tranquillitatis. The astronauts collected 21.7 kg of samples from the regolith within about 10 m of the lunar module during this remarkable first manned field trip to the lunar surface (LSPET, 1969; Schmitt et al., 1970). The material obtained was basaltic in character, consisting of individual clasts, breccias, and fines, the latter referring to fragments of less than about 10 mm diameter. These samples provided conclusive evidence that the maria were formed by basaltic lavas partly filling some of the large ringed basins, which had been excavated previously by major impacts with the lunar crust. This early formed lunar crust is partially preserved as the heavily cratered, light colored, highlands. Useful reviews of lunar geology were given by Taylor (1975, 1982).

Soon after arrival of the *Apollo 11* samples on Earth, several of the crystalline rocks were measured by the conventional K–Ar dating method. Apparent ages in the 3–4 Ga range were found (LSPET, 1969), indicating that these rocks were of considerable antiquity. Application of the ^{40}Ar/^{39}Ar step heating technique to whole rock samples of seven of the basalt clasts yielded a similar range of incremental total fusion ages, but the age spectra could be interpreted in terms of variable loss of radiogenic argon from the samples (Turner, 1970c,d). These results demonstrated the great power of the ^{40}Ar/^{39}Ar step heating method in detecting partial loss of radiogenic argon, and permitted more realistic estimates of cooling ages to be made compared with results obtained by the conventional K–Ar dating method. Subsequently, similar measurements have been carried out on many other lunar samples from the maria as well as from the highlands; for a comprehensive review of lunar geochronology by the ^{40}Ar/^{39}Ar technique the reader is referred to Turner (1977). As the results from the *Apollo 11* samples are of particular historic significance and provide excellent examples of ^{40}Ar/^{39}Ar age spectra exhibiting argon loss profiles, we will summarize these data here.

6.2.2.2 Apollo 11 *Basalts*

Sixteen basalt clasts of mass greater than 50 g were collected during the *Apollo 11* mission (Schmitt et al., 1970). These basalts show features in common with terrestrial basalts, including the presence of vesicles in some, typical igneous textures, and mineralogy that is familiar, but distinctive. Detailed characterization of the basalts was given by Beaty and Albee (1978). Compositionally, however, *Apollo 11* basalts are different from terrestrial basalts: they are essentially anhydrous, free of ferric iron, and have high TiO_2 (9–13%), high FeO (17.8–20.6%), low Al_2O_3 (7.1–10.5%), low Na_2O (0.3–0.7%), and relatively low SiO_2 (mainly 38–43.5%). Two distinct chemical types were recognized within this compositional range from the earliest investigations (Compston et al., 1970; Gast and Hubbard, 1970; Tera et al., 1970). These two types are now usually referred to as the low-K and high-K basalts (James and Wright, 1972; Beaty and Albee, 1978), characterized by potassium contents averaging ∼600 and 2600 parts per million (ppm), respectively. Within the low-K basalt suite, Beaty and Albee (1978) distinguished three textural, petrological, and compositional groups designated B1, B2, and B3, probably derived from separate lava flows. No subdivision of the high-K basalt group was made by Beaty and Albee (1978), who suggested that at least six of the samples may be from the same lava flow, and that several other samples are closely related, probably originating from a single chamber in which magma was differentiating during eruption.

Each of the larger basalt clasts has been measured by the ^{40}Ar/^{39}Ar step heating technique, and several smaller samples or fragments from the breccias and fines also have been dated. Results from more than 50 step heating experiments, performed on about 20 samples, are now available, including data on many mineral concentrates (Turner, 1970c,d; Stettler et al., 1973, 1974; Geiss et al., 1977; Guggisberg et al., 1979). Although these age spectra show a variety of patterns, they represent variations on essentially the same theme of loss of radiogenic argon, as will become evident. Several examples will be discussed here in relation to the inferred ages, the mineralogy of the rocks, and implications as to lunar history, dealing with the two major compositional suites separately.

6.2.2.3 *Age Spectra, Low-K Basalts*

In Figs. 6.1 and 6.2 age spectra are shown for a whole rock sample of basalt 10003, and plagioclase, clinopyroxene, and ilmenite mineral separates from the same rock (Turner 1970c,d; Stettler et al., 1974), recalculated to conform with the decay constants currently in use. All four age spectra exhibit low apparent ages of 1–2 Ga for the gas fraction released at the lowest temperature (400–500°C), with ages increasing in a regular manner in subsequent steps to a well-developed plateau segment comprising more than 60% of the ^{39}Ar release. For the clinopyroxene and the ilmenite separates, the highest temperature step, comprising 8–9% of the gas, yielded a lower and higher age, respectively, compared with the flat part of the spectrum. The plateau segments of all four spectra have ages that are concordant at 3.86 ± 0.01 Ga (1σ). In agreement with the views of the authors of the original papers, these remarkably consistent ages are most simply and readily interpreted as reflecting the time since crystallization and cooling of the basalt, following eruption on the lunar surface. The monotonic rise of the ages to the plateau conforms very well with the type predicted from the simple diffusion models of Turner (1968, 1969, 1970a), brought about by loss of radiogenic argon from the least retentive sites,

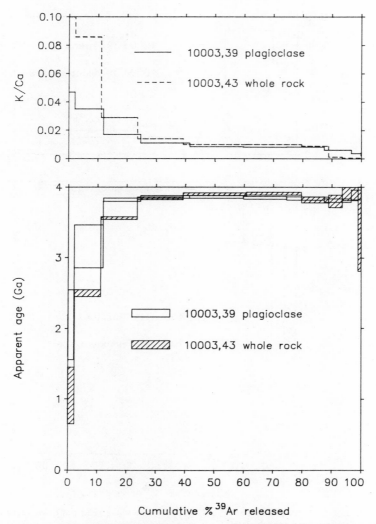

Fig. 6.1. Age spectra and K/Ca plots for whole rock and separated plagioclase from basalt sample 10003, collected on *Apollo 11* mission to Mare Tranquillitatis. Data from Turner (1970d) and Stettler et al. (1974), with permission; recalculated using $\lambda = 5.543 \times 10^{-10}/\text{a}$.

presumably owing to heating at a time or times younger than the youngest apparent ages.

Assuming that the high-temperature plateau ages indeed relate to the time of cooling of the basalt, the extent of subsequent loss of radiogenic argon is readily calculated as the fractional difference between the $^{40}\text{Ar}^*/^{39}\text{Ar}_K$ ratio corresponding to the plateau and the bulk sample ratio (Turner, 1970d). These fractional losses are 0.06, 0.10, 0.14, and 0.29 for the plagioclase, whole rock, clinopyroxene, and ilmenite, respectively. The K/Ca release patterns for each sample, also shown in Figs. 6.1 and 6.2, are all similar in form, decreasing essentially monotonically with increasing temperature of gas release. The higher temperature release, corresponding to the plateaus, is from sites with relatively low K/Ca ratios (<0.025), and the low-temperature gas release is from sites with substantially greater K/Ca ratios.

With this information it is instructive to consider the mineralogy and chemistry of the basalt, which is assigned to group B2 by Beaty and Albee (1978). They show that this rock consists of strongly zoned clinopyroxene (50 vol%), plagioclase (34%) of bytownite composition (An_{88-93}) but zoned to An_{70}, ilmenite (13%), cristobalite (1%), together with minor

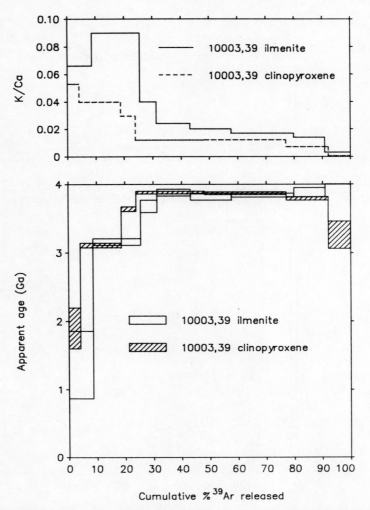

Fig. 6.2. Age spectra and K/Ca plots for ilmenite and clinopyroxene separated from basalt sample 10003. Data from Stettler et al. (1974), with permission; recalculated.

troilite, olivine, and a very small amount (0.03%) of K-rich (~4.4%K) glass. As basalt 10003 is very well crystallized, and the phases are anhydrous and were formed at high temperature, the sample should be particularly appropriate for a $^{40}Ar/^{39}Ar$ step heating experiment. This is because the minerals should remain stable throughout much of the extraction, meeting an important condition for the successful recovery of an argon diffusion gradient, if present. Indeed, this sample is about as close to ideal as it is possible to obtain. Thus, the interpretation of the age spectra for sample 10003 in terms of simple diffusion models seems entirely reasonable and appropriate.

From the $^{40}Ar/^{39}Ar$ analyses and much electron probe data it is clear that the bulk of the potassium (~70%±) in 10003 is carried within the plagioclase, particularly in the more sodic rims of the crystals (Lovering and Ware, 1970; Beaty and Albee, 1978). The close similarity of the age spectra obtained for the whole rock sample and the plagioclase separate is consistent with this view. As ilmenite and clinopyroxene do not readily accept the large cation, potassium, into their lattices, it is likely that the potassium found in these mineral separates resides mainly in small inclusions or composites of plagioclase and glass within the crystals (Albee et al., 1970; Philpotts and Schnetzler, 1970; Tera et al., 1970). The similarity of the age spectra and K/Ca plots may, thus, be explained in terms of release of the argon essentially from the same phase or phases, plagioclase and glass or mesostasis in each mineral fraction analyzed. Nevertheless, the excellent agreement between all the plateau ages is taken as strong evidence that the age relates to the crystallization and cooling of the basalt on the lunar surface, as the derived age is independent of the proportion of argon loss, which ranges from 0.06 to 0.29.

Argon loss observed in each age spectrum is associated with relatively high K/Ca ratios. This is interpreted as reflecting release of argon from glass, which has a K/Ca of ~ 3 (Beaty and Albee, 1978), and possibly also from the potassium-enriched outer zones of the plagioclase. Note, however, that the intersertal glass comprises only ~ 0.03 vol% of the rock, so that even with $\sim 4\%$ potassium (Beaty and Albee, 1978), the glass accounts for only a few percent of the total potassium.

The lower apparent ages, compared with the plateau age, for the gas released at the highest temperatures from pyroxene and some whole rock samples is a relatively common phenomenon in lunar rocks. Although several explanations have been advanced for this behavior (see summary in Turner, 1977), the consensus appears to be that in a significant number of cases it is related to recoil of ^{39}Ar during irradiation (see Section 4.9). The recoil distance of $\sim 0.1 \mu m$ (Turner and Cadogan, 1974) may result in a potassium-bearing phase developing a narrow zone at its margins somewhat depleted in ^{39}Ar, with a complementary enriched zone forming in an immediately adjacent potassium-poor phase, such as pyroxene or olivine. Different diffusion characteristics for argon in the two phases can result in variations in the ^{40}Ar*/^{39}Ar$_K$ ratio of the gas released during a step heating experiment. Thus, in the case of recoil of ^{39}Ar into pyroxene, a potassium-poor phase of relatively high argon retentivity, some of the ^{39}Ar may be retained until released near fusion of the mineral, at which temperature much of the radiogenic argon from the potassium-bearing material within or adjacent to the pyroxene may well have been released. Clearly such an explanation could apply to whole rock samples, but for it to apply also to pyroxene separates requires the presence of at least one other phase. It was suggested earlier that much of the potassium in the pyroxene separate may be carried in small inclusions or composites of plagioclase or glass.

Whether similar kinds of arguments can be advanced to explain the high apparent age for the last release of argon from 10003 ilmenite concentrate is less clear, but this kind of pattern is not commonly observed in lunar samples.

The question as to whether the apparent ages for the initial gas released in each experiment have any age significance in terms of a discrete geological event will be discussed after the age spectra from the *Apollo 11* high-K basalts have been considered.

A summary of the ^{40}Ar/^{39}Ar age results from the *Apollo 11* low-K basalts is given in Table 6.1. For simplicity the table includes only selected data, generally from the whole rock, but also from plagioclase, if available. Much of the remaining data on mineral separates is consistent with the whole rock and plagioclase results, as previously shown for sample 10003. The results are listed by groups using subdivisions proposed by Beaty and Albee (1978), based upon petrological and chemical characteristics of the basalts.

All the age spectra are similar in form to those for sample 10003, and, thus, can be interpreted in terms of relatively youthful loss of radiogenic argon, with the well-developed plateaus yielding estimates of the cooling age. For group B1 and B2 samples, the plateau age is only a few percent older than the total gas age, so that the indicated losses of argon are quite small. Note the excellent agreement between the plateau ages for samples within each group, providing good reason for having confidence in the results, and consistent with the view that samples in any given group may well be from the same lava flow. The cosmic ray exposure ages, based upon measurement of ^{38}Ar/^{37}Ar ratios in the argon released from the samples (Turner et al., 1971; Turner, 1977), are similar in each group. This might be expected if the samples in each group are derived from the same lava, possibly brought to the surface by the same event. For groups B1 and

TABLE 6.1. Summary of $^{40}Ar/^{39}Ar$ age data from Apollo 11 basalts[a]

Sample number	Material	$^{40}Ar/^{39}Ar$ total fusion age (Ga)	$^{40}Ar/^{39}Ar$ plateau age (Ga)	Span of ^{39}Ar release for plateau (%)	K (ppm)	Ca (wt%)	Exposure age (Ma)	Rb–Sr or Sm–Nd age (Ga)	References[b]
Low potassium basalts									
Group B1									
10044	Whole rock	3.56 ± 0.05	3.68 ± 0.05	21–98	780	8.6	70	3.63 ± 0.11	1, 2
10047	Whole rock	3.60 ± 0.02	3.69 ± 0.03	9–100	900	8.6	78	—	3
10058	Whole rock	3.54 ± 0.03	3.66 ± 0.04	35–74	620	8.6	71	3.55 ± 0.20	4, 2
Group B2									
10003	Whole rock	3.70 ± 0.05	3.86 ± 0.07	24–99	445	7.9	150	3.76 ± 0.08	1, 5
10003	Plagioclase	3.77 ± 0.03	3.85 ± 0.03	12–100	1110	11.6	120	—	3
10029	Whole rock	3.69 ± 0.02	3.81 ± 0.04	41–100	570	7.6	135	—	4
10029	Plagioclase	3.74 ± 0.02	3.83 ± 0.03	15–85	1220	12.3	133	—	4
Group B3									
10020	Clinopyroxene/ilmenite	3.32 ± 0.02	3.72 ± 0.04	30–70	440	8.1	122	—	4, 6
10045	Whole rock	3.39 ± 0.03	3.70 ± 0.03	53–78	480	7.8	114	—	4, 6
Unassigned									
10062	Whole rock	3.36 ± 0.05	3.77 ± 0.06	45–100	650	7.8	90	3.92 ± 0.11	1, 7
10050	Plagioclase	3.45 ± 0.03	3.70 ± 0.03	44–100	795	10.8	480	3.88 ± 0.06	4
High potassium basalts									
10017	Whole rock	2.30 ± 0.05	>3.20 ± 0.06	—	~2600	7.6	440	3.51 ± 0.05	1, 8
10022	Whole rock	3.22 ± 0.05	3.54 ± 0.06	55–99	2290	7.9	380	—	1
10024	Whole rock	2.56 ± 0.05	>3.44 ± 0.05	—	~3400	7.2	360	3.53 ± 0.07	1, 9
10031	Whole rock	3.48 ± 0.03	3.55 ± 0.08	24–100	2570	8.0	300	—	4
10032	Whole rock	3.19 ± 0.03	3.53 ± 0.06	54–100	2390	7.2	139	—	4
10057	Whole rock	2.45 ± 0.02	—	—	2340	7.2	52	3.55 ± 0.04	4, 2
10069	Whole rock	2.73 ± 0.02	—	—	2600	7.7	33	3.60 ± 0.06	4, 2
10071	Plagioclase	3.15 ± 0.02	3.46 ± 0.05	63–100	4550	9.0	390	3.60 ± 0.02	4, 2
10072	Plagioclase	3.43 ± 0.03	3.56 ± 0.06	30–100	3420	7.4	310	3.57 ± 0.05	4, 2
10072	Whole rock	3.26 ± 0.05	3.48 ± 0.05	53–99	2410	7.9	220	3.57 ± 0.03	1, 7

[a] $\lambda^{40}K = 5.543 \times 10^{-10}/a$; $\lambda^{87}Rb = 1.42 \times 10^{-11}/a$; $\lambda^{147}Sm = 6.54 \times 10^{-12}/a$. Errors quoted are as given in original publication, usually at the level of two standard deviations.

[b] (1) Turner (1970d); (2) Papanastassiou et al. (1970); (3) Stettler et al. (1974); (4) Guggisberg et al. (1979); (5) Papanastassiou and Wasserburg (1975); (6) Geiss et al. (1977); (7) Papanastassiou et al. (1977); (8) de Laeter et al. (1973); (9) Papanastassiou and Wasserburg (1971).

B2, Rb–Sr internal isochrons for three of the samples (Papanastassiou et al., 1970; Papanastassiou and Wasserburg, 1975), recalculated using $\lambda_{Rb} = 1.42 \times 10^{-11}/a$, yielded ages systematically younger than the $^{40}Ar/^{39}Ar$ plateau ages, but as the errors overlap the ages are statistically indistinguishable. This general consistency between results from two different decay schemes is most satisfactory, and serves to further increase confidence in the interpretation that these ages refer to the crystallization and cooling of the basalts.

For samples 10020, 10045, both of group B3, and samples 10062 and 10050, which are not assigned to a group, the differences between the plateau age in each case and the total gas age is somewhat greater, indicating radiogenic argon loss up to $\sim 25\%$, but the plateau segments are excellent. On the basis of Turner's (1968, 1969) diffusion models (see Section 4.2), the possibility that the plateau has been lowered somewhat, owing to the degree of younger outgassing, needs to be borne in mind. Internal isochrons utilizing the Rb–Sr and Sm–Nd dating methods on sample 10062 gave ages in agreement with one another at 3.9 Ga (Papanastassiou et al., 1977), a little older than the indicated $^{40}Ar/^{39}Ar$ plateau age of 3.77 ± 0.06 Ga. However, the errors are such that statistically the ages are not significantly different. If the $^{40}Ar/^{39}Ar$ age plateau has been lowered, owing to diffusion loss of radiogenic argon, the effect has been small, perhaps a few percent.

It should be emphasized that all the low-K basalts analyzed have chemistry and mineralogy similar to sample 10003, and are well crystallized, with the amount of glass not exceeding 0.7 vol% (Beaty and Albee, 1978).

Overall, these $^{40}Ar/^{39}Ar$ ages demonstrate, with little ambiguity, that the *Apollo 11* low-K basalts from several different lava flows were erupted in Mare Tranquillitatis over a time interval in the order of 0.1–0.2 Ga, from about 3.85 to about 3.7 Ga ago. Of particular note is that the results on plagioclase are very reliable, indicating that this mineral, which carries a large proportion of the potassium in these lunar rocks, has good retentivity for radiogenic argon under the conditions experienced on the lunar surface.

6.2.2.4 Age Spectra, High-K Basalts

The high-K basalts from the *Apollo 11* mission average about 0.26% potassium, a factor of four greater than found in the low-K basalts. Beaty and Albee (1978) provided detailed mineralogical and chemical documentation of these basalts, but preferred not to subdivide them. They suggested that most of the samples could be derived from a single lava flow that has undergone some crystal fractionation or from lavas derived from a single fractionating batch of magma.

The $^{40}Ar/^{39}Ar$ age spectra for the high-K basalts are quite similar to those found for the low-K basalts in that they show monotonically rising patterns, but in most cases a substantially greater degree of argon loss is indicated. In Fig. 6.3 patterns for two whole rock samples are illustrated. Sample 10072 shows a modest loss of radiogenic argon ($\sim 14\%$) with a well-developed plateau segment in its age spectrum, whereas sample 10017 exhibits a pattern of progressively increasing apparent age right through to the final gas release at the highest temperature. From this latter, markedly disturbed, age spectrum with $\sim 45\%$ loss of radiogenic argon, only a minimum age of 3.2 Ga for crystallization can be interpreted. Note that the K/Ca curves are similar to one another and also to those found for the low-K basalts, except that substantially greater ratios are observed over the first one-third of the gas release, related to the higher potassium content of these rocks.

A summary of the $^{40}Ar/^{39}Ar$ age data for the high-K basalts, given in Table 6.1, shows that there is considerable variability between the samples. In the case of sample 10072, the plateau ages for the whole rock and separated plagioclase agree to within the errors at 3.5 Ga, and only minor argon loss has occurred from the plagioclase. Four other samples show significant plateaus in their age spectra and yield ages that are concordant among themselves and agree with that derived for sample 10072 (Table 6.1). The remaining samples have more disturbed age spectra that can be interpreted as indicating cooling more than 3.2 Ga ago. All these data are consistent with the idea that the high-K basalts are either from the same lava

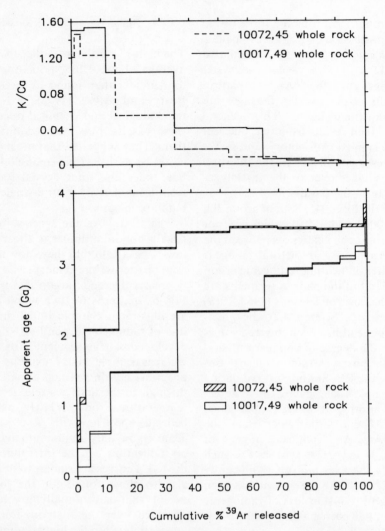

Fig. 6.3. Age spectra and K/Ca plots for *Apollo 11* whole rock basalt samples 10072 and 10017. After Turner (1970d), with permission; recalculated.

flow or are from closely related lavas that erupted 3.52 ± 0.05 Ga ago, the mean of results from six age spectra with plateaus, where the uncertainty quoted is one standard deviation.

Internal isochrons measured by the Rb–Sr method on six samples yielded concordant ages averaging 3.56 ± 0.04 Ga (Table 6.1). Note that several of these are on samples that show the highest degree of argon loss, such that no ^{40}Ar/^{39}Ar plateau age could be obtained. There is also a Sm–Nd internal isochron age of 3.57 ± 0.03 Ga for sample 10072. The remarkably good agreement between all these results, including the ^{40}Ar/^{39}Ar plateau ages, provides great confidence that the derived age is that of the crystallization and cooling of the lava flow(s), significantly younger than the low-K lavas. Note that the apparent systematic difference in age of ~100 Ma for the high-K basalts when measured by the ^{40}Ar/^{39}Ar and Rb–Sr methods, commented upon by Turner (1977), has disappeared as a result of using the currently recommended decay constants for the two systems (Steiger and Jäger, 1977).

As the *Apollo 11* low-K basalts have ages as great as 3.85 Ga, the results from the high-K basalts indicate that volcanism at this site in Mare Tranquillitatis extended over at least 0.3 Ga. The greater exposure ages generally found for the high-K basalts, compared with those found for the low-K basalts (Table 6.1), are consistent with the younger age inferred for the high-K basalts in that these basalts presumably overlie the low-K basalts and, thus, have been closer to the lunar surface for a longer time.

Mineralogically the high-K basalts are quite similar to the low-K basalts as they consist of abundant zoned clinopyroxene (45–53 vol%), plagioclase (22–24%) which is dominantly An_{79-82} but zoned to about An_{75}, ilmenite (14–16%), residual mesostasis (7–11%), cristobalite (1–2%), olivine (0–1%), troilite (0.4–0.7%), and a number of minor phases (Beaty and Albee, 1978). The major difference is the presence of considerable residual mesostasis in the high-K basalts, consisting of a mixture of plagioclase, cristobalite, and a potassic glass containing up to 5% potassium (Beaty and Albee, 1978). Although the plagioclase carries significant potassium, it accounts for less than half the total potassium in the high-K basalts; clearly the balance occurs mainly within the mesostasis. In view of the good retentivity of argon in plagioclase, and the high K/Ca ratios in the initial stages of gas release, the large losses of radiogenic argon interpreted from the age spectra are likely to be the result of argon leakage from the glass within the mesostasis. Especially if the glass was devitrified submicroscopically, the diffusion paths for argon to escape would be extremely short. The high degree of variability in argon loss from sample to sample is not readily explained in terms of mineralogical factors, except possibly by variable glass devitrification. The thermal history subsequent to crystallization may hold the key to the argon loss behavior.

6.2.2.5 Argon Loss Models

Finally, the cause of the variable loss of radiogenic argon from the *Apollo 11* basalts needs to be examined. From diffusion theory as applied to $^{40}Ar/^{39}Ar$ age spectra (Turner, 1968, 1969, 1970a), the apparent age for the initial gas release is expected to yield a maximum age for the reheating event that caused the argon loss. For *Apollo 11* basalts that have lost only a small proportion of their argon, the initial apparent ages generally are in the 1.0–2.0 Ga range. Samples that exhibit marked argon loss have apparent ages for the first gas release as low as ~200 Ma. Turner (1970d) inclined to the view that the argon loss was brought about by a discrete event such as meteorite impact. However, subsequently, Turner (1971) demonstrated that there was a remarkably good correlation between the observed natural loss of radiogenic argon from the lunar basalts and the ease with which argon was released from the samples at a given temperature in the laboratory. As recognized by Turner (1971), because the diffusion rate varies exponentially with temperature, this observation implies that all the samples have had similar thermal histories. On this basis Turner suggested that a more plausible explanation for the argon loss was thermal cycling at the lunar surface, where temperatures may exceed 100°C during the lunar day. Using diffusion data derived from the $^{40}Ar/^{39}Ar$ step heating experiments themselves to obtain estimates for the activation energies involved in argon loss from these lunar basalts (~30 kcal/mol), Turner (1971) showed that solar heating over periods of tens of million years could account for the argon loss interpreted from the age spectra. Indeed, as the cosmic ray exposure ages of *Apollo 11* rocks exceed 30 Ma and range up to nearly 500 Ma (Table 6.1), such an explanation seems quite tenable, although no direct correlation is found between degree of argon loss and exposure age. The reason the apparent ages for the gas released initially in the step heating experiments do not approach zero, as might be expected under the solar heating hypothesis, may be because of the limited resolution of the age spectra. Alternatively, variability in the grain size of the potassium-bearing phases, as well as variability of the diffusion coefficients for argon in the several phases, may be factors of importance in explaining the relatively high initial apparent ages.

6.2.3 Apollo 12 basalts

Two glassy basalts sampled by the *Apollo 12* mission from Oceanus Procellarum gave unusual $^{40}Ar/^{39}Ar$ age spectra, which are appropriate to discuss here. However, in order to place the results in context we shall first briefly consider $^{40}Ar/^{39}Ar$ age data available from well-crystallized *Apollo 12* mare basalts. These basalts are quite similar to those from the *Apollo 11* mission, but have much lower TiO_2 contents, in the 2.5–5% range (Papike et al., 1976). Potassium contents for the dated *Apollo 12* samples lie between 400 and 460 ppm (Turner, 1977), comparable with the abundances found in the low-K *Apollo 11* basalts. Olivine basalt, pigeonite basalt, and ilmenite basalt comprise the three major types of *Apollo 12* mare basalts (Papike et al., 1976; James and Wright, 1972). Seven basalts, including representatives from each of the three groups, dated as whole rocks by the $^{40}Ar/^{39}Ar$ method, yielded good high-temperature plateaus or correlations on three isotope plots (Turner, 1971; Alexander et al., 1972; Stettler et al., 1973; Turner, 1977), corresponding to ages in the 3.08 ± 0.06–3.23 ± 0.05 Ga range, with a mean of 3.14 ± 0.05 Ga. These age spectra exhibit typical argon loss patterns, similar to those found for the *Apollo 11* basalts. The seven basalts are essentially holocrystalline in clinopyroxene, plagioclase, olivine, opaques, and cristobalite (Warner, 1971; James and Wright, 1972; Papike et al., 1976), so that there is good reason for interpreting the ages found for the gas released at high temperature in terms of eruption, or, more strictly, cooling ages. The mean Rb–Sr age of 3.20 ± 0.08 Ga, based upon internal isochrons for four of these same samples (Papanastassiou and Wasserburg, 1970, 1971), is indistinguishable from the high-temperature $^{40}Ar/^{39}Ar$ ages, providing even greater confidence in the interpretation that the ages refer to the eruption and cooling of the basalts.

One olivine basalt sample (12018), however, showed extreme argon loss, although the highest temperature gas fractions in the monotonically rising age spectrum gave ages similar to those for the other *Apollo 12* basalts (Stettler et al., 1973). Assuming that this basalt is about the same age as the seven samples discussed previously, the loss of radiogenic argon is in the order of 70%. As this olivine basalt is well crystallized with less than 1% glass (Papike et al., 1976) and has an exposure age of ~ 150 Ma, comparable to that found for other *Apollo 12* basalts (Stettler et al., 1973), it would seem that the severe argon loss suffered by sample 12018 is related to the event or events (meteorite impact?) that brought the sample to the lunar surface. From the apparent age for the gas released initially in the step heating experiment, this outgassing occurred at or subsequent to 680 Ma ago.

The remaining two *Apollo 12* basalt samples that have been measured by the $^{40}Ar/^{39}Ar$ technique, and which yield unusual age spectra, are the olivine vitrophyres 12008 and 12009. These samples are dominated by a glassy to cryptocrystalline and microcrystalline mesostasis, comprising more than two-thirds of the rocks, with olivine as the phenocrystic phase (Brett et al., 1977; Warner, 1971; James and Wright, 1972; Papike et al., 1976). Clearly these basalts have crystallized rapidly, presumably owing to fast cooling. An age spectrum for sample 12008 is given in Fig. 6.4, after Stettler et al. (1973); that for sample 12009 is very similar. These spectra are characterized by high apparent ages for the gas initially released, showing a monotonic decrease in apparent age with increasing temperature of gas release. This type of pattern cannot be explained by the simple diffusion models. The incremental total fusion ages, based upon the average of two step heating experiments on each sample, are 3.10 and 3.19 Ga for samples 12008 and 12009, respectively. As noted by Stettler et al. (1973), these ages are indistinguishable from those obtained from the *Apollo 12* basalts that showed good plateaus in their age spectra. If the total fusion ages are significant, as seems likely, the implication is that these vitrophyres have not lost any radiogenic argon, despite their glassy to microcrystalline character. This is in marked contrast to the distinct but variable argon loss from the well-crystallized basalts, which are expected to retain argon much better than very fine grained or glassy rocks. If the total fusion ages are accepted as meaningful, then the ages derived from the individual gas fractions in the step heating experiment generally must be regarded as aberrant. As pointed out by Turner (1977), a possible explanation is that there has been closed system redistribution of ^{40}Ar or ^{39}Ar at

APPLICATIONS AND CASE HISTORIES

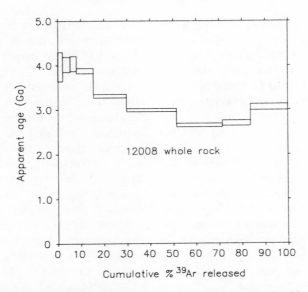

Fig. 6.4. Age spectrum for olivine vitrophyre basalt 12008 from the *Apollo 12* landing site. After Stettler et al. (1973), with permission; recalculated.

some time subsequent to the total argon age. Shock heating or melting might cause redistribution of ^{40}Ar. As discussed previously, ^{39}Ar redistribution can occur owing to recoil during irradiation in the nuclear reactor, in certain circumstances. Some actual loss of ^{39}Ar might have taken place during irradiation, from sites that also have poor retention for radiogenic argon, as found for some terrestrial basalts. Such a phenomenon might account for the apparent lack of overall argon loss from samples 12008 and 12009, in spite of their glassy character, and in contrast to the well-crystallized *Apollo 12* basalts, which show significant loss of radiogenic argon.

Similar age spectra have been observed in some fine grained lunar breccias (Turner et al., 1971), in several meteorites (Turner and Cadogan, 1974), and also in certain terrestrial basalts (Walker and McDougall, 1982). In virtually all cases of this kind of behavior, the samples contain abundant glass, in which most of the potassium resides, or are very fine grained. The lack of full understanding of the physical reason for this type of age spectrum highlights the difficulties inherent in interpreting age spectra that depart from the kind expected from the models based upon argon diffusion.

6.2.4 Significance of mare basalt dating

The ^{40}Ar/^{39}Ar dating of mare basalts from five sites on the Moon provided much important information relating to lunar history, well summarized by Turner (1977). The ^{40}Ar/^{39}Ar age spectra yielded extremely strong evidence for some diffusive loss of radiogenic argon from virtually all samples. The presence of plateaus in the age spectra often allowed crystallization and cooling ages to be established, in many cases amply supported by data from other decay schemes. Arguments presented by Turner (1971) indicated that the argon loss profiles generally could be accounted for in terms of diffusion losses owing to solar heating over a relatively long period on the lunar surface, rather than to a discrete event. The results showed that plagioclase was the most reliable mineral for dating lunar rocks, but that whole rock samples commonly yielded reliable ages, in part because much of the potassium is carried in plagioclase in these samples. The suitability of many of the mare basalts for ^{40}Ar/^{39}Ar dating is related to their high temperature, anhydrous mineralogy, and their high degree of crystallinity. These factors contributed significantly toward the recovery of gradients of radiogenic argon in the whole rocks and separated minerals, because of the stability of the minerals in the vacuum system during the step heating experiments. The great age of the samples, with their high content of radiogenic argon, made the technical measurements less difficult than would otherwise have been the case.

The $^{40}\text{Ar}/^{39}\text{Ar}$ age results provide convincing evidence that the mare basalts were erupted into some of the large impact basins on the moon over an interval extending from at least 3.85 to 3.15 Ga ago. These data also provide minimum ages for formation of the large ringed basins, and show that the major bombardment of the lunar surface ended earlier than ~3.8 Ga ago, as the mare surfaces have very low crater densities compared with the lunar highlands. The age results also contribute significantly toward providing constraints on the thermal history of the Moon, especially in relation to models of magma generation in the lunar interior.

6.2.5 Geochronology of the lunar highlands

The rugged, light-colored lunar highlands and the relatively smooth surfaces of the dark-colored maria are the most obvious first-order features of the Moon. The extremely dense cratering of the lunar highlands compared with the maria is universally interpreted as indicating a greater age for the highlands. The highlands generally are regarded as consisting essentially of crust produced by major differentiation early in lunar history, possibly from a magma ocean [see review by Warren (1985)]. Photogeological mapping and the vast amount of detailed information obtained on samples of highland rocks, collected on a number of the *Apollo* missions, have contributed greatly to the understanding of the origin and evolution of the highland regions, as well summarized by Taylor (1982).

The heavily cratered lunar highlands, covered by a megaregolith of brecciated material, clearly must be older than at least the younger great ringed basins excavated within the highlands. Many of the basins subsequently were partly filled by basaltic lavas to form the maria, the surfaces of which have a very much lower density of cratering. Thus, the rate of bombardment of the lunar crust appears to have decreased to near present-day levels by about 3.5 Ga ago, based upon measured ages of mare basalts.

Samples obtained on the *Apollo* missions show that the rocks comprising the lunar highlands are mainly anorthositic in character, dominated by calcic plagioclase, and ranging from anorthosite, through anorthositic norite, norite to troctolite, together with some basalt. Much of the recovered material is highly shocked, brecciated, and metamorphosed as shown by the presence of cataclastic and recrystallization textures. Indeed, the majority of the highlands samples are complex breccias, commonly exhibiting breccia-in-breccia textures. In addition, there is ample evidence of impact-induced melting, and many of the well-crystallized rocks have formed from such melts. Thus, samples from the lunar highlands often show considerable complexity best explained by multiple impact events. Because of the saturation cratering of the highlands it is doubtful that any undisturbed primary crust remains on the lunar surface, although about 100 so-called pristine rock fragments from the highlands have now been recognized (Warren et al., 1986). The composite nature of many rocks and the repeated impact mixing highlights the potential difficulties in interpretation of geochronological data obtained on rocks from the lunar highlands, contrasting with the relative simplicity of interpretation of data from the mare basalts. Nevertheless, application of the $^{40}\text{Ar}/^{39}\text{Ar}$ age spectrum method to rocks from the lunar highlands has been surprisingly successful.

Prior to discussing the $^{40}\text{Ar}/^{39}\text{Ar}$ age data it is important to note that other isotopic studies utilizing the Rb–Sr, U–Pb, and Sm–Nd systems provide strong evidence for major early differentiation of the Moon at about 4.5 Ga ago, soon after its formation, but possibly extending to about 4.3 Ga ago (see Carlson and Lugmair, 1981, for a useful summary). This early differentiation is thought to have produced the lunar crust now preserved in the lunar highlands, albeit much reworked by repeated impacts.

More than 100 samples from the lunar highlands have now been measured by the

^{40}Ar/^{39}Ar dating technique; a compilation and detailed discussion of results were given by Turner (1977), with much additional data on *Apollo 16* breccias in Maurer et al. (1978). These determinations have been made on a wide variety of samples, usually on small whole rock clasts from breccias or from the coarser fraction of lunar soils. However, plagioclase separates also have been used in several cases. Rock types dated include anorthosite, norite, basalt, polymict breccia, and recrystallized melt rock or glass. These samples exhibit a large range in potassium content from as low as 0.01% to ~0.5%, exceeding 1% in a few samples. As might be expected, the age spectra show considerable diversity. A large proportion (>80%) of the spectra have well-developed plateau segments, normally comprising more than half the gas release in a number of successive steps. Minor subsequent loss of radiogenic argon, usually <10%, is reflected in the results from the gas released in the early stages of the step heating experiments in the majority of cases. Several age spectra possess a good plateau at intermediate temperatures of gas release, but show a decrease in apparent age for the gas fractions released at higher temperatures; these are similar to the kind previously noted for some of the mare basalts, possibly recoil related. In a minority of cases the age spectra rise monotonically with increasing gas release, and there are several age spectra that are quite irregular and not readily interpretable.

For those age spectra that exhibit reasonable plateau segments the great majority yield ages (adjusted to the currently used decay constants) in the rather restricted 3.80–4.25 Ga range (Fig. 6.5), with few authenticated older plateau ages (cf. Dominik and Jessberger, 1978). Several samples gave plateau ages of ~3.5 Ga, and a number of glasses, some of which are devitrified, yielded still younger plateau ages (Turner, 1977; Maurer et al., 1978), possibly related to localized impact melting.

The widely accepted interpretation for the plateau ages is that they generally reflect resetting of the K–Ar system as a direct or indirect result of collisional events in the highlands, thus providing control on the timing of some of these events.

There are Rb–Sr age data, usually in the form of internal isochrons, available on 10 samples that have yielded good ^{40}Ar/^{39}Ar plateaus. In each case except one the ages agree to within the errors, and for eight of the intercomparisons the ages from the two systems are within 1% of one another. These remarkably concordant results on a range of samples, mainly basaltic fragments from breccias, together with knowledge of the petrography and geology, provide very strong evidence for the complete resetting of the Rb–Sr and K–Ar systems, and justification for accepting most of the ^{40}Ar/^{39}Ar plateau ages at face value. Several of the samples, however, are thought to be mare basalts that have reached the highlands owing to impact events, and the concordant ages in such cases may simply reflect original cooling. The U–Pb systematics of whole rock samples from the lunar

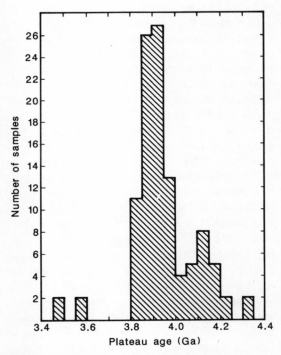

Fig. 6.5. Histogram of ^{40}Ar/^{39}Ar plateau ages for samples from the highlands regions of the Moon. Data from summary table in Turner (1977, see for primary sources), from Maurer et al. (1978), and from Dominik and Jessberger (1978). All ages recalculated using $\lambda = 5.543 \times 10^{-10}$/a. Results on several glassy samples with apparent ages in the 0.95–2.3 Ga range, reported in Maurer et al. (1978), are omitted from the plot.

highlands also provide powerful evidence for isotopic resetting at ~3.9 Ga ago (Tera et al., 1974). As summarized by Turner (1977), possible scenarios for resetting include direct melting owing to heating related to meteorite impact, heating within a hot ejecta blanket from a large impact crater or basin, or cooling from above the closure temperature for daughter products through excavation from depth as a result of a major impact.

On the basis of the remarkably high concentration of ages at ~3.9 Ga from the U–Pb, Rb–Sr, and ^{40}Ar/^{39}Ar data on highlands samples, Tera et al. (1974) proposed that the surface of the Moon suffered a terminal cataclysm at this time as a result of particularly intense bombardment by large objects, correlated with the younger major basin-forming events. It is widely accepted that the sharp cut-off of ages of highlands samples at ~3.8 Ga is because of a marked decrease in the rate of bombardment of the lunar surface at this time. However, it remains to be resolved whether this simply reflects the tail end of a progressive decrease in collisional events, or whether there was a period of enhanced bombardment over a short interval at ~3.9 Ga ago, as advocated by Tera et al. (1974). Although discussion continues on these questions, it is abundantly clear, however, that the majority of the ^{40}Ar/^{39}Ar results, and results from other isotopic systems on highlands samples, reflect the latter stages of intense bombardment of the Moon, providing important constraints on a number of aspects of lunar history.

Finally, we shall consider some results from breccia sample 65015 that possibly relate to the early history of the Moon. This sample, collected from the central highlands during the *Apollo 16* mission, was described in detail by Albee et al. (1973). They showed that sample 65015 is a metamorphosed clastic rock consisting of a variety of rock and mineral fragments, mainly angular plagioclase clasts, set in a fine grained matrix. The metamorphism caused extensive, but incomplete, intergranular recrystallization of the matrix without reconstitution of the clasts. Albee et al. (1973) suggested that the protolith of 65015 probably was a glassy agglutinate of mineral and lithic fragments cemented by high-potassium glass. The high temperature metamorphism was inferred to be related to a large impact event, with incorporation of the sample in an ejecta blanket. Thus, it is clear that this sample has had a complex history.

Papanastassiou and Wasserburg (1972) interpreted Rb–Sr data they obtained on sample 65015 as indicating that the age of the last major metamorphism was 3.85 ± 0.02 Ga. This conclusion was based upon analysis of two total rock samples and an alkali-rich separate ("quintessence" or mesostasis). However, they also found that measurements on plagioclase separates, including material from the clasts, showed that full isotopic equilibration did not occur during the metamorphism, as the plagioclase concentrates possessed significantly more primitive isotope compositions. For this reason, Jessberger et al. (1974a,b) chose to study whole rock and mineral separates from sample 65015 in considerable detail by means of ^{40}Ar/^{39}Ar age spectra.

In Fig. 6.6 the age spectrum for plagioclase separate B, previously analyzed by Papanastassiou and Wasserburg (1972), is shown plotted in the normal manner, and also against ^{37}Ar release, together with a K/Ca plot. Although the age spectrum clearly is complex, much useful information has been derived from it. There is a monotonic rise of the apparent age in the early stages of gas release to a maximum of 3.94 Ga, followed by a decrease to about 3.80 Ga, then a gentle rise to a semblance of a plateau at 3.92 Ga, finally rising steeply to 4.4 Ga in the last stages of gas release. Note that the K/Ca ratio decreases progressively over more than an order of magnitude with increasing temperature and gas release. Jessberger et al. (1974a,b) interpreted this as indicating that the bulk of potassium (~70%) is from potassium-rich mesostasis associated with the plagioclase, and that the gas released above ~750°C, exhibiting low K/Ca ratios, reflects argon originating from the plagioclase structure; this was emphasized by plotting apparent age versus ^{37}Ar release (Fig. 6.6), as the ^{37}Ar originates only from calcium, and the plagioclase is a high calcium variety.

On this basis we can interpret the early part

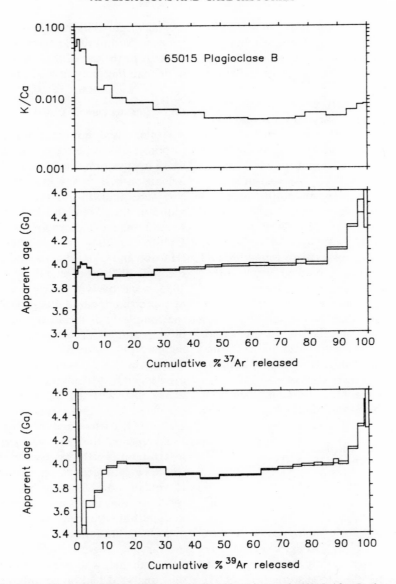

Fig. 6.6. Age spectrum for plagioclase B separated from *Apollo 16* breccia sample 65015. After Jessberger et al. (1974b), with permission; recalculated to currently used decay constants. Note that the age spectrum is plotted against ^{39}Ar release, and also against ^{37}Ar release, with the K/Ca similarly plotted relative to ^{37}Ar release.

of the age spectrum as a typical argon loss profile recorded in the potassium-rich mesostasis, with the loss occurring subsequent to 3.9 Ga ago. Note, however, that Jessberger et al. (1974a,b) proposed that this portion of the pattern was likely recoil related or due to grain size or activation energy effects. This argon loss profile from the high K/Ca material overlaps with a similar, but much broader loss profile from plagioclase, accounting for the decrease in apparent age around release temperatures of about 700°C, and the subsequent rise to a plateau at 3.91 ± 0.04 Ga. The latter age is reasonably interpreted as reflecting the major metamorphism, particularly as the age is similar to that derived from the Rb–Sr total rock and mesostasis data. The apparent ages rising to ~ 4.4 Ga in the last stages of gas release, together with the Rb–Sr results that show the plagioclase did not equilibrate with the groundmass during the metamorphic event, led Jessberger et al. to propose that this was strong evidence for much earlier events in the history of the Moon, soon after its formation.

This is a particularly good example as to how a complex age spectrum may reveal much detail, but it should be emphasized that the evidence remains permissive; care must be exercised to avoid the temptation of overinterpreting results from ^{40}Ar/^{39}Ar measurements. In this context, it is worth noting that two plagioclase clasts from another *Apollo 16* breccia (sample 67435) yielded well-developed ^{40}Ar/^{39}Ar plateaus with ages of 4.35 ± 0.05 and 4.33 ± 0.04 Ga, respectively (Dominik and Jessberger, 1978), thus providing further rather strong evidence favoring the view that some rocks or parts thereof on the lunar surface crystallized earlier than 4.3 Ga ago. Extremely strong additional evidence for the great antiquity of original crystallization of some material in the lunar highlands comes from U–Pb ages measured on zircons of 4.36 ± 0.02 Ga from lunar breccia 73217 by Compston et al. (1984) using a high-resolution ion microprobe.

6.3 AGE SPECTRA REFLECTING EPISODIC ARGON LOSS: EXAMPLES AND APPLICATIONS

6.3.1 Introduction

The near unique geochronological perspective available through ^{40}Ar/^{39}Ar age spectrum analyses makes it particularly attractive in the identification of discrete disturbances of the isotopic system. Indeed, the first application of the method (Turner et al., 1966) was for just this purpose—resolving the age of a thermal event experienced by a meteorite. Subsequent to this, several cases of episodic loss in terrestrial samples were documented by the ^{40}Ar/^{39}Ar approach, giving confidence that at least several minerals are capable of preserving naturally produced argon diffusion gradients as well as revealing these profiles during laboratory heating. The problem of argon loss from minerals during reheating is probably the single largest cause of uncertainty in interpreting K–Ar ages from terranes of deep crustal origin. The potential of the age spectrum method to reveal argon loss has given an impetus to the use of the ^{40}K decay scheme in understanding polythermal histories.

6.3.2 Contact aureole studies

An often used approach for examining the response of isotope systems to episodic loss is to sample within the thermal aureole of an igneous intrusion. Several classic areas of this sort were studied using the K–Ar dating technique in the 1960s; the Eldora Stock in the Front Ranges of Colorado (Hart, 1964) and the Duluth Gabbro in northern Minnesota (Hanson and Gast, 1967) were reinvestigated in the mid-1970s (Berger, 1975; Hanson et al., 1975, respectively) by means of the ^{40}Ar/^{39}Ar age spectrum method. Berger (1975) focused on the complexly disturbed age spectra found for biotites and alkali feldspars from rocks of the metamorphic aureole of the Eldora Stock. Hanson et al. (1975) concluded that a variety of experimental artifacts had obscured age gradients they believed to exist in the minerals measured from rocks adjacent to the Duluth Gabbro. Harrison and McDougall (1980a,b, 1982) repeated this kind of experiment utilizing a terrane in northwest Nelson, New Zealand, where a Cretaceous granitoid intrudes a Devonian hornblende gabbro. The thermal effect of the intrusion, pronounced at the margin but decaying rapidly with distance away from the contact, caused ^{40}Ar* to be lost in varying degrees from the hornblendes. Although many of the hornblende separates they analyzed had apparently acquired excess argon during the heating event, several samples did yield argon loss patterns that corresponded well with the simple predictions of Turner's (1968) model. For example, a hornblende from the gabbro 1 km from the contact with the granitoid revealed an age of 110 ± 2 Ma early in the gas release, with ages rising progressively to 310 ± 3 Ma in the latter stages of the release (Fig. 6.7). Given knowledge of the gabbro and granitoid intrusion ages of 367 ± 5 and 114 ± 1 Ma, respectively, the 228 ± 2 Ma total fusion age translates to 57% ^{40}Ar* loss. The synthetic spectrum for this degree of gas loss, shown as a smooth curve in Fig. 6.7, agrees

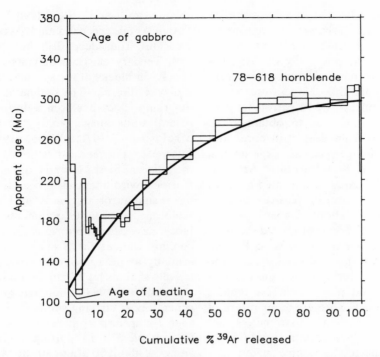

Fig. 6.7. Age spectrum for hornblende 78-618 from the Rameka Gabbro, Nelson, New Zealand. After Harrison and McDougall (1980b). This sample is 1 km from the intrusive contact with a younger granitic body, and the effects of this younger reheating are clearly evident in the age spectrum. The solid curve shows a theoretical age spectrum for a sample outgassed 57%, 114 Ma ago.

very well with the empirical data, providing clear support for the single-site diffusion model as well as testament to the fact that at least some amphiboles can reveal naturally imposed $^{40}Ar^*$ gradients during vacuum step heating experiments.

6.3.3 Sedimentary basin thermal histories

A common ingredient in evaluation of virtually all energy sources is some form of temperature measurement. A complete description of exploitable geothermal and fossil fuel resources requires an understanding of the temperature history of the system or source rocks over extremely long time periods. For example, the maturation kinetics of hydrocarbons to petroleum requires temperatures between about 100 and 150°C over million year time periods (Hunt, 1979). Temperatures in excess of this range cause rapid devolution of organic substances to methane, whereas temperatures significantly lower are insufficient to allow the reactions to proceed. Because of the economic importance and geodynamic significance of sedimentary basins, much thought has been directed at understanding the thermal evolution of these features. One result has been the development of an extraordinary array of thermometers utilizing such diverse temperature indicators as the optical reflectivity of a coal maceral (vitrinite reflectance) and the discoloration of a fossil group of unknown genesis (conodont color alteration) (Hunt, 1979).

The relatively poor retentivity of low-temperature alkali feldspars for $^{40}Ar^*$ was recognized during the early application of the K-Ar dating technique (e.g., Sardarov, 1957). Indeed, the retention characteristics of microcline were so variable and suspect that it became popular to believe that significant $^{40}Ar^*$ could be lost at room temperature (e.g., Faure, 1977). Foland (1974) predicted that the relatively poor retentivity of microcline was because perthite lamellae behave as channels of rapid argon diffusion, and that the closure temperature was of the order of 150°C. Harrison and McDougall (1982) reinvestigated the behavior of microcline in a slowly cooled terrane and estimated a similar closure temperature. These studies caused Harrison and Bé (1983) to explore the possibility of using detrital microcline to record the prograde basin heating history in the temperature range (~100–150°C) appropriate to petroleum generation. Their method was to obtain alkali feldspar separates from deep drill cores and perform ^{40}Ar/^{39}Ar age spectrum experiments. The appearance of characteristic ^{40}Ar* loss profiles (when compared to unheated samples higher up in the sequence) allows an estimate of temperature and duration of heating to be made, providing the temperature-dependent rate of argon diffusion in the phase is known. By calculating diffusion coefficients from the measured ^{39}Ar loss during the vacuum extraction experiment and plotting these estimates against the reciprocal absolute temperature of the heating step, Arrhenius parameters (activation energy and frequency factor/grain size parameter) unique to the sample in question could be obtained. Although poor experimental temperature control precluded Harrison and Bé (1983) from realizing this attribute, subsequent studies (e.g., Harrison and Burke, 1988) have borne out this promise.

Harrison and Bé (1983) performed ^{40}Ar/^{39}Ar age spectrum analysis on microcline separates from deep drill hole intersections with Tertiary clastic sediments of the Tejon and Basin blocks of the southern San Joaquin Valley, California. The sedimentary history of this ramp valley has been well documented in several studies (e.g., Bandy and Arnal, 1969; MacPherson, 1978; Webb, 1981). Briefly, Eocene to Pliocene sandstones derived from the adjacent Sierra Nevada and deposited in a marine environment have been further buried by a significant thickness of Pleistocene to Holocene terrigenous sediments. The Tejon block has experienced youthful uplift and/or cooling whereas the Basin block remains essentially at peak prograde conditions. In relatively shallow (≤ 3.5 km depth) samples Harrison and Bé (1983) found age spectra that, in general, bore striking similarity to alkali feldspar age spectra from the Sierra Nevada (see, for example, Fig. 5.15), rising in age from initial values of about 70 Ma to 80–85 Ma in the final portion of gas release. In deeper samples they observed progressively greater amounts of ^{40}Ar* loss, apparently in response to recent heating (Fig. 6.8). At a depth of 6.2 km, currently at a temperature of 157°C, 18% ^{40}Ar* loss

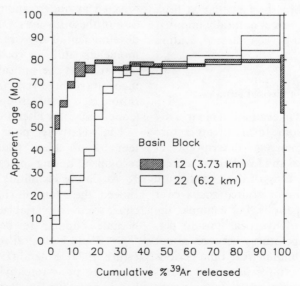

Fig. 6.8. Age spectra for microclines from sediments recovered from a borehole at depths of 3.7 and 6.2 km in the Basin Block, San Joaquin Valley, California. After Harrison and Bé (1983). The spectrum for microcline from sample 12 is similar to that found for microclines from the adjacent Sierra Nevada, whereas microcline from sample 22 shows significant loss of radiogenic argon (~18%).

was documented (Fig. 6.8). Using the Arrhenius parameters calculated for this sample by Harrison and Burke (1986), this ^{40}Ar* deficit was translated into a linear basin heating history proceeding at a rate of 0.03°C/ka over the past 2 Ma, a thermal evolution quite consistent with the stratigraphy. They used this information to estimate the perturbation in the equilibrium thermal gradient resulting from the transient effect of rapid sediment deposition.

Other studies have demonstrated the utility of this general approach in constraining the thermal history of extensional sedimentary basins (e.g., Harrison and Burke, 1988) and geothermal systems (e.g., Harrison et al., 1986).

6.3.4 Application of isochron analysis to partially outgassed xenoliths

Precise K-Ar age measurements of young (< 1 Ma) volcanic rocks often can be made as a result of the low level of trapped argon in fresh samples, typically 10^{-11}–10^{-12} mol/g (Fig. 2.6; McDougall, 1966; Dalrymple and Lanphere, 1969). But problems can arise if samples are altered, resulting in much higher atmospheric argon contamination, or if the potassium content is very low. In rare cases there is an alternative approach available. Thus, the ubiquitous occurrence of granitoid xenoliths in the Pleistocene basalts of the eastern Sierra Nevada, California, provided Gillespie (1982) with an opportunity to obtain age control on these geomorphologically important rocks. Gillespie surmised that upon being encompassed by molten basalt, at least a portion of the highly potassic mineral, microcline, would completely degas during the many hours to days that the lava remained hot. Although Dalrymple (1964) had previously shown that conventional K-Ar analyses of these xenoliths did not reveal meaningful geological ages, owing to the incomplete degassing of ^{40}Ar* during heating, the spatial dimension provided by ^{40}Ar/^{39}Ar dating might separate reservoirs of inherited ^{40}Ar* from a purely posteruption, *in situ* component.

The experiments of Gillespie et al. (1982, 1983, 1984) on partially degassed xenoliths tended to yield age spectra characterized in the first portion of gas release by a plateau at a relatively young age, presumably recording the basalt outgassing episode, followed in the latter portion by ages rising precipitously to older ages characteristic of Sierran granitic feldspars. This type of spectrum, indicative of two contrasting activation energies or grain sizes, is discussed in detail in Section 4.10. When plotted on an isochron diagram these data tend to

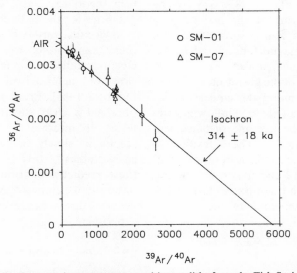

Fig. 6.9. Results from step heating experiments on two granitic xenoliths from the Fish Springs Cinder Cone in Owens Valley, California, plotted on a correlation diagram. After Martel et al. (1987). Data define an isochron, yielding an age of 314 ± 18 ka.

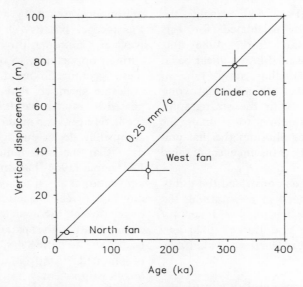

Fig. 6.10. Age of the Fish Springs Cinder Cone plotted against its displacement by the Fish Springs Fault, together with data from two alluvial fans, indicating an average vertical displacement of 0.25 mm/a for this fault. After Martel et al. (1987).

exhibit well-correlated arrays defining the age of degassing by the basalt and the composition of argon trapped at that time, before scattering below this line as the inherited ^{40}Ar* begins to degas. Perhaps the most immediate example of the geological application of the approach of Gillespie and co-workers is their dating of the Fish Springs Cinder Cone, Owens Valley, California (Martel et al., 1987). This cinder cone is vertically displaced 78 ± 6 m by the Fish Springs fault, part of the 300-km-long Owens Valley Fault zone. Surfaces of two distinctly different alluvial fans, dated by relative means, are also offset 31 ± 3 and 3.3 ± 0.3 m. Analysis of the granitoid xenoliths, found sporadically among the cinders, yields an age of 314 ± 18 ka (Fig. 6.9), allowing precise calculation of the average displacement rate of 0.25 ± 0.02 mm/a. These results together with the measured offsets of the two fans are shown in Fig. 6.10 and illustrate how this method can provide a quantitative framework for calibration of other kinds of data.

6.4 UPLIFT AND COOLING STUDIES

6.4.1 Introduction

The recognition that K–Ar ages in slowly cooled terranes reflect retrograde closure (e.g., Armstrong et al., 1966) means that all K–Ar measurements on crustal rocks are, in effect, uplift and/or cooling studies. Turner (1969) was the first to note that samples cooled slowly through their closure temperature interval (see Section 5.7), the relatively narrow temperature window over which daughter retention begins and becomes complete, should record ^{40}Ar* gradients within discrete diffusion domains. Dodson (1973) provided this concept with a firm theoretical basis, deriving analytical expressions for the bulk closure temperature (T_c) behavior of a variety of geometric forms. Later, these expressions were evaluated to yield closure profiles within mineral grains following slow cooling (Harrison and McDougall, 1982; Dodson, 1986). From these studies, we have obtained a clear theoretical framework with which to interpret ^{40}Ar/^{39}Ar age spectrum results in slowly cooled terranes. However, early studies of this kind were slow to reveal these predicted distributions, in part because naturally developed ^{40}Ar* gradients in biotite generally are not recovered in vacuum heating experiments.

6.4.2 Simple cooling of an igneous intrusion

Perhaps the clearest view of the response of ^{40}Ar/^{39}Ar age spectra to cooling is to examine an intrusive granitoid known to have been unaffected by any postemplacement thermal event. Because of the profound thermal contrast between magma and country rock during intrusion, cooling rates at near subsolidus temperatures are often relatively rapid ($\sim 100°C/Ma$), even for large plutons. Although this generally excludes highly retentive minerals such as hornblende from revealing slow cooling behavior, minerals with somewhat lower closure temperatures, such as biotite and alkali feldspar, should cool through their closure temperature interval over long enough time spans, say 2–20 Ma, that age gradients resulting from slow cooling may be analytically resolvable, at least in relatively young samples.

Harrison and McDougall (1980a, 1982) performed $^{40}Ar/^{39}Ar$, K–Ar, Rb–Sr, U–Pb, and fission track age measurements on minerals separated from the Separation Point Batholith, northwest Nelson, New Zealand, with the intent of resolving the cooling history of the pluton and examining the age spectra in this context. By plotting the measured mineral ages against an estimated or calculated value of the respective mineral closure temperatures, a simple cooling history emerged (Fig. 6.11). Beginning with crystallization at 114 Ma, as evidenced by a $^{206}Pb/^{238}U$ zircon date, the pluton initially cooled rapidly ($\sim 500°C/Ma$) for several hundred thousand years after which time the hornblende K–Ar system closed. The cooling rate, dropping asymptotically, slowed to a value of about 50°C/Ma during the closure interval for biotite, which gave a K–Ar age of 107.8 Ma, and was as low as $\sim 5°C/Ma$ at the time the microcline K–Ar systems closed at about 95 Ma.

These results (Fig. 6.11) are strong evidence that the intrusive experienced a simple thermal decay in response to emplacement and has not experienced any subsequent temperature perturbation. Furthermore, the solid curve shown in Fig. 6.11 is the prediction of a numerical heat flow model (see Section 5.8) using dimensions and thermal properties appropriate to the Separation Point Batholith. The close correspondence of the simulation with the isotopic results suggests that the mode and mechanism of cooling are well understood.

$^{40}Ar/^{39}Ar$ age spectra for microcline revealed substantial age gradients (Fig. 4.11) that were ascribed to long-term residence in the temperature transition between completely open and closed system behavior (Harrison and McDougall, 1982). Under ideal circumstances it may be possible to obtain cooling rate information directly from an age spectrum of this kind.

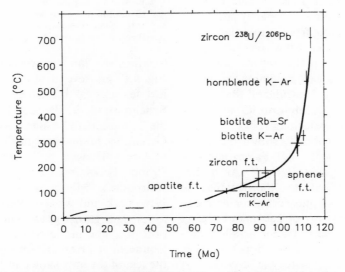

Fig. 6.11. Cooling history of Separation Point Batholith based upon mineral age data from sample 78-592. After Harrison and McDougall (1980a). Ages are plotted against estimated closure temperature for the particular isotopic system. The solid line is calculated cooling history for the sample based upon a conductive cooling model for the batholith subsequent to emplacement. Uncertainties are indicated at the level of one standard deviation.

6.4.3 Metamorphic cooling histories

^{40}Ar/^{39}Ar studies of metamorphic terranes (Pankhurst et al., 1973; Dallmeyer, 1975a,b, 1978, 1987; Dallmeyer et al., 1975; Dallmeyer and Sutter, 1976; Dallmeyer and Villeneuve, 1987; Roddick et al., 1980; Harrison and McDougall, 1981; Berger and York, 1981a) in general have provided incremental additions to our understanding, or at least our expectations, of the behavior of age spectra in a slow cooling environment. Most of these studies have been cited elsewhere in this book for specific contributions to age spectrum or isochron interpretation. This growth in interpretative skills is shown rather nicely in a sequence of papers spanning a dozen years relating to Appalachian geochronology.

The Appalachian Mountain Belt is perhaps one of the most intensively studied orogens in the world and has yielded many classic studies in stratigraphy, structure, and petrology (e.g., Zen et al., 1968). Field evidence and modern tectonic concepts have allowed recognition of several successive Paleozoic collision zones termed Taconian (mid-Ordovician), Acadian (early Devonian), and Alleghenian (Permo-Carboniferous). In many areas, stratigraphic relations are clear and observed metamorphism can be reliably ascribed to one or more orogenic events. However, ambiguity can exist in the higher grade regions, in part because of structural complexity and partly the result of uncertainties arising from extrapolations of stratigraphic relations from remote, lower grade sections. It is in this regard that geochronologic studies can be uniquely useful. Indeed, it may not be coincidental that nearly half of the ^{40}Ar/^{39}Ar laboratories in North America sit atop Appalachian rocks.

Figure 6.12 schematically shows the distribution of Proterozoic through Paleozoic rocks in central New England. Early K–Ar studies (e.g., Harper, 1968) confirmed the view that the westernmost metamorphosed Paleozoic rocks are exclusively the result of Taconian orogeny of late Proterozoic–early Paleozoic allochthonous sediments (e.g., Taconic klippe). Further east, questions arise as to the divisions of Taconic/Acadian/Alleghenian and the extent

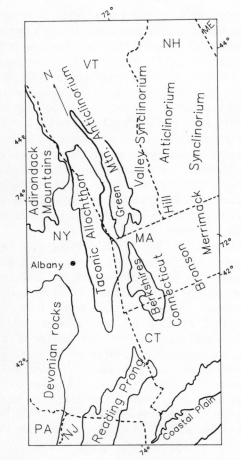

Fig. 6.12. Map of western New England, showing important geological features of the Paleozoic orogenies in this portion of Appalachia. The Adirondack, Berkshire, and Green Mountains as well as the Reading Prong are exposures of mid-Proterozoic basement cropping out amid late-Precambrian to late-Paleozoic sedimentary rocks.

to which overlap has occurred. Also shown in Fig. 6.12 are several of the significant geological features of the orogen; the Merrimack Synclinorium, the Bronson Hill Anticlinorium, the Connecticut Valley Synclinorium, the Green Mountain Anticlinorium, and exposures of Grenville aged basement in the Reading Prong, Berkshires, Green Mountains, and Adirondack Mountains. Metamorphism in New York and western Vermont is generally lower than garnet grade, and, with the exception of a low grade belt coincident with the Connecticut River, metamorphism throughout the rest of the area shown in Fig. 6.12 is garnet grade or higher. Barrovian style metamorphism is characteristic of western New England whereas Buchan style is prevalent to the east of the Bronson Hill Anticlinorium.

Dallmeyer et al. (1975) analyzed hornblende and biotite separates from the Grenville basement gneiss units of the Reading Prong in adjacent parts of New Jersey and New York (Fig. 6.12). They were interested in using the age spectrum approach to assess whether the pattern of K-Ar mineral ages found (i.e., hornblende about 900 Ma, biotite about 800 Ma) reflects cooling from the ~1.1 Ga Grenville Orogeny, or if these ages had perhaps been affected by Palaeozoic thermal disturbances. On the basis of essentially flat release patterns for the biotite, Dallmeyer et al. (1975) argued that this mitigated against the latter hypothesis in that characteristic staircase, episodic loss patterns were not observed. However, subsequent studies have cast doubt on the general ability of biotite to reveal diffusion gradients during vacuum extraction (see Section 4.12.5), making them less than ideal candidates for monitoring later disturbances. Curiously, coexisting altered amphiboles gave age spectra that tended to correspond to a simple model of argon loss during one or more of the Paleozoic orogenies impressed upon the Appalachians. Although the authors ruled out the likelihood of the rocks responding to a Paleozoic thermal pulse, the data suggest at least a low-grade alteration must have occurred at that time.

In a follow-up paper, Dallmeyer and Sutter (1976) investigated similar rocks of the Reading Prong, further to the northeast, in New York State, where variable retrogression was observed (Fig. 6.12). Several analyses of biotite and hornblende strengthened their view that the Late Proterozoic hornblende and biotite ages of about 900 and 800 Ma, respectively, resulted from slow cooling following the Grenville Orogeny. However, in the easternmost part of the Reading Prong, they recognized hornblende plateau ages of 420–440 Ma together with flat biotite age spectra at 380–390 Ma. They interpreted these ages as resulting from retrograde alteration during the Ordovician Taconic Orogeny, although the biotite ages likely represent cooling from the subsequent early Devonian Acadian Orogeny (e.g., Naylor, 1971). Lanphere and Albee (1974) and Sutter et al. (1985) used $^{40}Ar/^{39}Ar$ analyses of metamorphic minerals to investigate these overprinting events throughout western New England. Through $^{40}Ar/^{39}Ar$ hornblende analyses, Sutter et al. (1985) revealed that metamorphic recrystallization in western New England peaked at 465 ± 5 Ma, and that three separate metamorphic-structural domains of Taconic age could be recognized. They also delineated the western extent of Acadian overprinting of hornblende. However, $^{40}Ar/^{39}Ar$ biotite ages within the deformed Paleozoic rocks yielded ages of about 380 Ma, suggesting that Acadian unroofing/retrogression influenced rocks even further to the west.

Harrison et al. (1988) used $^{40}Ar/^{39}Ar$ analyses of hornblende, muscovite, biotite, and K-feldspar to study the post-Acadian unroofing in central New England. They found that following thrusting between 410 and 400 Ma ago, unroofing occurred in a complex but ultimately revealing fashion. Ages for all of these mineral systems revealed monotonically decreasing ages from the Merrimack Synclinorium to the Bronson Hill Anticlinorium. An abrupt increase in age was observed at the Vermont–New Hampshire border, including a narrow belt of rocks found to have experienced a Silurian metamorphism. These data argue strongly for a prolonged (200 Ma) hinged uplift of western New Hampshire about an axis in the Merrimack Synclinorium. In contrast, Vermont was unroofed rapidly, with most isostatic compensation occurring within about 50 Ma following thrusting.

Taken together, these studies reveal the distribution of metamorphism from the various collisional events and resolve the nature of postdeformation unroofing. Figure 6.13 shows the same area as Fig. 6.12 with $^{40}Ar/^{39}Ar$ age contours superimposed. West of the Bronson Hills, the contours are hornblende plateau ages whereas to the east, the contours are $^{40}Ar/^{39}Ar$ muscovite ages. The choice of muscovite for this portion of the map area results from the abundance of these data compared to hornblende. The western limit of Acadian overprinting is shown by a smooth north–

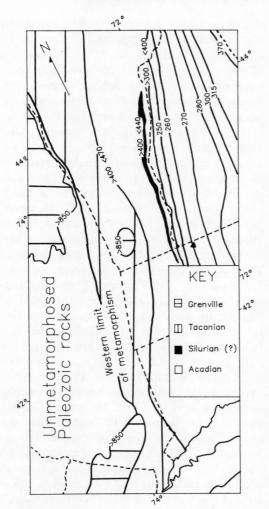

Fig. 6.13. Age contours from $^{40}Ar/^{39}Ar$ age spectrum results from a variety of studies (see text). These results reveal the areal extent of thermal metamorphism from each of three successive orogenic events. Rocks known to be affected by Alleghenian age deformation are shown as triangles.

south contour. Within the purely Taconic zone, windows of Precambrian cooling ages are preserved in the Reading Prong and southern Green Mountains. The belt of Silurian ages mentioned earlier is shown in black. Alleghenian metamorphism appears to be restricted to small areas, shown as triangles.

The geochronologic results for all minerals analyzed can be combined with closure temperature data to reconstruct the depth of the present exposure within the Earth at various times. Through this analysis, quantitative estimates of uplift rates can be made that ultimately bear on the mechanical behavior of the crust. Harrison et al. (1988) found that the prolonged and systematic post-Acadian uplift spanned almost 200 Ma. Because this duration is many times longer than the thermal relaxation time for any heat flow perturbation, isostatic rebound following erosion remains as the only mechanism to explain these results. Although it is well recognized that the stress distribution in the mantle lithosphere underlying an orogenic belt may take hundreds of millions of years to decay (e.g., Stephenson and Lambeck, 1985), thermochronometric studies are uniquely able to provide quantitative constraints on this problem. Central to this interpretation is the existence of a mountain range of initially Alpine elevation along the Bronson Hill Anticlinorium decaying to current Appalachian proportions with a time constant of about 100 Ma, throughout the Devonian and much of the Carboniferous.

These comprehensive results delineating the thermal extent of the successive Paleozoic orogenies of Appalachia demonstrate both the usefulness of the $^{40}Ar/^{39}Ar$ age spectrum approach in elucidating polymetamorphic histories and its potential for understanding the mechanisms behind orogeny.

6.5 GEOCHRONOLOGY AND PALEOMAGNETISM

The concept of blocking temperature was introduced into rock magnetism studies by Néel (1949, 1955). Similar ideas were developed in relation to geochronology by Armstrong (1966), Harper (1967), and Jäger et al. (1967) to help explain patterns of mineral ages in orogenic terranes. As discussed in Section 5.7, Dodson (1973) formally defined the term closure temperature for an isotope system in a cooling rock body, providing physical significance to the blocking temperature concept in the context of geochronology.

In rocks cooling from high temperature, an assemblage of ferromagnetic grains can acquire a magnetization parallel to the prevailing magnetic field as the temperature decreases below the Curie point for that mineral. During cooling a temperature known as the blocking temperature will be reached at which the magnetization becomes locked in, as illustrated in Fig. 6.14, after Dodson and McClelland-Brown (1985). Of particular importance, however, is that a magnetic mineral does not have a unique blocking temperature; this varies widely with the shape, size, and chemistry of the grains within a rock sample. Thus, a sample may record within its magnetic minerals a history of the direction of the external magnetic field over an extended period of time (Fig. 6.14), and subsequent thermal events may cause partial or complete overprinting of the earlier magnetic record.

As magnetization is a thermally activated process, Arrhenius-type relationships apply. Indeed, the equation derived by Dodson (1973) for closure temperature in a cooling isotopic system is directly applicable to the blocking temperature of magnetic minerals. Thus, after Dodson and McClelland-Brown (1980, 1985) [see also York (1978a,b)]

$$(6.1) \qquad T_B = \frac{E}{R \ln(A\tau k_0)}$$

where E approximates the activation energy, R is the gas constant, T_B is the blocking temperature, A is a parameter relating to the geometry of the grains, k_0 is the frequency factor, and τ is a characteristic cooling time constant, the time required for the value of $\exp(-E/RT_B)$ to decrease by a factor of e (2.718...) in the transitional temperature range.

The application of this equation to paleomagnetic studies is rather more complicated than in the case of isotopic systems, because of the variability of blocking temperatures, and the fact that the activation energy is itself a function of temperature. In addition, magnetization may be carried in a rock in both single and multidomain grains, as well as in different mineral phases. The reader is referred to Dodson and McClelland-Brown (1980, 1985) for more detailed discussion of these matters. In addition, the question of relating laboratory-determined estimates of magnetic blocking temperatures to geologically realistic values has received a good deal of attention (Pullaiah et al., 1975; Dunlop and Hale, 1977; McFadden, 1977; McClelland Brown, 1981).

In the case of rapidly cooled igneous rocks, such as lavas or shallow intrusives, it has been common practice over the last 25 years to combine geochronological data with paleomagnetic results from the same rocks, to document the changes of the geomagnetic field with time. The direction of primary thermoremanent magnetization normally can be determined readily in igneous rocks not subsequently reheated, and an isotopic age on the same rock is likely to provide a good measure

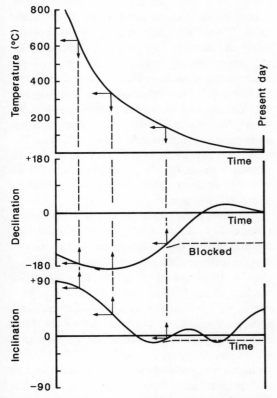

Fig. 6.14. Diagrammatic representation of the concepts of magnetic blocking temperature in a cooling system. After Dodson and McClelland-Brown (1985). The top curve shows cooling of a rock body from high temperature, and the lower curves show the variation of declination and inclination of the magnetic field during the cooling to ambient temperature. During cooling, grains with successively lower blocking temperatures become blocked, recording the two components of the magnetic field at the time. The blocking behavior for three grains is shown. For one blocking temperature, the horizontal dashed lines illustrate how the direction of magnetization may differ considerably from that of the magnetic field at the present time. Reproduced by permission of the Geological Society, London, from "Isotopic and palaeomagnetic evidence for rates of cooling, uplift and erosion" by Dodson and McClelland-Brown, *The Chronology of the Geological Record*, Memoir 10, 1985.

of the time since the magnetization was acquired. This type of approach was very successful in delineating the geomagnetic polarity time scale for the last 5 Ma of geological time (for summaries see McDougall, 1979; Maniken and Dalrymple, 1979). Similarly, numerical age calibration of apparent polar wander paths commonly has been attempted in relation to studies of lithospheric plate motions and continental drift (McElhinny, 1973).

It is now widely recognized from detailed paleomagnetic studies that plutonic igneous and metamorphic rocks, especially those from orogenic terranes, may record several different components of remanent magnetization. Techniques of measurement and analysis have been developed to assist in isolating the component magnetizations and their relative ages. Clearly, it is also desirable to provide numerical age control for the various magnetic directions derived from the paleomagnetic analysis. Thus, it is useful to combine age data and isotopic closure temperature information with magnetic blocking temperature estimates from the same rock body to gain a more complete picture of the variation of magnetization directions through time. Attempts to do just this have been documented in several papers by workers from Toronto (Buchan et al., 1977; Berger et al., 1979; Berger and York, 1979, 1981a,b). They have studied rocks from Precambrian shield areas in Canada that have experienced a rather complex history, using ^{40}Ar/^{39}Ar age spectra and detailed paleomagnetic analysis techniques.

In each of these studies more than one direction of magnetization was identified and estimates of geological blocking temperatures were made; the latter have physical significance provided that the magnetizations are thermoremanent rather than chemical in origin. The results from the Thanet gabbro, Ontario, however, were equivocal, in part because of the difficulty of distinguishing between thermoremanent and chemical remanent magnetizations (Berger and York, 1981b). Greater success was claimed by Berger et al. (1979) and Berger and York (1981a) in matching magnetizations and isotopic ages determined on several mafic igneous intrusions, subsequently metamorphosed, from the Haliburton Highlands within the Grenville Province of the Canadian Shield. These rocks were metamorphosed at upper amphibolite to granulite facies, at temperatures of the order 600–700°C, so that cooling, following the peak of the metamorphism, would likely result in progressive closure of the various isotopic and magnetic systems. Using ^{40}Ar/^{39}Ar dating of hornblende, biotite, plagioclase, and alkali feldspar, Berger et al. (1979) and Berger and York (1981a) derived a generalized cooling curve for the terrane. This cooling curve was obtained mainly using closure temperatures calculated from ^{40}Ar/^{39}Ar step heating data on the minerals themselves, a procedure about which we have strong reservations for hydrous minerals such as hornblende and biotite (see Section 5.9.2). Nevertheless, there is little doubt that the cooling curve for the orogen has the form given, using appropriate closure temperatures for these minerals (Fig. 6.15). On this basis, the particular segment of the Grenville Province under study probably cooled from $\sim 550°C$ at about 1000 Ma ago to $\sim 200°C$ by about 800 Ma ago.

Paleomagnetic data on these Haliburton Highlands rocks were interpreted as indicating the presence of three distinct natural remanent magnetizations that yield paleopoles extending over $\sim 60°$ of arc (Buchan and Dunlop, 1976; Buchan et al., 1977). From laboratory experiments, Berger et al. (1979) were able to estimate geological blocking temperatures in the 520–650°C range for the presumed earliest preserved magnetization, and $\sim 250°C$ for the youngest magnetization. Combining the age and paleomagnetic data, they proposed that the highest temperature magnetic component was acquired at 900 ± 10 Ma ago, as the closure temperature for the hornblende K–Ar system and the blocking temperature for the magnetization are in the same general range. The paleopole corresponding to this magnetization is situated at 142°E, 40°S (Buchan and Dunlop, 1976). By similar reasoning the magnetic component with the lowest blocking temperature was estimated to have become locked in at about 820 Ma ago on the basis of the feldspar ages. The paleopole for this component is located at 172°E, 24.5°N (Buchan and Dunlop, 1976), about 60° of arc from the

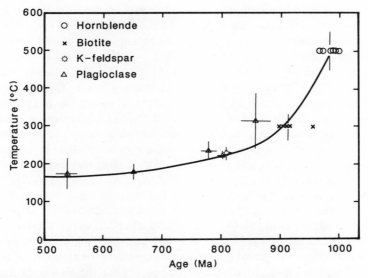

Fig. 6.15. Cooling curve for the Haliburton Highlands, Grenville Province, Ontario, derived from $^{40}Ar/^{39}Ar$ geochronological data. After Berger and York (1981a), with permission. Ages are plotted against estimated average closure temperatures for argon in the case of hornblende and biotite with notional uncertainties ($\pm 50°C$) indicated. For K-feldspar and plagioclase, closure temperature estimates are as reported by Berger and York (1981a), with uncertainties at the level of one standard deviation.

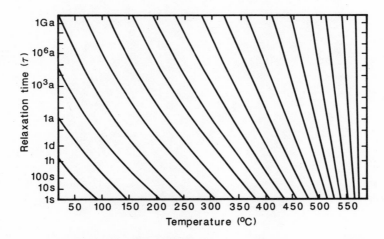

Fig. 6.16. Theoretical magnetic blocking curves for magnetite (Curie temperature = 580°C), relative to the relaxation time. From Pullaiah et al. (1975) and Dunlop and Hale (1977). For the temperature-time conditions illustrated, grains that plot to the left of their respective curves are magnetically blocked, whereas those plotting to the right are unblocked and do not contribute to the remanence. As an example, particles with a blocking temperature of ~400°C, as determined in a laboratory stepwise thermal demagnetization experiment of duration, say, 300 s (= relaxation time), heated to 250°C for times exceeding 10^4 years will be unblocked, acquiring a new magnetization on subsequent cooling.

paleopole previously discussed. The paleopoles lie within an apparent polar wander path determined from rocks of the Grenville Province, and, thus, help to provide some numerical age control for the curve.

Although methods are available whereby laboratory-determined blocking temperatures are extrapolated to geological blocking temperatures (Pullaiah et al., 1975), the question of relaxation time of the magnetization and its relation to the characteristic time constant in isotopic dating systems has not been fully addressed. Because of the strong temperature dependence of spontaneous magnetization, the relaxation of a magnetic domain (that is the time constant of the decay from the magnetized state) changes dramatically as the magnetic blocking temperature decreases below the Curie temperature (Pullaiah et al., 1975; Dunlop and Hale, 1977). And for a given ferromagnetic mineral species a range of blocking temperatures is expected because of shape and magnetocrystalline anisotropy. A consequence of these factors is that magnetization acquired at a temperature significantly below that of the Curie temperature has a strikingly different response to a later thermal excursion, compared with the effect of the same thermal event on the loss of radiogenic argon by diffusion from, say, biotite. Consider the case of a rock sample in which magnetite grains become magnetically blocked at 400°C, followed shortly thereafter by the isotopic closure of the biotite for argon, cooling to an ambient temperature of 150°C. Under the latter conditions the magnetization would be stable for time scales exceeding the age of the Earth (Fig. 6.16).

If, however, a temperature increase of only 100°C occurred for, say, 10^4 years, the magnetization would be completely unblocked (Fig. 6.16), whereas the biotite would experience negligible argon loss in response to the heating pulse. This difference in behavior reflects, in effect, the temperature dependence of the activation energy for the decay of magnetization in magnetic domains, and also is a consequence of the variability of magnetic blocking temperatures.

Despite the complexities, continued interdisciplinary research in geochronology and paleomagnetism on the same rocks holds considerable promise for the future in the elucidation and time calibration of apparent polar wander paths for segments of continental crust, especially for the Precambrian. The approach appears to be potentially the most promising way of providing the documentation necessary to map large scale motions of crustal segments of the Earth in the geological past.

REFERENCES

Adams, C. J. D., Harper, C. T., and Laird, M. G. (1975). K-Ar ages of low grade metasediments of the Greenland and Waiuta Groups in Westland and Buller, New Zealand. *N.Z. J. Geol. Geophys.* **18**, 39–48.

Albarede, F. (1978). The recovery of spatial isotope distributions from stepwise degassing data. *Earth Planet. Sci. Lett.* **39**, 387–397.

Albarède, F., Feraud, G., Kaneoka, I., and Allègre, C. J. (1978). ^{39}Ar-^{40}Ar dating: The importance of K-feldspars on multi-mineral data of polyorogenic areas. *J. Geol.* **86**, 581–598.

Albee, A. L., Burnett, D. S., Chodos, A. A., Eugster, O. J., Huneke, J. C., Papanastassiou, D. A., Podosek, F. A., Price Russ II, G., Sanz, H. G., Tera, F., and Wasserburg, G. J. (1970). Ages, irradiation history, and chemical composition of lunar rocks from the Sea of Tranquillity. *Science* **167**, 463–466.

Albee, A. L., Gancarz, A. J., and Chodos, A. A. (1973). Metamorphism of Apollo 16 and 17 and Luna 20 metaclastic rocks at about 3.95 AE: Samples 61156, 64424, 14-2, 65015, 67483, 15-2, 76055, 22006 and 22007. *Geochim. Cosmochim. Acta* Suppl. **4** (Proceedings of the Fourth Lunar Science Conference), 569–595.

Aldrich, L. T., and Nier, A. O. (1948). Argon 40 in potassium minerals. *Phys. Rev.* **74**, 876–877.

Alexander, E. C., Jr., and Davis, P. K. (1974). ^{40}Ar-^{39}Ar ages and trace element contents of Apollo 14 breccias; an interlaboratory cross-calibration of ^{40}Ar-^{39}Ar standards. *Geochim. Cosmochim. Acta* **38**, 911–928.

Alexander, E. C., Jr., Davis, P. K., and Reynolds, J. H. (1972). Rare-gas analyses on neutron irradiated Apollo 12 samples. *Geochim. Cosmochim. Acta* Suppl. **3** (Proceedings of the Third Lunar Science Conference), 1787–1795.

Alexander, E. C., Jr., Mickelson, G. M., and Lanphere, M. A. (1978). MMhb-1: A new ^{40}Ar-^{39}Ar dating standard. *U.S. Geol. Surv. Open-File Rept.* **78-701**, 6–8.

Amirkhanov, Kh. I., Bartnitskii, E. N., Brandt, S. B., and Voitkevich, G. V. (1960). The migration of argon and helium in certain rocks and minerals. *Acad. Nauk. S.S.S.R. Dokl.* **126**, 394–395.

Aprub, S. V., Levskiy, L. K., and Fedorova, I. V. (1973). Discordant age values and the reconstruction of the thermal rock evolution in the south-western Pamirs. *Akad. Nauk. S.S.S.R. Izv. Ser. Geol.* no. 3 for 1973, 3–16.

Armstrong, R. L. (1966). K-Ar dating of plutonic and volcanic rocks in orogenic belts. In *Potassium-argon dating* (ed. O. A. Schaeffer and J. Zähringer), pp. 117–133. Springer-Verlag, New York.

Armstrong, R. L., Jäger, E., and Eberhardt, P. (1966). A comparison of K-Ar and Rb-Sr ages on Alpine biotites. *Earth Planet. Sci. Lett.* **1**, 13–19.

Aston, F. W. (1921). The mass spectra of the alkali metals. *Phil. Mag. Ser. 6* **42**, 436–441.

Baksi, A. K. (1973). Quantitative unspiked argon runs in K-Ar dating. *Can. J. Earth Sci.* **10**, 1415–1419.

———. (1974). Isotopic fractionation of a loosely held atmospheric argon component in the Picture Gorge Basalts. *Earth Planet. Sci. Lett.* **21**, 431–438.

Baksi, A. K., and Farrar, E. (1973). The orifice correction for K-Ar dating. *Can. J. Earth Sci.* **10**, 1410–1414.

Baksi, A. K., York, D., and Watkins, N. D. (1967). Age of the Steens Mountain geomagnetic polarity transition. *J. Geophys. Res.* **72**, 6299–6308.

Bandy, O. L., and Arnal, R. E. (1969). Middle Tertiary basin development, San Joaquin Valley, California. *Geol. Soc. Am. Bull.* **80**, 783–819.

Barnard, G. P. (1953). *Modern mass spectrometry*. Institute of Physics, London.

Bass, R., Häenni, H. P., Bonner, T. W., and Gabbard, F. (1961). Disintegration of K^{39} by fast neutrons. *Nuclear Phys.* **28**, 478–493.

Bass, R., Fanger, U., and Saleh, F. M. (1964). Cross sections for the reactions ^{39}K(n,p)^{39}A and ^{39}K(n,α)^{36}Cl. *Nuclear Phys.* **56**, 569–576.

Batyrmurzayev, A. S., and Voronovskiy, S. N. (1976). Comparative study of desorption and diffusion of radiogenic Ar in some minerals. In: *Aklual'nyye voprosy sovremennoy Geokhrona-lagii* (ed. A. P. Vinogradov), pp. 157–168. Moscow.

Batyrmurzayev, A. S., Salautidinova, B.Sh., and Gargatsev, I. O. (1971). On the radiogenic argon migration in the hornblendes. *Akad. Nauk. S.S.S.R. Izv. Ser. Geol.* no. 2 for 1971, 128–131.

Baur, H. (1980). Numerische Simulation und praktische Erprobung einer rotationssymmetrischen Ionenquelle fuer Gasmassenspektrometer. Ph.D. Diss. in Natural Science, Eidgenoessischen Technischen Hochschule, Zurich.

Beaty, D. W., and Albee, A. L. (1978). Comparative petrology and possible genetic relations among the Apollo 11 basalts. *Geochim. Cosmochim. Acta* Suppl. **10** (Proceedings of the Ninth Lunar and Planetary Science Conference), 359–463.

Beckinsale, R. D., and Gale, N. H. (1969). A reappraisal of the decay constants and branching ratio of ^{40}K. *Earth Planet. Sci. Lett.* **6**, 289–294.

Becquerel, H. (1896a). Sur les radiations émises par phosphorescence. *Compt. Rend. Acad. Sci. Paris* **122**, 420–421.

———. (1896b). Sur les radiations invisibles émises par les corps phosphorescents. *Compt. Rend. Acad. Sci. Paris* **122**, 501–503.

———. (1896c). Sur les radiations invisibles émises par les sels d'uranium. *Compt. Rend. Acad. Sci. Paris* **122**, 689–694.

Berger, G. W. (1975). ^{40}Ar/^{39}Ar step heating of thermally overprinted biotite, hornblende and potassium feldspar from Eldora, Colorado. *Earth Planet. Sci. Lett.* **26**, 387–408.

Berger, G. W., and York, D. (1970). Precision of the ^{40}Ar/^{39}Ar dating technique. *Earth Planet. Sci. Lett.* **9**, 39–44.

———. (1979). ^{40}Ar-^{39}Ar dating of multicomponent magnetizations in the Archean Shelly Lake granite, northwestern Ontario. *Can. J. Earth Sci.* **16**, 1933–1941.

——— (1981a). Geothermometry from ^{40}Ar/^{39}Ar dating experiments. *Geochim. Cosmochim. Acta* **45**, 795–811.

———. (1981b). ^{40}Ar/^{39}Ar dating of the Thanet gabbro, Ontario: Looking through the Grenvillian metamorphic veil and implications for paleomagnetism. *Can. J. Earth Sci.* **18**, 266–273.

Berger, G. W., York, D., and Dunlop, D. J. (1979). Calibration of Grenvillian palaeopoles by ^{40}Ar/^{39}Ar dating. *Nature (London)* **277**, 46–48.

Bernatowicz, T. J., and Fahey, A. J. (1986). Xe isotopic fractionation in a cathodeless glow discharge. *Geochim. Cosmochim. Acta* **50**, 445–452.

Bernatowicz, T. J., Hohenberg, C. M., Hudson, B., Kennedy, B. M., and Podosek, F. A. (1978). Argon ages for lunar breccias 14064 and 15405. *Geochim. Cosmochim. Acta* Suppl. **10** (Proceedings of the Ninth Lunar and Planetary Science Conference), 905–919.

Berry, R. F. and McDougall, I. (1986). Interpretation of ^{40}Ar/^{39}Ar and K/Ar dating evidence from the Aileu Formation, East Timor. *Chem. Geol. (Isot. Geosci. Sect.)* **59**, 43–58.

Binns, R. A. (1964). Zones of progressive regional metamorphism in the Willyama Complex, Broken Hill district, New South Wales. *Geol. Soc. Aust. J.* **11**, 283–330.

Bleakney, W. (1930). Ionization potentials and probabilities for the formation of multiply charged ions in helium, neon and argon. *Phys. Rev.* **36**, 1303–1308.

Bogard, D. D., Husain, L., and Wright, R. J. (1976). ^{40}Ar-^{39}Ar dating of collisional events in chondrite parent bodies. *J. Geophys. Res.* **81**, 5664–5678.

Boltwood, B. B. (1907). On the ultimate disintegration products of the radio-active elements. Part II. The disintegration products of uranium. *Am. J. Sci.* **173**, 77–88.

Bottomley, R. J., and York, D. (1976). ^{40}Ar-^{39}Ar age determinations on the Owyhee basalt of the Columbia Plateau. *Earth Planet. Sci. Lett.* **31**, 75–84.

Boyce, W. E., and DiPrima, R. C. (1965). *Elementary differential equations and boundary value problems.* Wiley, New York.

Bramley, A. (1937). The potassium-argon transformation. *Science* **86**, 424–425.

Brandt, S. B., and Voronovskiy, S. N. (1967). Dehydration and diffusion of radiogenic argon in micas. *Int. Geol. Rev.* **9**, 1504–1507.

Brandt, S. B., Smirnov, V. N., Lapides, I. L., Volkova, N. V., and Kovalenko, V. I. (1967). Radiogenic argon as geochemical indicator of hydrothermal stability of some minerals. *Geochem. Int.* **4**, 826–829.

Brereton, N. R. (1970). Corrections for interfering isotopes in the ^{40}Ar/^{39}Ar dating method. *Earth Planet. Sci. Lett.* **8**, 427–433.

———. (1972). A reappraisal of the ^{40}Ar/^{39}Ar stepwise degassing technique. *Geophys. J. R. Astron. Soc.* **27**, 449–478.

Brereton, N. R., Hooker, P. J., and Miller, J. A. (1976). Some conventional potassium-argon and ^{40}Ar/^{39}Ar age studies of glauconite. *Geol. Mag.* **113**, 329–340.

Brett, R., Butler, P., Jr., Meyer, C., Jr., Reid, A. M., Takeda, H., and Williams, R. (1971). Apollo 12 igneous rocks 12004, 12008, 12009, and 12022: A mineralogical and petrological study. *Geochim. Cosmochim. Acta* Suppl. **2** (Proceedings of the Second Lunar Science Conference), 301–317.

Brewer, A. K. (1935). Further evidence for the existence of K^{40}. *Phys. Rev.* **48**, 640.

Brewer, M. S. (1969). Excess radiogenic argon in metamorphic micas from the Eastern Alps, Austria. *Earth Planet. Sci. Lett.* **6**, 321–331.

Brookins, D. G., Register, J. K., and Krueger, H. W.

(1980). Potassium-argon dating of polyhalite in southeastern New Mexico. *Geochim. Cosmochim. Acta* **44**, 635–637.

Brooks, C., Hart, S. R., and Wendt, I. (1972). Realistic use of two-error regression treatments as applied to rubidium-strontium data. *Rev. Geophys. Space Phys.* **10**, 551–577.

Brown, W. L., (ed.) (1984). *Feldspars and feldspathoids: Structures, properties and occurrences.* Reidel, Dordrecht. (NATO ASI series C, Maths. Phys. Sciences, Vol. 137.)

Buchan, K. L., and Dunlop, D. J. (1976). Paleomagnetism of the Haliburton intrusions: Superimposed magnetizations, metamorphism, and tectonics in the late Precambrian. *J. Geophys. Res.* **81**, 2951–2967.

Buchan, K. L., Berger, G. W., McWilliams, M. O., York, D., and Dunlop, D. J. (1977). Thermal overprinting of natural remanent magnetization and K/Ar ages in metamorphic rocks. *J. Geomag. Geoelectr.* **29**, 401–410.

Buckley, H. A., Bevan, J. C., Brown, K. M., Johnson, L. R., and Farmer, V. C. (1978). Glauconite and celadonite: Two separate mineral species. *Mineral. Mag.* **42**, 373–382.

Bukowinski, M. S. T. (1979). Theoretical estimate of compressional changes of decay constant of ^{40}K. *Geophys. Res. Lett.* **6**, 697–699.

Burbridge, E. M., Burbridge, G. R., Fowler, W. A., and Hoyle, F. (1957). Synthesis of the elements in stars. *Rev. Mod. Phys.* **29**, 547–650.

Burchfield, J. D. (1975). *Lord Kelvin and the age of the Earth.* Macmillan, London.

Burnett, D. S., Lippolt, H. J., and Wasserburg, G. J. (1966). The relative isotopic abundance of K^{40} in terrestrial and meteoritic samples. *J. Geophys. Res.* **71**, 1249–1269.

Campbell, N. (1908). The radioactivity of potassium, with special reference to solutions of its salts. *Proc. Camb. Phil. Soc.* **14**, 557–567.

Campbell, N. R., and Wood, A. (1906). The radioactivity of the alkali metals. *Proc. Camb. Phil. Soc.* **14**, 15–21.

Carlson, R. W., and Lugmair, G. W. (1981). Time and duration of lunar highlands crust formation. *Earth Planet. Sci. Lett.* **52**, 227–238.

Carslaw, H. S., and Jaeger, J. C. (1959). *Conduction of heat in solids*, 2nd ed. Clarendon, Oxford.

Cassignol, C., and Gillot, P.-Y. (1982). Range and effectiveness of unspiked potassium-argon dating: Experimental groundwork and applications. In *Numerical dating in stratigraphy* (ed. G. S. Odin), pp. 159–179. Wiley, Chichester.

Cerling, T. E., Brown, F. H., and Bowman, J. R. (1985). Low-temperature alteration of volcanic glass: Hydration, Na, K, ^{18}O and Ar mobility. *Isot. Geosci.* **52**, 281–293.

Chopin, C., and Maluski, H. (1980). ^{40}Ar-^{39}Ar dating of high pressure metamorphic micas from the Gran Paradiso area (Western Alps): Evidence against the blocking temperature concept. *Contrib. Mineral. Petrol.* **74**, 109–122.

Claesson, S., and Roddick, J. C. (1983). ^{40}Ar/^{39}Ar data on the age and metamorphism of the Ottfjället dolerites, Särv Nappe, Swedish Caledonides. *Lithos* **16**, 61–73.

Clague, D. A., Dalrymple, G. B., and Moberly, R. (1975). Petrography and K-Ar ages of dredged volcanic rocks from the western Hawaiian Ridge and the southern Emperor Seamount Chain. *Geol. Soc. Am. Bull.* **86**, 991–998.

Clauer, N., Giblin, P., and Lucas, J. (1984). Sr and Ar isotope studies of detrital smectites from the Atlantic Ocean (D.S.D.P., Legs 43, 48 and 50). *Isot. Geosci.* **2**, 141–151.

Coleman, R. G., and Lanphere, M. A. (1971). Distribution and age of high-grade blueschists, associated eclogites, and amphibolites from Oregon and California. *Geol. Soc. Am. Bull.* **82**, 2397–2412.

Compston, W., Chappell, B. W., Arriens, P. A., and Vernon, M. J. (1970). The chemistry and age of Apollo 11 lunar material. *Geochim. Cosmochim. Acta* Suppl. **1** (Proceedings of the Apollo 11 Lunar Science Conference), 1007–1027.

Compston, W., Williams, I. S., and Meyer, C. (1984). U-Pb geochronology of zircons from lunar breccia 73217 using a sensitive high mass-resolution ion microprobe. *J. Geophys. Res.* **89**, Suppl. B525–B534 (Proceedings of the Fourteenth Lunar and Planetary Science Conference).

Cox, A., and Dalrymple, G. B. (1967). Statistical analysis of geomagnetic reversal data and the precision of potassium-argon dating. *J. Geophys. Res.* **72**, 2603–2614.

Crank, J. (1975). *The mathematics of diffusion*, 2nd ed. Clarendon, Oxford.

Cross, W. G. (1951). Two-directional focusing of charged particles with a sector-shaped, uniform magnetic field. *Rev. Sci. Inst.* **22**, 717–722.

Dallmeyer, R. D. (1974). ^{40}Ar/^{39}Ar incremental release ages of biotite and hornblende from pre-Kenoran gneisses between the Matagami-Chibougamau and Frotet-Troilus greenstone belts, Quebec. *Can. J. Earth Sci.* **11**, 1586–1593.

———. (1975a). Incremental ^{40}Ar/^{39}Ar ages of biotite and hornblende from retrograded basement gneisses of the southern Blue Ridge: Their bearing on the age of Paleozoic metamorphism. *Am. J. Sci.* **275**, 444–460.

———. (1975b). ^{40}Ar/^{39}Ar ages of biotite and hornblende from a progressively remetamorphosed basement terrane: Their bearing on interpretation of release spectra. *Geochim. Cosmochim. Acta* **39**, 1655–1669.

———. (1978). $^{40}Ar/^{39}Ar$ incremental-release ages of hornblende and biotite across the Georgia Inner Piedmont: Their bearing on Late Paleozoic–Early Mesozoic tectonothermal history. *Am. J. Sci.* **278**, 124–149.

———. (1987). $^{40}Ar/^{39}Ar$ mineral age record of variably superimposed Proterozoic tectonothermal events in the Grenville Orogen, central Labrador. *Can. J. Earth Sci.* **24**, 314–333.

Dallmeyer, R. D., and Rivers, T. (1983). Recognition of extraneous argon components through incremental-release $^{40}Ar/^{39}Ar$ analysis of biotite and hornblende across the Grenvillian metamorphic gradient in southwestern Labrador. *Geochim. Cosmochim. Acta* **47**, 413–428.

Dallmeyer, R. D., and Sutter, J. F. (1976). $^{40}Ar/^{39}Ar$ incremental-release ages of biotite and hornblende from variably retrograded basement gneisses of the northeasternmost Reading Prong, New York: Their bearing on early Paleozoic metamorphic history. *Am. J. Sci.* **276**, 731–747.

Dallmeyer R. D., and Villeneuve M. (1987). $^{40}Ar/^{39}Ar$ mineral age record of polyphase tectonothermal evolution in the southern Mauritanide orogen, southeastern Senegal. *Geol. Soc. Am. Bull.* **98**, 602–611.

Dallmeyer, R. D., Sutter, J. F., and Baker, D. J. (1975). Incremental $^{40}Ar/^{39}Ar$ ages of biotite and hornblende from the northeastern Reading Prong: Their bearing on late Proterozoic thermal and tectonic history. *Geol. Soc. Am. Bull.* **86**, 1435–1443.

Dalrymple, G. B. (1964). Argon retention in a granitic xenolith from a Pleistocene basalt, Sierra Nevada, California. *Nature (London)* **201**, 282.

———. (1967). Potassium-argon ages of Recent rhyolites of the Mono and Inyo Craters, California. *Earth Planet. Sci. Lett.* **3**, 289–298.

———. (1969). $^{40}Ar/^{36}Ar$ analyses of historic lava flows. *Earth Planet. Sci. Lett.* **6**, 47–55.

Dalrymple, G. B., and Clague, D. A. (1976). Age of the Hawaiian-Emperor bend. *Earth Planet. Sci. Lett.* **31**, 313–329.

Dalrymple, G. B., and Lanphere, M. A. (1969). *Potassium-argon dating.* Freeman, San Francisco.

———. (1971). $^{40}Ar/^{39}Ar$ technique of K-Ar dating: A comparison with the conventional technique. *Earth Planet. Sci. Lett.* **12**, 300–308.

———. (1974). $^{40}Ar/^{39}Ar$ age spectra of some undisturbed terrestrial samples. *Geochim. Cosmochim. Acta* **38**, 715–738.

Dalrymple, G. B., and Moore, J. G. (1968). Argon-40: Excess in submarine pillow basalts from Kilauea Volcano, Hawaii. *Science* **161**, 1132–1135.

Dalrymple, G. B., Grommé, C. S., and White, R. W. (1975). Potassium-argon age and paleomagnetism of diabase dikes in Liberia: Initiation of central Atlantic rifting. *Geol. Soc. Am. Bull.* **86**, 399–411.

Dalrymple, G. B., Lanphere, M. A., and Clague, D. A. (1980). Conventional and $^{40}Ar/^{39}Ar$ K-Ar ages of volcanic rocks from the Ojin (site 430), Nintoku (site 432), and Suiko (site 433) seamounts and the chronology of volcanic propagation along the Hawaiian-Emperor chain. *Deep Sea Dril. Proj., Initial Rep.* **55**, 659–676.

Dalrymple, G. B., Alexander, E. C., Jr., Lanphere, M. A., and Kraker, G. P. (1981). Irradiation of samples for $^{40}Ar/^{39}Ar$ dating using the Geological Survey TRIGA reactor. *U.S. Geol. Surv., Prof. Paper* **1176**.

Damon, P. E. (1968). Potassium-argon dating of igneous and metamorphic rocks with applications to the Basin ranges of Arizona and Sonora. In *Radiometric dating for geologists* (ed. E. I. Hamilton and R. M. Farquhar), pp. 1–71. Interscience, London.

Damon, P. E., Laughlin, A. W., and Percious, J. K. (1967). Problem of excess argon-40 in volcanic rocks. In *Radioactive dating and methods of low-level counting*, pp. 463–481. International Atomic Energy Agency, Vienna.

Davies, J. A., Brown, F., and McCargo, M. R., Jr. (1963). Range of Xe^{133} and Ar^{41} ions of kiloelectron volt energies in aluminum. *Can. J. Phys.* **41**, 829–843.

Davy, H. (1808). The Bakerian Lecture, on some new phenomena of chemical changes produced by electricity, particularly the decomposition of the fixed alkalies, and the exhibition of the new substances which constitute their bases; and on the general nature of alkaline bodies. *R. Soc. Lond. Phil. Trans.* **98**, 1–44.

Deer, W. A., Howie, R. A., and Zussman, J. (1962; 1963). *Rock-forming minerals*, 5 volumes, Longmans, London.

de Laeter, J. R., Vernon, M. J., and Compston, W. (1973). Revision of lunar Rb-Sr ages. *Geochim. Cosmochim. Acta* **37**, 700–702.

DeLong, S. E., Mitchell, D. W., Cherniak, D., and Harrison, T. M. (1988). Z-axis oscillation sidebands in FT/ICR mass spectra. *Internat. J. Mass Spec. Ion Processes*, submitted.

Dodson, M. H. (1973). Closure temperature in cooling geochronological and petrological systems. *Contrib. Mineral. Petrol.* **40**, 259–274.

———. (1982). Closure profiles in cooling mineral grains. *Fifth Int. Conf. Geochron. Cosmochron.*

Isotope Geol., Nikko, Japan (Post-program abstract).

———. (1986). Closure profiles in cooling systems. In *Materials Science Forum*, Vol. 7, pp. 145–153. Trans Tech Publications, Aedermannsdorf, Switzerland.

Dodson, M. H., and McClelland-Brown, E. (1980). Magnetic blocking temperatures of single-domain grains during slow cooling. *J. Geophys. Res.* **85**, 2625–2637.

———. (1985). Isotopic and palaeomagnetic evidence for rates of cooling, uplift and erosion. In *The chronology of the geological record* (ed. N. J. Snelling), pp. 315–325. Geol. Soc. London, Memoir 10, Blackwell, Oxford.

Dodson, M. H., and Rex, D. C. (1971). Potassium-argon ages of slates and phyllites from south-west England. *Geol. Soc. Lond., Qt. J.* **126**, 465–499.

Dominik, B., and Jessberger, E. K. (1978). Early lunar differentiation: 4.42-AE-old plagioclase clasts in Apollo 16 breccia 67435. *Earth Planet. Sci. Lett.* **38**, 407–415.

Drake, R. E., Curtis, G. H., Cerling, T. E., Cerling, B. W., and Hampel, J. (1980). KBS Tuff dating and geochronology of tuffaceous sediments in the Koobi Fora and Shungura Formations, East Africa. *Nature (London)* **283**, 368–372.

Duckworth, H. E. (1958). *Mass spectroscopy*. Cambridge University Press, Cambridge.

Duckworth, H. E., Barber, R. C., and Venkatasubramanian, V. S. (1986). *Mass spectroscopy*, 2nd ed. Cambridge University Press, Cambridge.

Duncan, R. A., and Clague, D. A. (1985). Pacific plate motion recorded by linear volcanic chains. In *The ocean basins and margins* (ed. A. E. M. Nairn, F. G. Stehli, and S. Uyeda), Vol. 7A, pp. 89–121. Plenum, New York.

Duncan, R. A., and McDougall, I. (1976) Linear volcanism in French Polynesia. *J. Volcanol. Geotherm. Res.* **1**, 197–227.

Duncan, R. A., and Staudigel, H. (1986). K/Ar and Rb/Sr ages from celadonites and their application to dating seafloor hydrothermal circulation. *Terra Cognita* **6**, 214 (Abstract).

Dunham, K. C., Fitch, F. J., Ineson, P. R., Miller, J. A., and Mitchell, J. G. (1968). The geochronological significance of argon-40/argon-39 age determinations on White Whin from the northern Pennine orefield. *R. Soc. Lond. Proc.* **A307**, 251–266.

Dunlop, D. J., and Hale, C. J. (1977). Simulation of long-term changes in the magnetic signal of the oceanic crust. *Can. J. Earth Sci.* **14**, 716–744.

Dymond, J. (1970). Excess argon in submarine basalt pillows. *Geol. Soc. Am. Bull.* **81**, 1229–1232.

Dymond, J., and Hogan, L. (1973). Noble gas abundance patterns in deep-sea basalts—primordial gases from the mantle. *Earth Planet. Sci. Lett.* **20**, 131–139.

Eichelberger, J. C., Lysne, P. C., and Younker, L. W. (1984). Research drilling at Inyo Domes, Long Valley Caldera, California. *EOS Trans. Am. Geophys. Un.* **65**, 721–725.

Eichhorn, G., McGee, J. J., James, O. B., and Schaeffer, O. A. (1979). Consortium breccia 73255: Laser ^{39}Ar-^{40}Ar dating of aphanite samples. *Geochim. Cosmochim. Acta* Suppl. **11** (Proceedings of the Tenth Lunar and Planetary Science Conference), 763–788.

Engels, J. C., and Ingamells, C. O. (1970). Effect of sample inhomogeneity in K-Ar dating. *Geochim. Cosmochim. Acta* **34**, 1007–1017.

Evans, B. W. (1965). Application of reaction-rate method to the breakdown equilibria of muscovite plus quartz. *Am. J. Sci.* **263**, 647–667.

Evernden, J. F., and Curtis, G. H. (1965). The potassium-argon dating of late Cenozoic rocks in East Africa and Italy. *Curr. Anthropol.* **6**, 343–385.

Evernden, J. F., Curtis, G. H., Kistler, R. W., and Obradovich, J. (1960). Argon diffusion in glauconite, microcline, sanidine, leucite and phlogopite. *Am. J. Sci.* **258**, 583–604.

Evernden, J. F., Curtis, G. H., Obradovich, J., and Kistler, R. (1961). On the evaluation of glauconite and illite for dating sedimentary rocks by the potassium-argon method. *Geochim. Cosmochim. Acta* **23**, 78–99.

Farver, J. R., and Giletti, B. J. (1985). Oxygen diffusion in amphiboles. *Geochim. Cosmochim. Acta* **49**, 1403–1411.

Faure, G. (1977). *Principles of isotope geology*. Wiley, New York.

———. (1986). *Principles of isotope geology*, 2nd ed. Wiley, New York.

Fechtig, H., and Kalbitzer, S. (1966). The diffusion of argon in potassium-bearing solids. In *Potassium-argon dating* (ed. O. A. Schaeffer and J. Zähringer), pp. 68–107. Springer-Verlag, New York.

Féraud, G., Gastaud, J., Auzende, J.-M., Olivet, J.-L., and Cornen, G. (1982). ^{40}Ar/^{39}Ar ages for the alkaline volcanism and the basement of Gorringe Bank, North Atlantic Ocean. *Earth Planet. Sci. Lett.* **57**, 211–226.

Fick, A. (1855). Ueber diffusion. *Ann. Phys. Chem.* **94**, 59–86.

Fitch, F. J., Miller, J. A., and Mitchell, J. G. (1969). A new approach to radio-isotopic dating in orogenic belts. In *Time and place in orogeny* (eds. P. E. Kent, G. E. Satterthwaite, and A. M. Spencer), pp. 157–195. Geol. Soc. London Spec. Publ. 3.

Fitch, F. J., Hooker, P. J., Miller, J. A., and Brereton, N. R. (1978). Glauconite dating of Palaeocene-Eocene rocks from East Kent and the time-scale of Palaeogene volcanism in the North Atlantic region. *Geol. Soc. Lond. J.* **135**, 499–512.

Fleck, R. J., Sutter, J. F., and Elliot, D. H. (1977). Interpretation of discordant $^{40}Ar/^{39}Ar$ age-spectra of Mesozoic tholeiites from Antarctica. *Geochim. Cosmochim. Acta* **41**, 15–32.

Fleischer, R. L., Price, P. B., and Walker, R. M. (1975). *Nuclear tracks in solids.* University of California Press, Berkeley.

Flisch, M. (1982). Potassium-argon analysis. In *Numerical dating in stratigraphy* (ed. G. S. Odin), pp. 151–158. Wiley, Chichester.

Foland, K. A. (1974). ^{40}Ar diffusion in homogeneous orthoclase and an interpretation of Ar diffusion in K-feldspar. *Geochim. Cosmochim. Acta* **38**, 151–166.

———. (1983). $^{40}Ar/^{39}Ar$ incremental heating plateaus for biotites with excess argon. *Isot. Geosci.* **1**, 3–21.

Foland, K. A., Linder, J. S., Laskowski, T. E., and Grant, N. K. (1984). $^{40}Ar/^{39}Ar$ dating of glauconites: Measured ^{39}Ar recoil loss from well-crystallized specimens. *Isot. Geosci.* **2**, 241–264.

Fourier, J. B. J. (1820). Extrait d'un mémoire sur le refroidissement séculaire du globe terrestre. *Bull. Sci. Soc. Philomathique Paris Ser.* **3**, 58–70.

Fourier, J. B. J. (1822). *Theorie analytique de la chaleur* (English translation by A. Freeman). Dover, New York, 1955.

Frank, E., and Stettler, A. (1979). K-Ar and ^{39}Ar-^{40}Ar systematics of white K-mica from an Alpine metamorphic profile in the Swiss Alps. *Schweiz. Mineral. Petrog. Mitt.* **59**, 375–394.

Friedlander, G., Kennedy, J. W., Macias, E. S., and Miller, J. M. (1981). *Nuclear and radiochemistry*, 3rd ed., Wiley, New York.

Funkhouser, J. G., Barnes, I. L., and Naughton, J. J. (1966). Problems in the dating of volcanic rocks by the potassium-argon method. *Bull. Volcanol.* **29**, 709–716.

Funkhouser, J. G., Fisher, D. E., and Bonatti, E. (1968). Excess argon in deep-sea rocks. *Earth Planet. Sci. Lett.* **5**, 95–100.

Garner, E. L., Machlan, L. A., and Barnes, I. L. (1975a). The isotopic composition of lithium, potassium, and rubidium in some Apollo 11, 12, 14, 15, and 16 samples. *Geochim. Cosmochim. Acta* Suppl. **6** (Proceedings of the Sixth Lunar Science Conference), 1845–1855.

Garner, E. L., Murphy, T. J., Gramlich, J. W., Paulsen, P. J., and Barnes, I. L. (1975b). Absolute isotopic abundance ratios and the atomic weight of a reference sample of potassium. *J. Res. Natl. Bureau Stand.* **79A**, 713–725.

Gast, P. W., and Hubbard, N. J. (1970). Abundance of alkali metals, alkaline and rare earths, and rubidium-87/strontium-86 ratios in lunar samples. *Science* **167**, 485–487.

Geiss, J., Eberhardt, P., Grögler, N., Guggisberg, S., Maurer, P., and Stettler, A. (1977). Absolute time scale of lunar mare formation and filling. *R. Soc. Lond. Phil. Trans. Ser A* **285**, 151–158.

Gentner, W., and Zähringer, J. (1960). Das Kalium-Argon-Alter von Tektiten. *Z. Naturforsch.* **15a**, 93–99.

Gentner, W., Lippolt, H. J., and Schaeffer, O. A. (1963). Argonbestimmungen an Kaliummineralien-XI. Die Kalium-Argon-Alter der Gläser des Nördlinger Rieses und der böhmisch-mährischen Tektite. *Geochim. Cosmochim. Acta* **27**, 191–200.

Gerling, E. K., and Morozova, I. M. (1957). Determination of activation energy of argon liberation from micas. *Geochemistry* for 1957, 359–367.

Gerling, E. K., Kol'tsova, T. V., Petrov, B. V., and Zul'fikarova, Z. K. (1965). On the suitability of amphiboles for age determination by the K-Ar method. *Geochem. Int.* **2**, 148–154.

Gerling, E. K., Petrov, B. V., and Kol'tsova, T. V. (1966). A comparative study of the activation energy of argon liberation and dehydration energy in amphiboles and biotites. *Geochem. Int.* **3**, 295–305.

Gerling, E. K., Morozova, I. M., Sprintsson, V. D., and Vinogradov, D. P. (1969). Use of nepheline in K-Ar dating. *Geochem. Int.* **6**, 1105–1108.

Giletti, B. J. (1974a). Diffusion related to geochronology. In *Geochemical transport and kinetics* (ed. A. W. Hofmann, B. J. Giletti, H. S. Yoder, Jr., and R. A. Yund), pp. 61–76. Carnegie Inst. of Wash. Publ. 634.

———. (1974b). Studies in diffusion I: Argon in phlogopite mica. In *Geochemical transport and kinetics* (ed. A. W. Hofmann, B. J. Giletti, H. S. Yoder, Jr., and R. A. Yund), pp. 107–115. Carnegie Inst. of Wash. Publ. 634.

Giletti, B. J., and Nagy, K. L. (1981). Grain boundary diffusion of oxygen along lamellar boundaries in perthitic feldspar. *EOS, Trans. Am. Geophys. Un.* **62**, 428 (Abstract).

Giletti, B. J., and Tullis, J. (1977). Studies in diffusion. IV. Pressure dependence of Ar diffusion in phlogopite mica. *Earth Planet. Sci. Lett.* **35**, 180–183.

Gillespie, A. R. (1982). Quaternary glaciation and tectonism in the southeastern Sierra Nevada, Inyo County, California. Ph.D. Diss., California Institute of Technology, Pasadena.

Gillespie, A. R., Huneke, J. C., and Wasserburg, G. J. (1982). An assessment of ^{40}Ar-^{39}Ar dating of incompletely degassed xenoliths. *J. Geophys. Res.* **87**, 9247–9257.

———. (1983). Eruption age of a Pleistocene basalt from ^{40}Ar-^{39}Ar analysis of partially degassed xenoliths. *J. Geophys. Res.* **88**, 4997–5008.

———. (1984). Eruption age of ~100,000-year-old basalt from ^{40}Ar-^{39}Ar analysis of partially degassed xenoliths. *J. Geophys. Res.* **89**, 1033–1048.

Glass, B. P. (1982). *Introduction to planetary geology.* Cambridge University Press, Cambridge.

Glasstone, S., and Edlund, M. C. (1952). *The elements of nuclear reactor theory.* Van Nostrand, New York.

Glen, W. (1982). *The road to Jaramillo.* Stanford University Press, Stanford.

Goldsmith, J. R., and Laves, F. (1954). The microcline-sanidine stability relations. *Geochim. Cosmochim. Acta* **5**, 1–19.

Grasty, R. L., and Mitchell, J. G. (1966). Single sample potassium-argon ages using the omegatron. *Earth Planet. Sci. Lett.* **1**, 121–122.

Guggisberg, S., Eberhardt, P., Geiss, J., Grögler, N., Stettler, A., Brown, G. M., and Peckett, A. (1979). Classification of the Apollo-11 mare basalts according to Ar^{39}-Ar^{40} ages and petrological properties. *Geochim. Cosmochim. Acta* Suppl. 11 (Proceedings of the Tenth Lunar and Planetary Science Conference), 1–39.

Gulson, B. L. (1984). Uranium-lead and lead-lead investigations of minerals from the Broken Hill lodes and mine sequence rocks. *Econ. Geol.* **79**, 476–490.

Hall, C. M., and York, D. (1978). K-Ar and ^{40}Ar/^{39}Ar age of the Laschamp geomagnetic polarity reversal. *Nature (London)* **274**, 462–464.

Hallam, A. (1983). *Great geological controversies.* Oxford University Press, Oxford.

Halliday, A. N. (1978). ^{40}Ar-^{39}Ar stepheating studies of clay concentrates from Irish orebodies. *Geochim. Cosmochim. Acta* **42**, 1851–1858.

Hanes, J. A., York, D., and Hall, C. M. (1985). An ^{40}Ar/^{39}Ar geochronological and electron microprobe investigation of an Archean pyroxenite and its bearing on ancient atmospheric compositions. *Can. J. Earth Sci.* **22**, 947–958.

Hanson, G. N. (1971). Radiogenic argon loss from biotites in whole rock heating experiments. *Geochim. Cosmochim. Acta* **35**, 101–107.

Hanson, G. N., and Gast, P. W. (1967). Kinetic studies in contact metamorphic zones. *Geochim. Cosmochim. Acta* **31**, 1119–1153.

Hanson, G. N., Simmons, K. R., and Bence, A. E. (1975). ^{40}Ar/^{39}Ar spectrum ages for biotite, hornblende and muscovite in a contact metamorphic zone. *Geochim. Cosmochim. Acta* **39**, 1269–1277.

Harper, C. T. (1964). Potassium-argon ages of slates and their geological significance. *Nature (London)* **203**, 468–470.

———. (1966). Potassium-argon ages of slates from the southern Caledonides of the British Isles. *Nature (London)* **212**, 1339–1341.

———. (1967). The geological interpretation of potassium-argon ages of metamorphic rocks from the Scottish Caledonides. *Scott. J. Geol.* **3**, 46–66.

———. (1968). Isotopic ages from the Appalachians and their tectonic significance. *Can. J. Earth Sci.* **5**, 49–59.

Harrison, T. M. (1980). Thermal histories from the ^{40}Ar/^{39}Ar age spectrum method. Ph.D. Diss., Australian National University, Canberra.

———. (1981). Diffusion of ^{40}Ar in hornblende. *Contrib. Mineral. Petrol.* **78**, 324–331.

———. (1983). Some observations on the interpretation of ^{40}Ar/^{39}Ar age spectra. *Isot. Geosci.* **1**, 319–338.

———. (1987). Comment on "Kelvin and the age of the earth." *J. Geol.* **95**, 725–727.

Harrison, T. M., and Bé, K. (1983). ^{40}Ar/^{39}Ar age spectrum analysis of detrital microclines from the southern San Joaquin Basin, California: An approach to determining the thermal evolution of sedimentary basins. *Earth Planet. Sci. Lett.* **64**, 244–256.

Harrison, T. M., and Burke, K. (1988). ^{40}Ar/^{39}Ar thermochronology of sedimentary basins using detrital K-feldspar: Examples from the San Joaquin Valley, California, Rio Grande Rift, New Mexico and North Sea. In *Thermal histories of sedimentary basins* (ed. N. Naeser, and T. McCulloh) Springer-Verlag.

Harrison, T. M., and Clarke, G. K. C. (1979). A model of the thermal effects of igneous intrusion and uplift as applied to Quottoon Pluton, British Columbia. *Can. J. Earth Sci.* **16**, 411–420.

Harrison, T. M., and Fitz Gerald, J. D. (1986). Exsolution in hornblende and its consequences for ^{40}Ar/^{39}Ar age spectra and closure temperature. *Geochim. Cosmochim. Acta* **50**, 247–253.

Harrison, T. M., and McDougall, I. (1980a). Investigations of an intrusive contact, northwest Nelson, New Zealand-I. Thermal, chronological and isotopic constraints. *Geochim. Cosmochim. Acta* **44**, 1985–2003.

———. (1980b). Investigations of an intrusive contact, northwest Nelson, New Zealand-II. Diffusion of radiogenic and excess ^{40}Ar in hornblende revealed by ^{40}Ar/^{39}Ar age spectrum analysis. *Geochim. Cosmochim. Acta* **44**, 2005–2020.

———. (1981). Excess ^{40}Ar in metamorphic rocks from Broken Hill, New South Wales: Implications for ^{40}Ar/^{39}Ar age spectra and the thermal history of the region. *Earth Planet. Sci. Lett.* **55**, 123–149.

———. (1982). The thermal significance of potassium feldspar K-Ar ages inferred from ^{40}Ar/^{39}Ar age spectrum results. *Geochim. Cosmochim. Acta* **46**, 1811–1820.

Harrison, T. M., and Wang, S. (1981). Further ^{40}Ar/^{39}Ar evidence for the multi-collisional heating of the Kirin meteorite. *Geochim. Cosmochim. Acta* **45**, 2513–2517.

Harrison, T. M., and Watson, E. B. (1983). Kinetics of zircon dissolution and zirconium diffusion in granitic melts of variable water content. *Contrib. Mineral. Petrol.* **84**, 66–72.

Harrison, T. M., Duncan, I., and McDougall, I. (1985). Diffusion of ^{40}Ar in biotite: Temperature, pressure and compositional effects. *Geochim. Cosmochim. Acta* **49**, 2461–2468.

Harrison, T. M., Morgan, P., and Blackwell, D. D. (1986). Constraints on the age of heating at the Fenton Hill Site, Valles Caldera, New Mexico. *J. Geophys. Res.* **91**, 1899–1908.

Harrison T. M., Spear F. S., and Heizler M. T. (1988). Geochronologic studies in central New England, II: Post-Acadian hinged and differential uplift. *Geology*, submitted.

Hart, S. R. (1960). Some diffusion measurements relating to the K-Ar dating method. In *Variations in isotopic abundances of Sr, Ca, and Ar and related topics*, pp. 87–129. USAEC Contract AT(30-1)-1381. NYO-3941, 8th Ann. Rept. for 1960, Department of Geology and Geophysics, Massachusetts Institute of Technology, Cambridge.

———. (1961). The use of hornblendes and pyroxenes for K-Ar dating. *J. Geophys. Res.* **66**, 2995–3001.

———. (1964). The petrology and isotopic-mineral age relations of a contact zone in the Front Range, Colorado. *J. Geol.* **72**, 493–525.

———. (1981). Diffusion compensation in natural silicates. *Geochim. Cosmochim. Acta* **45**, 279–291.

Hart, S. R., and Dodd, R. T. (1962). Excess radiogenic argon in pyroxenes. *J. Geophys. Res.* **67**, 2998–2999.

Harvey, B. G. (1962). *Introduction to nuclear physics and chemistry*. Prentice-Hall, Englewood Cliffs, New Jersey.

Hayatsu, A., and Waboso, C. E. (1985). The solubility of rare gases in silicate melts and implications for K-Ar dating. *Chem. Geol.* **52**, 97–102.

Heaviside, O. (1899). *Electromagnetic theory*, Vol. II. Van Nostrand, New York.

Heizler, M. T., and Harrison, T. M. (1988). Multiple trapped argon isotope components revealed by ^{40}Ar/^{39}Ar isochron analysis. *Geochim. Cosmochim. Acta*, **52**, 1295–1303.

Henshaw, D. G. (1958). Atomic distribution in liquid and solid neon and solid argon by neutron diffraction. *Phys. Rev.* **111**, 1470–1475.

Herzog, R. (1934). Ionen- und electronenoptische Zylinderlinsen und Prismen. I. *Zeitschr. Phys.* **89**, 447–473.

Hess, J. C., and Lippolt, H. J. (1986). ^{40}Ar/^{39}Ar ages of tonstein and tuff sanidines: New calibration points for the improvement of the Upper Carboniferous time scale. *Chem. Geol. (Isot. Geosci. Sect.)* **59**, 143–154.

Hevesy, G. v. (1935). Die Radioaktivität des Kaliums. *Naturwissenschaften* **34**, 583–585.

Heymann, D. (1967). On the origin of hypersthene chondrites: Ages and shock effects of black chondrites. *Icarus* **6**, 189–221.

Hickman, B. S. (1958). Reactor irradiation techniques. In *Australian Atomic Energy Symposium - 1958-*, pp. 726–730. Melbourne University Press, Melbourne.

Hiyagon, H., and Ozima, M. (1982). Noble gas distribution between basalt melt and crystals. *Earth Planet. Sci. Lett.* **58**, 255–264.

———. (1986). Partition of noble gases between olivine and basalt melt. *Geochim. Cosmochim. Acta* **50**, 2045–2057.

Hohenberg, C. M. (1980). High sensitivity pulse-counting mass spectrometer system for noble gas analysis. *Rev. Sci. Instrum.* **51**, 1075–1082.

Hohenberg, C. M., Hudson, B., Kennedy, B. M., and Podosek, F. A. (1981). Noble gas retention chronologies for the St. Séverin meteorite. *Geochim. Cosmochim. Acta* **45**, 535–546.

Holmes, A. (1928). Radioactivity and earth movements. *Trans. Geol. Soc. Glasgow* **18**, 559–606.

Honda, M., Bernatowicz, T. J., and Podosek, F. A. (1983). ^{129}Xe-^{128}Xe and ^{40}Ar-^{39}Ar chronology of two Antarctic enstatite meteorites. *Memoirs Natl. Inst. Polar Res., Tokyo, Special Issue* **30**, 275–291.

Horn, P., Jessberger, E. K., Kirsten, T., and Richter, H. (1975). ^{39}Ar-^{40}Ar dating of lunar rocks: Effects of grain size and neutron irradiation. *Geochim. Cosmochim. Acta* Suppl. **6** (Proceedings of Sixth Lunar Science Conference), 1563–1591.

Huneke, J. C. (1976). Diffusion artifacts in dating by stepwise thermal release of rare gases. *Earth Planet. Sci. Lett.* **28**, 407–417.

———. (1978). ^{40}Ar-^{39}Ar microanalysis of single 74220 glass balls and 72435 breccia clasts. *Geochim. Cosmochim. Acta* Suppl. **10** (Proceedings of the Ninth Lunar and Planetary Science Conference), 2345–2362.

Huneke, J. C., and Smith, S. P. (1976). The realities of recoil: ^{39}Ar recoil out of small grains and anomalous patterns in ^{39}Ar-^{40}Ar dating. *Geochim. Cosmochim. Acta* Suppl. 7 (Proceedings of the Seventh Lunar Science Conference), 1987–2008.

Hunt, J. M. (1979). *Petroleum geochemistry and geology.* Freeman, San Francisco.

Hunziker, J. C., Frey, M., Clauer, N., Dallmeyer, R. D., Friedrichsen, H., Flehmig, W., Hochstrasser, K., Roggwiler, P., and Schwander, H. (1986). The evolution of illite to muscovite: Mineralogical and isotopic data from the Glarus Alps, Switzerland. *Contrib. Mineral. Petrol.* **92**, 157–180.

Hurley, P. M., Brookins, D. G., Pinson, W. H., Hart, S. R., and Fairbairn, H. W. (1961). K-Ar age studies of Mississippi and other river sediments. *Geol. Soc. Am. Bull.* **72**, 1807–1816.

Hurley, P. M., Hughes, H., Pinson, W. H., and Fairbairn, H. W. (1962). Radiogenic argon and strontium diffusion parameters in biotite at low temperatures obtained from Alpine Fault uplift in New Zealand. *Geochim. Cosmochim. Acta* **26**, 67–80.

Hurley, P. M., Hunt, J. M., Pinson, W. H., and Fairbairn, H. W. (1963). K-Ar age values on the clay fractions in dated shales. *Geochim. Cosmochim. Acta* **27**, 279–284.

Husain, L. (1974). ^{40}Ar-^{39}Ar chronology and cosmic ray exposure ages of the Apollo 15 samples. *J. Geophys. Res.* **79**, 2588–2606.

Ingamells, C. O., and Engels, J. C. (1976). Preparation, analysis and sampling constants for a biotite. *Accuracy in trace analysis: sampling, sample handling, and analysis.* U.S. National Bureau of Standards, Special Publication 422, pp. 401–419.

Jaeger, J. C. (1957). The temperature in the neighborhood of a cooling intrusive sheet. *Am. J. Sci.* **255**, 306–318.

———. (1959). Temperatures outside a cooling intrusive sheet. *Am. J. Sci.* **257**, 44–54.

———. (1961). The cooling of irregularly shaped igneous bodies. *Am. J. Sci.* **259**, 721–734.

———. (1964). Thermal effects of intrusions. *Rev. Geophys.* **2**, 443–466.

Jäger, E. (1979). Introduction to geochronology. In *Lectures in isotope geology* (ed. E. Jäger and J. C. Hunziker), pp. 1–12. Springer-Verlag, Berlin.

Jäger, E., Niggli, E., and Wenk, E. (1967). Rb-Sr Altersbestimmungen an Glimmern der Zentralalpen. *Beiträge zur geol. Karte der Schweiz*, NF 134, Lieferungen, Kümmerly and Frey, Bern.

Jain, S. C. (1958). Simple solutions of the partial differential equation for diffusion (or heat conduction). *R. Soc. Lond., Proc.* **A243**, 359–374.

Jambon, A., Weber, H., and Braun, O. (1986). Solubility of He, Ne, Ar, Kr and Xe in a basalt melt in the range 1250–1600°C. Geochemical implications. *Geochim. Cosmochim. Acta* **50**, 401–408.

James, O. B., and Wright, T. L. (1972). Apollo 11 and 12 mare basalts and gabbros: Classification, compositional variations, and possible petrogenetic relations. *Geol. Soc. Am. Bull.* **83**, 2357–2382.

Jeffery, P. M., and Reynolds, J. H. (1961). Origin of excess Xe^{129} in stone meteorites. *J. Geophys. Res.* **66**, 3582–3583.

Jessberger, E. K. (1977). Comment on "Identification of excess ^{40}Ar by the ^{40}Ar/^{39}Ar age spectrum technique" by M. A. Lanphere and G. B. Dalrymple. *Earth Planet. Sci. Lett.* **37**, 167–168.

Jessberger, E. K., Huneke, J. C., and Wasserburg, G. J. (1974a). Evidence for a $\sim$4.5 aeon age of plagioclase clasts in a lunar highland breccia. *Nature (London)* **248**, 199–202.

Jessberger, E. K., Huneke, J. C., Podosek, F. A., and Wasserburg, G. J. (1974b). High resolution argon analysis of neutron-irradiated Apollo 16 rocks and separated minerals. *Geochim. Cosmochim. Acta* Suppl. 5 (Proceedings of the Fifth Lunar Science Conference), 1419–1449.

Jost, W. (1960). *Diffusion in solids, liquids, gases.* Academic Press, New York.

Kaneoka, I. (1972). The effect of hydration on the K/Ar ages of volcanic rocks. *Earth Planet. Sci. Lett.* **14**, 216–220.

———. (1980). Rare gas isotopes and mass fractionation: An indicator of gas transport into or from a magma. *Earth Planet. Sci. Lett.* **48**, 284–292.

———. (1983). Investigation of the weathering effect on the ^{40}Ar-^{39}Ar ages of Antarctic meteorites. *Memoirs Natl. Inst. Polar Res., Tokyo, Special Issue* 30 (Proceedings of the Eighth Symposium on Antarctic Meteorites), pp. 259–274.

Kaneoka, I., and Aoki, K.-I. (1978). ^{40}Ar/^{39}Ar analyses of phlogopite nodules and phlogopite-bearing peridotites in South African kimberlites. *Earth Planet. Sci. Lett.* **40**, 119–129.

Kaneoka, I., Ozima, M., and Yanagisawa, M. (1979). ^{40}Ar-^{39}Ar age studies of four Yamato-74 meteorites. *Memoirs Natl. Inst. Polar Res., Tokyo, Special Issue* 12 (Proceedings of the Third Symposium on Antarctic Meteorites), pp. 186–206.

Kellas, A. (1895). On the percentage of argon in atmospheric and in respired air. *R. Soc. London Proc.* **59**, 66–68.

Kelley, S., Turner, G., Butterfield, A. W., and Shepherd, T. J. (1986). The source and significance of argon isotopes in fluid inclusions from areas of mineralization. *Earth Planet. Sci. Lett.* **79**, 303–318.

Kelvin, Lord (1899). The age of the earth as an abode fitted for life. *Phil. Mag. Ser. 5* **47**, 66–90.

Kendall, B. R. F. (1960). Isotopic composition of potassium. *Nature (London)* **186**, 225–226.

Kirsten, T., Deubner, J., Horn, P., Kaneoka, I., Kiko, J., Schaeffer, O. A., and Thio, S. K. (1972). The rare gas record of Apollo 14 and 15 samples. *Geochim. Cosmochim. Acta* Suppl. **3** (Proceedings of the Third Lunar Science Conference), 1865–1889.

Kirsten, T., Horn, P., and Heymann, D. (1973a). Chronology of the Taurus-Littrow region 1: Ages of two major rock types from the Apollo 17-site. *Earth Planet. Sci. Lett.* **20**, 125–130.

Kirsten, T., Horn, P., and Kiko, J. (1973b). ^{39}Ar-^{40}Ar dating and rare gas analysis of Apollo 16 rocks and soils. *Geochim. Cosmochim. Acta* Suppl. **4** (Proceedings of the Fourth Lunar Science Conference), 1757–1784.

Kistler, R. W., Obradovich, J. D., and Jackson, E. D. (1969). Isotopic ages of rocks and minerals from the Stillwater Complex, Montana. *J. Geophys. Res.* **74**, 3226–3237.

Klemperer, O. (1935). On the radioactivity of potassium and rubidium. *R. Soc. London Proc.* **A148**, 638–648.

Kolhörster, W. von (1930). Gammastrahlen an Kaliumsalzen. *Zeitschr. Geophy.* **6**, 341–357.

Kotlovskaya, F. I. (1964). Retention of radiogenic argon by hornblende. *Geochem. Int.* for 1964, 812–815.

———. (1975). Estimation of activation energy levels for the diffusion of Ar deterioration and oxidation of some minerals. In *Sostoyaniye methodicheskikh is slededovaniy V. oblasti absyutnoy geokhronologii* (ed. G. D. Afanas'ev), pp. 85–88. Nauka Press, Moscow.

Kotlovskaya, F. I., and Tovarenko, K. S. (1974). Study of the mechanism of radiogenic argon loss by amphiboles and biotite with heating. *Geol. Zh. Ukrain. S.S.S.R.* **24**, 52–58.

Krummenacher, D. (1970). Isotopic composition of argon in modern surface volcanic rocks. *Earth Planet. Sci. Lett.* **8**, 109–117.

Lanford, W. A., Davies, K., LaMarche, P., Laursen, T., Groleau, R., and Doremus, R. H. (1979). Hydration of soda-lime glass. *J. Non-Crystalline Solids* **33**, 249–266.

Lanphere M. A., and Albee A. L. (1974). ^{40}Ar/^{39}Ar age measurements in the Worcester Mountains: Evidence of Ordovician and Devonian metamorphic events in northern Vermont. *Am. J. Sci.* **274**, 545–555.

Lanphere, M. A., and Dalrymple, G. B. (1971). A test of the ^{40}Ar/^{39}Ar age spectrum technique on some terrestrial materials. *Earth Planet. Sci. Lett.* **12**, 359–372.

———. (1976). Identification of excess ^{40}Ar by the ^{40}Ar/^{39}Ar age spectrum technique. *Earth Planet. Sci. Lett.* **32**, 141–148.

———. (1978). The use of ^{40}Ar/^{39}Ar data in evaluation of disturbed K-Ar systems. *U.S. Geological Survey Open-file Report* **78-701**, 241–243.

Lasaga, A. C., Richardson, S. M., and Holland, H. D. (1977). The mathematics of cation diffusion and exchange between silicate minerals during retrograde metamorphism. In *Energetics of geological processes* (ed. S. K. Saxena and S. Bhattacharji), pp. 353–388. Springer-Verlag, New York.

Layer, P. W., Hall, C. M., and York, D. (1987). The derivation of ^{40}Ar/^{39}Ar age spectra of single grains of hornblende and biotite by laser step-heating. *Geophys. Res. Lett.* **14**, 757–760.

Lederer, C. M., and Shirley, V. S. (eds.) (1978). *Table of isotopes*, 7th ed. Wiley, New York.

Leitch, E. C., and McDougall, I. (1979). The age of orogenesis in the Nambucca Slate Belt: A K-Ar study of low-grade regional metamorphic rocks. *Geol. Soc. Aust. J.* **26**, 111–119.

Lippolt, H. J., and Oesterle, F.-P. (1977). Argon retentivity of the mineral langbeinite. *Naturwissenschaften* **64**, 90–91.

Livingstone, D. E., Damon, P. E., Mauger, R. L., Bennett, R., and Laughlin, A. W. (1967). Argon 40 in cogenetic feldspar-mica mineral assemblages. *J. Geophys. Res.* **72**, 1361–1375.

Lovering, J. F., and Richards, J. R. (1964). Potassium-argon age study of possible lower-crust and upper-mantle inclusions in deep-seated intrusions. *J. Geophys. Res.* **69**, 4895–4901.

Lovering, J. F., and Ware, N. G. (1970). Electron probe microanalyses of minerals and glasses in Apollo 11 lunar samples. *Geochim. Cosmochim. Acta* Suppl. **1** (Proceedings of the Apollo 11 Lunar Science Conference), 633–654.

Lovering, T. S. (1935). Theory of heat conduction applied to geological problems. *Geol. Soc. Am. Bull.* **46**, 69–94.

———. (1936). Heat conduction in dissimilar rocks and the use of thermal models. *Geol. Soc. Am. Bull.* **47**, 87–100.

LSPET (The Lunar Sample Preliminary Examination Team). (1969). Preliminary examination of lunar samples from Apollo 11. *Science* **165**, 1211–1227.

Lux, G. (1987). The behavior of noble gases in silicate liquids: Solution, diffusion, bubbles and surface effects, with applications to natural samples. *Geochim. Cosmochim. Acta* **51**, 1549–1560.

Macintyre, R. M., York, D., and Gittins, J. (1966). Argon retentivity of nephelines. *Nature (London)* **209**, 702–703.

MacPherson, B. A. (1978). Sedimentation and trapping mechanism in Upper Miocene Stevens and older turbidite fans of southeastern San Joaquin Valley, California. *Am. Assoc. Pet. Geol. Bull.* **62**, 2243–2274.

Mahon, K. I., Harrison, T. M., and Drew, D. A. (1988). Ascent of a granitoid diapir in a temperature varying medium. *J. Geophys. Res.* **93**, 1174–1188.

Mak, E. K., York, D., Grieve, R. A. F., and Dence, M. R. (1976). The age of the Mistastin Lake crater, Labrador, Canada. *Earth Planet. Sci. Lett.* **31**, 345–357.

Maluski, H. (1978). Behaviour of biotites, amphiboles, plagioclases and K-feldspars in response to tectonic events with the ^{40}Ar-^{39}Ar radiometric method. Example of Corsican granite. *Geochim. Cosmochim. Acta* **42**, 1619–1633.

Maluski, H., and Schaeffer, O. A. (1982). ^{39}Ar-^{40}Ar laser probe dating of terrestrial rocks. *Earth Planet. Sci. Lett.* **59**, 21–27.

Mankinen, E. A., and Dalrymple, G. B. (1972). Electron microprobe evaluation of terrestrial basalts for whole-rock K-Ar dating. *Earth Planet. Sci. Lett.* **17**, 89–94.

———. (1979). Revised geomagnetic polarity time scale for the interval 0–5 m.y. B.P. *J. Geophys. Res.* **84**, 615–626.

Marshall, B. D., and DePaolo, D. J. (1982). Precise age determinations and petrogenetic studies using the K-Ca method. *Geochim. Cosmochim. Acta* **46**, 2537–2545.

Martel, S. J., Harrison, T. M., and Gillespie, A. R. (1987). Late Quaternary vertical displacement rate across the Fish Springs Fault, Owens Valley Fault Zone, California. *Quat. Res.*, **27**, 113–129.

Martinez, M. L., York, D., Hall, C. M., and Hanes, J. A. (1984). Oldest reliable ^{40}Ar/^{39}Ar ages for terrestrial rocks: Barberton Mountain komatiites. *Nature (London)* **307**, 352–354.

Mattauch, J. (1934). Zur Systematik der Isotopen. *Zeitschr. Phys.* **91**, 361–371.

Maurer, P., Eberhardt, P., Geiss, J., Grögler, N., Stettler, A., Brown, G. M., Peckett, A., and Krähenbühl, U. (1978). Pre-Imbrian craters and basins: Ages, compositions and excavation depths of Apollo 16 breccias. *Geochim. Cosmochim. Acta* **42**, 1687–1720.

McClelland-Brown, E. (1981). Paleomagnetic estimates of temperatures reached in contact metamorphism. *Geology* **9**, 112–116.

McConville, P., Kelley, S., and Turner, G. (1985). Laser probe ^{40}Ar-^{39}Ar studies of the Peace River L6 chondrite. *Meteoritics* **20**, 707 (Abstract).

McDougall, I. (1961). Determination of the age of a basic igneous intrusion by the potassium-argon method. *Nature (London)* **190**, 1184–1186.

———. (1964). Potassium-argon ages from lavas of the Hawaiian Islands. *Geol. Soc. Am. Bull.* **75**, 107–128.

———. (1966). Precision methods of potassium-argon isotopic age determination on young rocks. In *Methods and techniques in geophysics* (ed. S. K. Runcorn), Vol. 2, pp. 279–304. Interscience, London.

———. (1971). The geochronology and evolution of the young volcanic island of Reunion, Indian Ocean. *Geochim. Cosmochim. Acta* **35**, 261–288.

———. (1974). The ^{40}Ar/^{39}Ar method of K-Ar age determination of rocks using HIFAR reactor. *Aust. Atomic Energy Comm. J.* **17**(3), 3–12.

———. (1979). The present status of the geomagnetic polarity time scale. In *The Earth, its origin, structure and evolution* (ed. M. W. McElhinny), pp. 543–565. Academic Press, London.

———. (1981). ^{40}Ar/^{39}Ar age spectra from the KBS Tuff, Koobi Fora Formation. *Nature (London)* **294**, 120–124.

———. (1985). K-Ar and ^{40}Ar/^{39}Ar dating of the hominid-bearing Pliocene-Pleistocene sequence at Koobi Fora, Lake Turkana, northern Kenya. *Geol. Soc. Am. Bull.* **96**, 159–175.

McDougall, I., and Duncan, R. A. (1980). Linear volcanic chains—recording plate motions? *Tectonophysics* **63**, 275–295.

McDougall, I., and Green, D. H. (1964). Excess radiogenic argon in pyroxene and isotopic ages on minerals from Norwegian eclogites. *Norsk Geol. Tidss.* **44**, 183–196.

McDougall, I., and Lovering, J. F. (1969). Apparent K-Ar dates on cores and excess Ar in flanges of australites. *Geochim. Cosmochim. Acta* **33**, 1057–1070.

McDougall, I., and Roksandic, Z. (1974). Total fusion ^{40}Ar/^{39}Ar ages using HIFAR reactor. *Geol. Soc. Aust. J.* **21**, 81–89.

McDougall, I., and Schmincke, H.-U. (1977). Geochronology of Gran Canaria, Canary Islands: Age of shield building volcanism and other magmatic phases. *Bull. Volcanol.* **40**, 57–77.

McDougall, I., Polach, H. A., and Stipp, J. J. (1969). Excess radiogenic argon in young subaerial basalts from the Auckland volcanic field, New Zealand. *Geochim. Cosmochim. Acta* **33**, 1485–1520.

McDougall, I., Watkins, N. D., and Kristjansson, L. (1976). Geochronology and paleomagnetism of a Miocene-Pliocene lava sequence at Bessastadaa, eastern Iceland. *Am. J. Sci.* **276**, 1078–1095.

McDougall, I., Maier, R., Sutherland-Hawkes, P., and Gleadow, A. J. W. (1980). K-Ar age estimate for the KBS Tuff, East Turkana, Kenya. *Nature (London)* **284**, 230–234.

McDougall, I., Kristjansson, L., and Saemundsson, K. (1984). Magnetostratigraphy and geochronology of northwest Iceland. *J. Geophys. Res.* **82**, 7029–7060.

McDowell, F. W. (1983). K-Ar dating: Incomplete extraction of radiogenic argon from alkali feldspar. *Isot. Geosci.* **1**, 119–126.

McDowell, F. W., Lehman, D. H., Gucwa, P. R., Fritz, D., and Maxwell, J. C. (1984). Glaucophane schists and ophiolites of the northern California Coast Ranges: Isotopic ages and their tectonic implications. *Geol. Soc. Am. Bull.* **95**, 1373–1382.

McElhinny, M. W. (1973). *Palaeomagnetism and plate tectonics*. Cambridge University Press, Cambridge.

McFadden, P. L. (1977). A palaeomagnetic determination of the emplacement temperature of some South African kimberlites. *Geophys. J. R. Astr. Soc.* **50**, 587–604.

McIntyre, G. A., Brooks, C., Compston, W., and Turek, A. (1966). The statistical assessment of Rb-Sr isochrons. *J. Geophys. Res.* **71**, 5459–5468.

McKenzie, D. P. (1978). Some remarks on the development of sedimentary basins. *Earth Planet. Sci. Lett.* **40**, 25–32.

Megrue, G. H. (1967). Isotopic analysis of rare gases with a laser microprobe. *Science* **157**, 1555–1556.

———. (1971). Distribution and origin of helium, neon, and argon isotopes in Apollo 12 samples measured by in situ analysis with a laser-probe mass spectrometer. *J. Geophys. Res.* **76**, 4956–4968.

———. (1973). Spatial distribution of $^{40}Ar/^{39}Ar$ ages in lunar breccia 14301. *J. Geophys. Res.* **78**, 3216–3221.

Melenevskiy, V. N., Morozova, I. M., and Yurgina, Ye. K. (1978). The migration of radiogenic argon and biotite dehydroxylation. *Geochem. Int.* **15**(6), 7–16.

Merrihue, C. (1965). Trace-element determinations and potassium-argon dating by mass spectroscopy of neutron-irradiated samples. *Trans. Am. Geophys. Un.* **46**, 125 (Abstract).

Merrihue, C., and Turner, G. (1966) Potassium-argon dating by activation with fast neutrons. *J. Geophys. Res.* **71**, 2852–2857.

Merrill, R. T., and McElhinny, M. W. (1983). *The Earth's magnetic field*. Academic Press, London.

Miller, J. A., Mitchell, J. G., and Evans, A. L. (1970). The argon-40/argon-39 dating method applied to basic rocks. In *Palaeogeophysics* (ed. S. K. Runcorn), pp. 481–489. Academic Press, London.

Mitchell, J. G. (1968a). The argon-40/argon-39 method for potassium-argon age determination. *Geochim. Cosmochim. Acta* **32**, 781–790.

———. (1968b). Ph.D. Diss., University of Cambridge.

Morgan, K. J., and Turner, J. E. (eds.) (1967). *Principles of radiation protection*. Wiley, New York.

Morozova, I. M., and Alferovskiy, A. A. (1974). Fractionation of lithium and potassium isotopes in geological processes. *Geochem. Int.* **11**, 17–25.

Müller, H. W., Plieninger, T., James, O. B., and Schaeffer, O. A. (1977). Laser probe ^{39}Ar-^{40}Ar dating of materials from consortium breccia 73215. *Geochim. Cosmochim. Acta* Suppl. **8** (Proceedings of the Eighth Lunar Science Conference), 2551–2565.

Müller, N., and Jessberger, E. K. (1985). Dating Jilin and constraints on its temperature history. *Earth Planet. Sci. Lett.* **72**, 276–285.

Mussett, A. E. (1969). Diffusion measurements and the potassium-argon method of dating. *Geophys. J. R. Astr. Soc.* **18**, 257–303.

———. (1986). ^{40}Ar-^{39}Ar step-heating ages of the Tertiary igneous rocks of Mull, Scotland. *J. Geol. Soc. Lond.* **143**, 887–896.

Mussett, A. E., Ross, J. G., and Gibson, I. L. (1980). $^{40}Ar/^{39}Ar$ dates of eastern Iceland lavas. *Geophys. J. R. Astr. Soc.* **60**, 37–52.

Naylor, R. S. (1971). Acadian Orogeny: An abrupt and brief event. *Science* **172**, 558–559.

Néel, L. (1949). Théorie du traînage magnétique des ferromagnétiques en grains fins avec applications aux terres cuites. *Ann. Geophys.* **5**, 99–136.

———. (1955). Some theoretical aspects of rock-magnetism. *Phil. Mag. Suppl. Adv. Phys.* **4**, 191–243.

Newman, F. H., and Walke, H. J. (1935). The radioactivity of potassium and rubidium. *Phil. Mag. Ser. 7* **19**, 767–773.

Nier, A. O. (1935). Evidence for the existence of an isotope of potassium of mass 40. *Phys. Rev.* **48**, 283–284.

———. (1940). A mass spectrometer for routine isotope abundance measurements. *Rev. Sci. Inst.* **11**, 212–216.

———. (1947). A mass spectrometer for isotope and gas analysis. *Rev. Sci. Inst.* **18**, 398–411.

———. (1950). A redetermination of the relative abundances of the isotopes of carbon, nitrogen, oxygen, argon, and potassium. *Phys. Rev.* **77**, 789–793.

Noble, C. S., and Naughton, J. J. (1968). Deep-ocean basalts: Inert gas content and uncertainties in age dating. *Science* **162**, 265–267.

Norwood, C. B. (1974). Radiogenic argon diffusion in the biotite micas. M.S. thesis, Brown University, Providence, RI.

Obradovich, J. D., Tatsumoto, M., Manuel, O. K., Mehnert, H., Domenick, M., and Wildman, T. (1982). K-Ar and K-Ca dating of sylvite from the late Permian Salado Formation, New Mexico. Implications regarding stability of evaporite minerals. *Fifth Int. Conf. Geochronol. Cosmochronol. Isotope Geol., Japan*, 283–284 (Abstract).

Odin, G. S. (ed.) (1982a). *Numerical dating in stratigraphy*. Wiley, New York.

Odin, G. S., and 35 collaborators (1982b). Interlaboratory standards for dating purposes. In *Numerical dating in stratigraphy* (ed. G. S. Odin), pp. 123–149. Wiley, Chichester.

O'Nions, R. K., Smith, D. G. W., Baadsgaard, H., and Morton, R. D. (1969). Influence of chemical composition on argon retentivity in metamorphic calcic amphiboles from south Norway. *Earth Planet. Sci. Lett.* **5**, 339–345.

Onstott, T. C., and Peacock, M. W. (1987). Argon retentivity of hornblendes: A field experiment in a slowly cooled metamorphic terrane. *Geochim. Cosmochim. Acta* **51**, 2891–2903.

Ozima, M., and Podosek, F. A. (1983). *Noble gas geochemistry*. Cambridge University Press, Cambridge.

Pankhurst, R. J., Moorbath, S., Rex, D. C., and Turner, G. (1973). Mineral age patterns in ca. 3700 my old rocks from West Greenland. *Earth Planet. Sci. Lett.* **20**, 157–170.

Papanastassiou, D. A., and Wasserburg, G. J. (1970). Rb-Sr ages from the Ocean of Storms. *Earth Planet. Sci. Lett.* **8**, 269–278.

———. (1971). Lunar chronology and evolution from Rb-Sr studies of Apollo 11 and 12 samples. *Earth Planet. Sci. Lett.* **11**, 37–62.

———. (1972). Rb-Sr systematics of Luna 20 and Apollo 16 samples. *Earth Planet. Sci. Lett.* **17**, 52–63.

———. (1975). A Rb-Sr study of Apollo 17 Boulder 3: Dunite clast, microclasts, and matrix. *Lunar Sci.* **VI**, 631–633.

Papanastassiou, D. A., Wasserburg, G. J., and Burnett, D. S. (1970). Rb-Sr ages of lunar rocks from the Sea of Tranquillity. *Earth Planet. Sci. Lett.* **8**, 1–19.

Papanastassiou, D. A., DePaolo, D. J., and Wasserburg, G. J. (1977). Rb-Sr and Sm-Nd chronology and genealogy of mare basalts from the Sea of Tranquillity. *Geochim. Cosmochim. Acta* Suppl. **8** (Proceedings of the Eighth Lunar Science Conference), 1639–1672.

Papike, J. J., Hodges, F. N., Bence, A. E., Cameron, M., and Rhodes, J. M. (1976). Mare basalts: Crystal chemistry, mineralogy, and petrology. *Rev. Geophys. Space Phys.* **14**, 475–540.

Parker, R. L. (1977). Understanding inverse theory. *Annu. Rev. Earth Planet. Sci.* **5**, 35–64.

Parrish, R. R. (1982). Cenozoic thermal and tectonic history of the Coast Mountains of British Columbia as revealed by fission track and geological data and quantitative thermal models. Ph.D. Diss., University of British Columbia, Vancouver.

Perry, J. (1895a). On the age of the earth. *Nature (London)* **51**, 224–227.

———. (1895b). On the age of the earth. *Nature (London)* **51**, 341–342.

———. (1895c). The age of the earth. *Nature (London)* **51**, 582–585.

Phillips, D., and Onstott, T. C. (1986). Application of $^{36}Ar/^{40}Ar$ versus $^{39}Ar/^{40}Ar$ correlation diagrams to the $^{40}Ar/^{39}Ar$ spectra of phlogopites from southern African kimberlites. *Geophys. Res. Lett.* **13**, 689–692.

Philpotts, J. A., and Schnetzler, C. C. (1970). Apollo 11 lunar samples: K, Rb, Sr, Ba and rare-earth concentrations in some rocks and separated phases. *Geochim. Cosmochim. Acta* Suppl. **1** (Proceedings of the Apollo 11 Lunar Science Conference), 1471–1486.

Plieninger, T., and Schaeffer, O. A. (1976). Laser probe ^{39}Ar-^{40}Ar ages of individual mineral grains in lunar basalt 15607 and lunar breccia 15465. *Geochim. Cosmochim. Acta* Suppl. **7** (Proceedings of the Seventh Lunar Science Conference), 2055–2066.

Podosek, F. A., and Huneke, J. C. (1973). Argon 40-argon 39 chronology of four calcium-rich achondrites. *Geochim. Cosmochim. Acta* **37**, 667–684.

Podosek, F. A., Huneke, J. C., Gancarz, A. J., and Wasserburg, G. J. (1973). The age and petrography of two Luna 20 fragments and inferences for widespread lunar metamorphism. *Geochim. Cosmochim. Acta* **37**, 887–904.

Pullaiah, G., Irving, E., Buchan, K. L., and Dunlop, D. J. (1975). Magnetization changes caused by burial and uplift. *Earth Planet. Sci. Lett.* **28**, 133–143.

Purdy, J. W., and Jäger, E. (1976). K-Ar ages on rock-forming minerals from the Central Alps. *Mem. Ist. Geol. Min. Univ. Padova* **30**, 31 pp.

Radicati di Brozolo, F., Huneke, J. C., Papanastassiou, D. A., and Wasserburg, G. J. (1981). ^{40}Ar-^{39}Ar and Rb-Sr age determinations on Quaternary volcanic rocks. *Earth Planet. Sci. Lett.* **53**, 445–456.

Rakovic, M. (1970). *Activation analysis*. CRC Press, Cleveland.

Rayleigh, Lord, and Ramsay, W. (1895). Argon, a new constituent of the atmosphere. *R. Soc. Lond. Phil. Trans. Ser. A* **186**, 187–241.

Reichenberg, D. (1953). Properties of ion-exchange resins in relation to their structure. III. Kinetics of exchange. *Am. Chem. Soc. J.* **75**, 589–597.

Reynolds, J. H. (1956). High sensitivity mass spectrometer for noble gas analysis. *Rev. Sci. Inst.* **27**, 928–934.

———. (1960). Rare gases in tektites. *Geochim. Cosmochim. Acta* **20**, 101–114.

———. (1963). Xenology. *J. Geophys. Res.* **68**, 2939–2956.

Reynolds, P. H., and Muecke, G. K. (1978). Age studies on slates: Applicability of the $^{40}Ar/^{39}Ar$ stepwise outgassing method. *Earth Planet. Sci. Lett.* **40**, 111–118.

Ribando, R. J., Torrance, K. E., and Turcotte, D. L. (1978). Numerical calculations of the convective cooling of an infinite sill. *Tectonophysics* **50**, 337–347.

Ribbe, P. H. (ed.) (1975). Feldspar mineralogy. *Mineral. Soc. Am. Short Course Notes* **2**.

Richards, J. R., and Pidgeon, R. T. (1963). Some age measurements on micas from Broken Hill, Australia. *Geol. Soc. Aust. J.* **10**, 243–260.

Richter, F. M. (1986). Kelvin and the age of the earth. *J. Geol.* **94**, 395–401.

Robbins, G. A. (1972). Radiogenic argon diffusion in muscovite under hydrothermal conditions. M.S. thesis, Brown University. Providence.

Roberts, W. H. (1958). The design and construction of HIFAR. *Aust. Atomic Energy Symp. 1958*, pp. 363–379. Melbourne University Press, Melbourne.

Roboz, J. (1968). *Introduction to mass spectrometry: Instrumentation and techniques*. Interscience, New York.

Roddick, J. C. (1978). The application of isochron diagrams in ^{40}Ar-^{39}Ar dating: A discussion. *Earth Planet. Sci. Lett.* **41**, 233–244.

———. (1983). High precision intercalibration of ^{40}Ar-^{39}Ar standards. *Geochim. Cosmochim. Acta* **47**, 887–898.

———. (1987). Generalized numerical error analysis with applications to geochronology and thermodynamics. *Geochim. Cosmochim. Acta* **51**, 2129–2135.

Roddick, J. C., Cliff, R. A., and Rex, D. C. (1980). The evolution of excess argon in alpine biotites—A ^{40}Ar-^{39}Ar analysis. *Earth Planet. Sci. Lett.* **48**, 185–208.

Royden, L., and Hodges, K. V. (1984). A technique for analyzing the thermal and uplift histories of eroding orogenic belts: A Scandinavian example. *J. Geophys. Res.* **89**, 7091–7106.

Rutherford, E. (1906). *Radioactive transformations*. Scribners, New York.

Rutherford, E., and Soddy, F. (1903). Radioactive change. *Phil. Mag. Ser. 6* **5**, 576–591.

Ryerson, F., Harrison, T. M., and Heizler, M. (1985). Thermal constraints on the emplacement of the rhyolite conduit at Inyo Domes, Long Valley Caldera, California. *EOS, Trans. Am. Geophys. Un.* **66**, 1125 (Abstract).

Samson, S. D., and Alexander, E. C., Jr. (1987). Calibration of the interlaboratory ^{40}Ar-^{39}Ar dating standard, MMhb-1. *Chem. Geol. (Isot. Geosci. Section)*, **66**, 27–34.

Sardarov, S. S. (1957). Retention of radiogenic argon in microcline. *Geochemistry* for 1957, No. 3, 233–237.

———. (1961). Bond energy and retention of radiogenic argon in micas. *Geochemistry* for 1961, 33–44.

Schaeffer, O. A. (1982). Laser microprobe argon-39-argon-40 dating of individual mineral grains. In *Nuclear and chemical dating techniques* (ed. L. A. Currie), pp. 139–148. American Chemical Society Symposium Series 176, Washington, D.C.

Schaeffer, O. A., and Zähringer, J. (eds.) (1966). *Potassium argon dating*. Springer-Verlag, New York.

Schaeffer, O. A., Müller, H. W., and Grove, T. L. (1977). Laser ^{39}Ar-^{40}Ar study of Apollo 17 basalts. *Geochim. Cosmochim. Acta* Suppl. **8** (Proceedings of the Eighth Lunar Science Conference), 1489–1499.

Schilling, J. H. (ed.) (1973). K-Ar dates on Permian potash minerals from southeastern New Mexico. *Isochron/West* **6**, 37.

Schmitt, H. H., Lofgren, G., Swann, G. A., and Simmons, G. (1970). The Apollo 11 samples: Introduction. *Geochim. Cosmochim. Acta* Suppl. **1** (Proceedings of the Apollo 11 Lunar Science Conference), 1–54.

Schwartzman, D. W., and Giletti, B. J. (1977). Argon diffusion and absorption studies of pyroxenes from the Stillwater Complex, Montana. *Contrib. Mineral. Petrol.* **60**, 143–159.

Seidemann, D. E. (1977). Effects of submarine alteration on K-Ar dating of deep-sea igneous rocks. *Geol. Soc. Am. Bull.* **88**, 1660–1666.

———. (1978). $^{40}Ar/^{39}Ar$ studies of deep-sea igneous rocks. *Geochim. Cosmochim. Acta* **42**, 1721–1734.

Shanin, L. L., Kononova, V. A., and Ivanov, I. B. (1967). On the use of nepheline in K-Ar dating. *Akad. Nauk. U.S.S.R. Izv. Ser. Geol.* no. 5, for 1967, 19–30.

Sigurgeirsson, T. (1962). Age dating of young basalts with the potassium argon method (in Icelandic). Unpublished report Physics Laboratory, University of Iceland. (English translation by L. Kristjansson, University of Iceland, 1973.)

Sisson, V. B., and Onstott, T. C. (1986). Dating blueschist metamorphism: A combined ^{40}Ar/^{39}Ar and electron microprobe approach. *Geochim. Cosmochim. Acta* **50**, 2111–2117.

Sitte, K. v. (1935). Über die Radioaktivität von Kalium und Rubidium. *Zeitschr. Phys.* **96**, 593–599.

Smith, J. V. (1974). *Feldspar minerals*, 2 volumes. Springer-Verlag, Berlin.

Smits, F., and Gentner, W. (1950). Argonbestimmungen an Kalium-Mineralien I. Bestimmungen an tertiären Kalisalzen. *Geochim. Cosmochim. Acta* **1**, 22–27.

Smythe, W. R., and Hemmendinger, A. (1937). The radioactive isotope of potassium. *Phys. Rev.* **51**, 178–182.

Snee, L. W. (1982). Determination of thermal histories in complex plutonic terranes by use of details of ^{40}Ar/^{39}Ar age-spectrum diagrams. *EOS, Trans. Am. Geophys. Un.* **63**, 453 (Abstract).

Stacey, J. S., Sherrill, N. D., Dalrymple, G. B., Lanphere, M. A., and Carpenter, N. V. (1981). A five-collector system for the simultaneous measurement of argon isotope ratios in a static mass spectrometer. *Int. J. Mass Spectrom. Ion Phys.* **39**, 167–180.

Staudacher, Th., Jessberger, E. K., Dörflinger, D., and Kiko, J. (1978). A refined ultrahigh-vacuum furnace for rare gas analysis. *J. Phys. E: Sci. Instrum.* **11**, 781–784.

Staudacher, Th., Jessberger, E. K., Dominik, B., Kirsten, T., and Schaeffer, O. A. (1982). ^{40}Ar-^{39}Ar ages of rocks and glasses from the Nördlinger Ries Crater and the temperature history of impact breccias. *J. Geophys.* **51**, 1–11.

Steiger, R. H., and Jäger, E. (1977). Subcommission on geochronology: Convention on the use of decay constants in geo- and cosmochronology. *Earth Planet. Sci. Lett.* **36**, 359–362.

Stephenson, R., and Lambeck, K. (1985). Erosion-isostatic rebound models for uplift: An application to south-eastern Australia. *Geophys. J.R. Astr. Soc.* **82**, 31–55.

Stettler, A., Eberhardt, P., Geiss, J., Grögler, N., and Maurer, P. (1973). Ar^{39}-Ar^{40} ages and Ar^{37}-Ar^{38} exposure ages of lunar rocks. *Geochim. Cosmochim. Acta* Suppl. **4** (Proceedings of the Fourth Lunar Science Conference), 1865–1888.

———. (1974). On the duration of lava flow activity in Mare Tranquillitatis. *Geochim. Cosmochim. Acta* Suppl. **5** (Proceedings of the Fifth Lunar Science Conference), 1557–1570.

Stevens, W. L. (1951). Asymptotic regression. *Biometrics* **7**, 247–267.

Stoenner, R. W., Schaeffer, O. A., and Katcoff, S. (1965). Half-lives of argon-37, argon-39, and argon-42. *Science* **148**, 1325–1328.

Strutt, R. J. (1906). On the distribution of radium in the earth's crust and on the earth's internal heat. *R. Soc. London Proc.* **77A**, 472–485.

Suppe, J., and Armstrong, R. L. (1972). Potassium-argon dating of Franciscan metamorphic rocks. *Am. J. Sci.* **272**, 217–233.

Sutter, J. F., and Hartung, J. B. (1984). Laser microprobe ^{40}Ar/^{39}Ar dating of mineral grains *in situ*. *Soc. Elec. Micros.* **4**, 1525–1529.

Sutter, J. F., Radcliffe, N. M., and Mukasa, S. B. (1985). ^{40}Ar/^{39}Ar and K-Ar data bearing on the metamorphic and tectonic history of western New England. *Geol. Soc. Am. Bull.* **96**, 123–136.

Tagai, T., and Korekawa, M. (1981). Crystallographic investigation of the Huttenlocher exsolution at high temperature. *Phys. Chem. Miner.* **7**, 77–81.

Takeda, H., and Morosin, B. (1975). Comparison of observed and predicted structural parameters of mica at high temperature. *Acta Crystallogr.* **B31**, 2444–2452.

Taylor, S. R. (1975). *Lunar science: A post-Apollo view*. Pergamon Press, New York.

———. (1979). Chemical composition and evolution of the continental crust: The rare earth element evidence. In *The Earth: Its origin, structure and evolution* (ed. M. W. McElhinny), pp. 353–376. Academic Press, London.

———. (1982). *Planetary science: A lunar perspective*. Lunar and Planetary Institute, Houston.

Taylor, S. R., and McLennan, S. M. (1985). *The continental crust: Its composition and evolution*. Blackwell, Oxford.

Tera, F., Eugster, O., Burnett, D. S., and Wasserburg, G. J. (1970). Comparative study of Li, Na, K, Rb, Cs, Ca, Sr and Ba abundances in achondrites and in Apollo 11 lunar samples. *Geochim. Cosmochim. Acta* Suppl **1** (Proceedings of the Apollo 11 Lunar Science Conference), 1637–1657.

Tera, F., Papanastassiou, D. A., and Wasserburg, G. J. (1974). Isotopic evidence for a terminal lunar cataclysm. *Earth Planet. Sci. Lett.* **22**, 1–21.

Tetley, N. W. (1978). Geochronology by the ^{40}Ar/^{39}Ar technique using HIFAR reactor. Ph.D. Diss., Australian National University, Canberra.

Tetley, N., and McDougall, I. (1978). Anomalous ^{40}Ar/^{39}Ar release spectra for biotites from the Berridale Batholith, New South Wales, Australia. *U.S. Geol. Surv. Open-file Rept.* **78-701**, 427–430.

Tetley, N., McDougall, I., and Heydegger, H. R. (1980). Thermal neutron interferences in the ^{40}Ar/^{39}Ar dating technique. *J. Geophys. Res.* **85**, 7201–7205.

Thompson, F. C., and Rowlands, S. (1943). Dual decay of potassium. *Nature (London)* **152**, 103.

Thomson, J. J. (1905). On the emission of negative corpuscles by the alkali metals. *Phil. Mag. Ser. 6* **10**, 584–590.

Thomson, W. (1863). On the secular cooling of the earth. *Phil. Mag. Ser. 4* **25**, 1–14.

Tingey, R. J., McDougall, I., and Gleadow, A. J. W. (1983). The age and mode of formation of Gaussberg, Antarctica. *Geol. Soc. Aust. J.* **30**, 241–246.

Turner, G. (1968). The distribution of potassium and argon in chondrites. In *Origin and distribution of the elements* (ed. L. H. Ahrens), pp. 387–398. Pergamon, London.

———. (1969). Thermal histories of meteorites by the ^{39}Ar-^{40}Ar method. In *Meteorite research* (ed. P. M. Millman), pp. 407–417. Reidel, Dordrecht.

———. (1970a). Thermal histories of meteorites. In *Palaeogeophysics* (ed. S. K. Runcorn), pp. 491–502. Academic Press, London.

———. (1970b). ^{40}Ar-^{39}Ar age determination of lunar rock 12013. *Earth Planet. Sci. Lett.* **9**, 177–180.

———. (1970c). Argon-40/argon-39 dating of lunar rock samples. *Science* **167**, 466–468.

———. (1970d). Argon-40/argon-39 dating of lunar rock samples. *Geochim. Cosmochim. Acta*, Suppl. 1 (Proceedings of the Apollo 11 Lunar Science Conference), 1665–1684.

———. (1971a). Argon 40-argon 39 dating: The optimization of irradiation parameters. *Earth Planet. Sci. Lett.* **10**, 227–234.

———. (1971b). ^{40}Ar-^{39}Ar ages from the lunar maria. *Earth Planet. Sci. Lett.* **11**, 169–191.

———. (1972). ^{40}Ar-^{39}Ar age and cosmic ray irradiation history of the Apollo 15 anorthosite, 15415. *Earth Planet. Sci. Lett.* **14**, 169–175.

———. (1977). Potassium-argon chronology of the moon. *Phys. Chem. Earth* **10**, 145–195.

Turner, G., and Cadogan, P. H. (1974). Possible effects of ^{39}Ar recoil in ^{40}Ar-^{39}Ar dating. *Geochim. Cosmochim. Acta* Suppl. 5 (Proceedings of the Fifth Lunar Science Conference), 1601–1615.

Turner, G., Miller, J. A., and Grasty, R. L. (1966). The thermal history of the Bruderheim meteorite. *Earth Planet. Sci. Lett.* **1**, 155–157.

Turner, G., Huneke, J. C., Podosek, F. A., and Wasserburg, G. J. (1971). ^{40}Ar-^{39}Ar ages and cosmic ray exposure age of Apollo 14 samples. *Earth Planet. Sci. Lett.* **12**, 19–35.

———. (1972). Ar^{40}-Ar^{39} systematics in rocks and separated minerals from Apollo 14. *Geochim. Cosmochim. Acta* Suppl. 3 (Proceedings of the Third Lunar Science Conference), 1589–1612.

Turner, G., Cadogan, P. H., and Yonge, C. J. (1973). Argon selenochronology. *Geochim. Cosmochim. Acta* Suppl. 4 (Proceedings of the Fourth Lunar Science Conference), 1889–1914.

Verbeek, A. A., and Schreiner, G. D. L. (1967). Variations in ^{39}K:^{41}K ratio and movement of potassium in a granite-amphibolite contact region. *Geochim. Cosmochim. Acta* **31**, 2125–2133.

Villa, I. M. (1986). ^{40}Ar$_{xs}$ in K-rich magmas: What can it mean? *Terra Cognita* **6**, 149 (Abstract).

Villa, I. M., Huneke, J. C., and Wasserburg, G. J. (1983). ^{39}Ar recoil losses and presolar ages in Allende inclusions. *Earth Planet. Sci. Lett.* **63**, 1–12.

Von Weizsäcker, C. F. (1937). Über die Möglichkeit eines dualen β^- Zerfalls von Kalium. *Phys. Zeitschr.* **38**, 623–624.

Walker, D. A., and McDougall, I. (1982). ^{40}Ar/^{39}Ar and K-Ar dating of altered glassy volcanic rocks: The Dabi Volcanics, P.N.G. *Geochim. Cosmochim. Acta* **46**, 2181–2190.

Walker, F. W., Miller, D. G., and Feiner, F. (1983). *Chart of the nuclides*, 13th ed. General Electric Company, San Jose.

Walker, K. L., and Homsy, G. M. (1978). Convection in a porous cavity. *J. Fluid Mech.* **87**, 449–474.

Wang, S., McDougall, I., Tetley, N., and Harrison, T. M. (1980). ^{40}Ar/^{39}Ar age and thermal history of the Kirin chondrite. *Earth Planet. Sci. Lett.* **49**, 117–131.

Wang, S., Sang, H., Hu, S., and Qiu, J. (1985). ^{40}Ar/^{39}Ar age determination using 49-2 reactor and ^{40}Ar/^{39}Ar age spectrum for amphibolite from Qianan, China. *Acta Petrol. Sinica* **1**, 35–44 (in Chinese).

Wang, S., Zai, M., Hu, S., Sang, H., and Qiu, J. (1986). ^{40}Ar/^{39}Ar age spectrum for biotite separated from Qingyuan Tonalite, NE China. *Scientia Geol. Sinica* for 1986, 97–100 (in Chinese).

Wänke, H., and König, H. (1959). Eine neue Methode zur Kalium-Argon-Altersbestimmung und ihre Anwendung auf Steinmeteorite. *Z. Naturforsch.* **14a**, 860–866.

Warner, J. L. (1971). Lunar crystalline rocks: Petrology and geology. *Geochim. Cosmochim. Acta* Suppl. 2 (Proceedings of the Second Lunar Science Conference), 469–480.

Warren, P. H. (1985). The magma ocean concept and lunar evolution. *Annu. Rev. Earth Planet. Sci.* **13**, 201–240.

Warren, P. H., Shirley, D. N., and Kallemeyn, G. W. (1986). A potpourri of pristine moon rocks, including a VHK mare basalt and a unique, augite-rich Apollo 17 anorthosite. *J. Geophys. Res.* **91**, B4, D319-D330 (Proceedings of the Sixteenth Lunar and Planetary Science Conference).

Watt, B. E. (1952). Energy spectrum of neutrons from thermal fission of U^{235}. *Phys. Rev.* **87**, 1037–1041.

Webb, A. W., and McDougall, I. (1967). A comparison of mineral and whole rock potassium-argon ages of Tertiary volcanics from Central Queensland, Australia. *Earth Planet. Sci. Lett.* **3**, 41–47.

Webb, G. W. (1981). Stevens and earlier Miocene turbidite sandstones, southern San Joaquin Valley, California. *Am. Assoc. Pet. Geol. Bull.* **65**, 438–465.

Weertman, J. (1970). The creep strength of the mantle. *Rev. Geophys. Space Phys.* **8**, 145–168.

Wellman, P. (1973). Early Miocene potassium-argon age for the Fitzroy Lamproites of Western Australia. *Geol. Soc. Aust. J.* **19**, 471–474.

Wellman, P., Cundari, A., and McDougall, I. (1970). Potassium-argon ages for leucite-bearing rocks from New South Wales, Australia. *R. Soc. N.S.W. J. Proc.* **103**, 103–107.

Wells, P. R. A. (1980). Thermal models for the magmatic accretion and subsequent metamorphism of continental crust. *Earth Planet. Sci. Lett.* **46**, 253–265.

Wenk, H.-R. (1979). An albite-anorthite assemblage in low-grade amphibolite facies rocks. *Am. Mineral.* **64**, 1294–1299.

Westcott, M. R. (1966). Loss of argon from biotite in a thermal metamorphism. *Nature (London)* **210**, 83–84.

White, F. A., and Wood, G. M. (1986). *Mass spectrometry.* Wiley, New York.

Wijbrans, J. R. (1985). Geochronology of metamorphic terrains by the $^{40}Ar/^{39}Ar$ age spectrum method. Ph.D. Diss., Australian National University, Canberra.

Wijbrans, J. R., and McDougall, I. (1986). $^{40}Ar/^{39}Ar$ dating of white micas from an Alpine high-pressure metamorphic belt on Naxos (Greece): The resetting of the argon isotopic system. *Contrib. Mineral. Petrol.* **93**, 187–194.

Williams, I. S., Tetley, N. W., Compston, W., and McDougall, I. (1982). A comparison of K-Ar and Rb-Sr ages of rapidly cooled igneous rocks: Two points in the Palaeozoic time scale re-evaluated. *J. Geol. Soc. London* **139**, 557–568.

Wones, D. R. (1967). A low pressure investigation of the stability of phlogopite. *Geochim. Cosmochim. Acta* **31**, 2248–2253.

Wones, D. R., and Eugster, H. P. (1965). Stability of biotite: Experiment, theory, and application. *Am. Mineral.* **50**, 1228–1272.

Yanase, Y., Wampler, J. M., and Dooley, R. E. (1975). Recoil-induced loss of ^{39}Ar from glauconite and other clay minerals. *Trans. Am. Geophys. Un.* **56**, 472 (Abstract).

York, D. (1966). Least-squares fitting of a straight line. *Can. J. Phys.* **44**, 1079–1086.

———. (1969). Least squares fitting of a straight line with correlated errors. *Earth Planet. Sci. Lett.* **5**, 320–324.

———. (1978a). A formula describing both magnetic and isotopic blocking temperatures. *Earth Planet. Sci. Lett.* **39**, 89–93.

———. (1978b). Magnetic blocking temperature. *Earth Planet. Sci. Lett.* **39**, 94–97.

———. (1984). Cooling histories from $^{40}Ar/^{39}Ar$ age spectra: Implications for Precambrian plate tectonics. *Annu. Rev. Earth Planet. Sci.* **12**, 383–409.

York, D., and Berger, G. W. (1970). $^{40}Ar/^{39}Ar$ age determinations on nepheline and basic whole rocks. *Earth Planet. Sci. Lett.* **7**, 333–336.

York, D., and Farquhar, R. M. (1972). *The Earth's age and geochronology.* Pergamon, Oxford.

York, D., Yanase, Y., and Berger, G. W. (1971). Determination of geological time with a nuclear reactor and a mass spectrometer. In *Activation analysis in geochemistry and cosmochemistry* (ed. A. O. Brunfelt and E. Steinnes), pp. 419–422. Universitetsforlaget, Oslo.

York, D., Hall, C. M., Yanase, Y., Hanes, J. A., and Kenyon, W. J. (1981). $^{40}Ar/^{39}Ar$ dating of terrestrial minerals with a continuous laser. *Geophys. Res. Lett.* **8**, 1136–1138.

York, D., Hall, C. M., Gaspar, M. J., and Lynch, M. (1984). Laser-probe $^{40}Ar/^{39}Ar$ dating with ultrasensitive mass spectrometer. *EOS Trans. Am. Geophys. Un.* **65**, 303 (Abstract).

Zähringer, J. (1963). K-Ar measurements of tektites. In *Radioactive dating*, pp. 289–305. International Atomic Energy Agency, Vienna.

Zartman, R. E., Brock, M. R., Heyl, A. V., and Thomas, H. H. (1967). K-Ar and Rb-Sr ages of some alkalic intrusive rocks from central and eastern United States. *Am. J. Sci.* **265**, 848–870.

Zeitler, P. K. (1985). Closure temperature implications of concordant $^{40}Ar/^{39}Ar$ potassium feldspar and zircon fission-track ages from high grade terranes. *Nucl. Tracks* **10**, 441–442.

———. (1988). The geochronology of metamorphic processes. In *Spec. Pub. Geol. Soc. London.*

Zeitler, P. K., and Fitz Gerald, J. D. (1986). Saddle-shaped $^{40}Ar/^{39}Ar$ age spectra from young, microstructurally complex potassium feldspars. *Geochim. Cosmochim. Acta* **50**, 1185–1199.

Zen, E-an, White, W. S., Hadley, J. B., and Thompson, J. B. (eds) (1968). *Studies of Appalachian Geology: Northern and Maritime.* Wiley, New York.

Zimmerman, J.-L. (1972). L'eau et les gaz dans les principales familles de silicates. *Sci. Terre Mem.* **22**, 1–188.

INDEX

Abundance sensitivity. *See* Mass spectrometer
Actinolite. *See* Amphiboles
Activity, sample, 67
Activation energy, 91, 134, 144–45
Age of Earth, debate about, 3
Age equations
  of ^{40}Ar/^{39}Ar, 18–19
  general, 17–18
  for K-Ar, 18
Age spectra
  definition of, 13, 86
  disturbed, 13–14, 87–89
  flat, undisturbed, 13, 88–89
  grain size effects on, 112
  from lunar samples, 165–66, 170, 173, 177
  from microcline in sediments, 180
  from mixed phases, 99–106
  model, 13
  numerical simulations of, 97–99
  phase change effects on, 113–14
  plateau in, 13, 88–89
  recoil effects on, 110–12
  resolution within, 106
  saddle-shaped, 102–3, 108–10
  from samples with exsolution, 105–6
  from slowly cooled samples, 96–97
  theoretical diagrams of, 88, 90, 93, 97–98
  Turner model of, 86–88
Alkali feldspars
  anorthoclase, 22, 114
  closure temperature of, 23, 116, 153
  deciphering thermal histories, use in, 115, 179–81
  diffusion of argon in, 23, 153
  excess argon in, 23
  exsolution in, 22, 114
  high-temperature, 22, 114
  homogenization of, 114
  low-temperature, 23, 115
  microcline, 115, 179–80
  orthoclase, 114
  perthite, 23, 114–16
  sanidine, 22, 38, 114
Amphiboles, 27–28, 117. *See also* Hornblende
  Ar uptake profiles in, 28, 107–9
  closure temperature for Ar, 28, 152
  excess Ar in, 28, 107–9
  exsolution in, 28, 106
Anorthoclase. *See* Alkali feldspars
Anorthosite, lunar sample 15415, 94

Apollo 11, Moon, ^{40}Ar/^{39}Ar dating of samples. *See* Lunar geochronology
Apollo 12, Moon, ^{40}Ar/^{39}Ar dating of samples. *See* Lunar geochronology
Appalachian Mountain Belt, cooling histories of, 184–86
^{36}Ar
  atmospheric, 19–20
  correcting for nonradiogenic Ar, use in, 5, 12, 20
  neutron-induced interferences, minimization of, 60–63
  production from Ca, 51–55, 60–62
  production from Cl, 52, 57, 62
^{37}Ar
  half life of, 52, 54, 67
  production from ^{40}Ca, 51–55
^{38}Ar
  atmospheric, 19–20
  production from Ca, 52–53
  production from Cl, 52, 57
^{39}Ar
  half life of, 5, 67
  minimization of interference, 62
  production from ^{39}K, 3, 5–8, 12, 47–50
  production from Ca, 51–55
  production rates of, 48–49
  recoil of, 110–12
  sufficient production of, 59–60
^{40}Ar. *See also* Argon
  atomic size of, 9
  neutron induced, 55, 60
  product of ^{40}K decay, 4–5, 9, 15–17
  transformation to ^{41}Ar, 5
^{41}Ar
  half life of, 5
  production from ^{40}Ar, 5
^{40}Ar*/^{39}Ar$_K$, calculation of, 83–85
Argon
  abundance in air, 19
  atmospheric, 11, 19–20
  cosmogenic, 11
  discovery of, 4
  excess, 11, 106–10
  extraction systems, 69–73
  extraneous, 11
  isotope dilution, measurement by 10, 69
  measurement, 10
  neutron-induced, 11
  nomenclature, 11
  radiogenic, 11
  tracer, 69

  trapped, 11, 20, 120–26
Argon loss, episodic
  in contact aureoles, 178
  effect on age spectrum, 87–96
  expressions for, 132
  from lunar samples, 171
  from meteorites, 8
Arrhenius parameters, 91, 94, 144–45, 147
Arrhenius relationship, 94, 105, 116, 134, 144–45, 148–50
  for Ar in biotite, 150
  for Ar in feldspars, 147
  for Ar in hornblende, 152
  for Ar in phlogopite, 149
  for Ar in plagioclase, 104
Assumptions, in K-Ar and ^{40}Ar/^{39}Ar dating, 11–12
Atmospheric argon, 19–21
  abundance in atmosphere, 19
  calibration, use in, 19, 80
  corrections for, 20
  fractionation of, 29
  isotopic composition of, 20
  solubility in melts, 36
  in various minerals, 35–37
Avogadro's number, 17

Barberton Mountains, South Africa, ^{40}Ar/^{39}Ar dating, 31
Barkevikite, *See* Amphiboles
Basalt, lunar, ^{40}Ar/^{39}Ar dating, 163–74
Basalt, terrestrial. *See* Volcanic rocks
Biotite, 25–26, 116
  age spectra from, 25, 116–17
  behavior *in vacuo*, 26, 113
  closure temperature for Ar, 25, 151
  composition of, 25
  diffusion of Ar in, 148–51
Blanks, in extraction systems, 73, 122
Blocking temperature. *See* Magnetization
Bjurböle, meteorite, 6
Branched decay, ^{40}K, 15–17
Broken Hill, N.S.W., Australia, age spectra at, 102–3, 108–9
Bruderheim, meteorite, ^{40}Ar/^{39}Ar dating of, 7–8, 86–87, 93, 163

Cadmium
  control of nuclear reactors, 46
  shielding, 50, 56–57, 65–66, 68
Calcium

K-Ca method of dating, 15
  neutron reactions on, 51–54
  product of ^{40}K decay, 4, 15–17
  production of Ar isotopes from, 51–54
Calibration of mass spectrometer, 19, 80–81
Capture cross section, neutron, 19, 44–45
Carnallite, 35
Celadonite, 27, 34
Chain reaction, nuclear fission, 45
Chart of isotopes, 16
Chlorine, neutron reactions on, 51–52, 57
Clay minerals, 34
  ^{39}Ar recoil effects, 112
  argon loss from, 34
  dating of, 34
  unsuitability for ^{40}Ar/^{39}Ar dating, 22
  in volcanic rocks, alteration product, 30
Closed system, assumption, 12
Closure temperature, 22, 96
  calculation of, 138
  definition of, 134
  diagram of closure model, 133
  for first-order loss, 134–36, 160–62
  with position in crystals, 136–38
  in slowly cooled systems, 133–38
Cooling, conductive
  of dike-like body, 140–42
  from simple uplift, 139–40
Cooling history
  and closure temperature, 133–38
  effect on age spectrum, 96–97
  of Haliburton Highlands, 188–89
  of metamorphic terranes, 184–86, 188
  of pluton, 182–83
  of Separation Point Batholith, 183
Correction factors, in ^{40}Ar/^{39}Ar dating, 8, 51–57
Correlation diagrams, 6, 85, 120–26
  for anorthoclase, 123–24
  for biotite, 123–25
  for Bjurböle meteorite, 122
  Fish Springs cinder cone, 181
  schematic, 121, 123
Cosmogenic argon, definition of, 11
Cross section, nuclear, 44
  for ^{40}Ca(n, α)^{37}Ar reaction, 61
  for ^{39}K(n,p)^{39}Ar reaction, 47, 50

Daughter product, 3, 15, 18
Diagenesis, effect on ^{40}Ar retention, 118–20
Decay constants
  for ^{39}Ar, 67
  for ^{37}Ar, 67
  definition of, 18
  for ^{40}K, 16
Decay factors, 67
Decay scheme, ^{40}K, 15–17
Diffusion, 86–99, 127–62
  of Ar from biotite-phlogopite, 148–51
  of Ar from feldspars, 153
  of Ar from hornblende, 151–53
  of Ar from muscovite, 153
  and closure temperature, 133
  coefficient, D, 130, 144–45
  from a cylinder, 131–32
  from a cube, 132
  criteria for successful experiments, 145–47
  during slow cooling, 133–38
  effective radius, 147
  episodic Ar loss by, 132
  grain size effect on, 112
  and heat flow, 154–56
  laboratory studies of, 147–54
  mechanisms of, 143–45
  from mixed phases, 99–106
  numerical models of, 97–99
  phenomenological basis of, 128–30
  process of, 127
  sample calculations, 162
  single-site, 87–96
  from a slab or plane sheet, 93, 130–32
  from a sphere, 89–93, 131–32
Discrimination, mass, 19, 80
Duluth Gabbro, Minnesota, age spectra, 178

Eldora Stock, Colorado, age spectra, 100, 178
Electron capture, by ^{40}K, 5, 16–17
Electron multiplier, 79
Error magnification in Ar measurement, 35–36
Errors
  in calculated age, 85
  estimates from mass spectrometry, 85
Evaporites, dating by K-Ar, 35
Excess argon
  in alkali feldspar, 105
  in amphibole, 107–9
  in biotite, 25, 110
  definition of, 11
  in plagioclase, Broken Hill, 102–5, 108
  in plagioclase, Liberia, 109
  in submarine basalts, 31
Extraneous argon, 11
  in biotite, 25
  definition of, 11

Faraday Cup, 79, 82
Fast neutrons, 12, 44–48
Feldspars, 22–24. See also Alkali feldspars; Plagioclase
Feldspathoids, 24. See also Leucite; Nepheline
Fick's First Law, 128
Fick's Second Law, 89, 130, 134, 146, 155
Fission, nuclear, 45
  energy of neutrons produced by, 45–46
Flux gradient, neutron, 47, 63–66
Flux monitors, 6, 12, 19, 41–43, 48
  intercalibration of, 43
Frequency factor, D_O, 91, 134, 144–45
Furnace, argon extraction, 70–72
  laser microprobe, 71
  radiofrequency heating, 70
  resistance heating, 70–71
  temperature control, 71

Gas, purification, 72
Geomagnetic polarity time scale, 9, 30, 188
Gettering, 73
Glass
  dating of, 33–34
  excess Ar in, 31, 33
  in igneous rocks, 29
  lunar, 34, 171–73
  tektites, 33–34
  trapped Ar in, 31
Glauconite, 26–27. See also Glaucony
Glaucony
  Ar diffusion from, 27
  dating of, 27
  general, 26–27
  inherited Ar in, 27
  loss of ^{39}Ar by recoil, 27
  unsuitability for ^{40}Ar/^{39}Ar dating, 22, 27
Glaucophane. See Amphiboles
Grain size effects on age spectra, 112–13

Half life
  of ^{37}Ar, 52, 54, 67
  of ^{39}Ar, 5, 52, 67
  of ^{36}Cl, 52, 57
  of ^{38}Cl, 52, 57
  definition of, 18
  of ^{40}K, 9, 16
Hazards, radioactive, 68
Heat transfer, conductive, 128–30. See also Diffusion
  and Ar diffusion, 154–56
  cooling of pluton by, 140–41

INDEX

in dike intrusion and uplift, 141–42
effect of uplift on, 139–40
Fourier's Law of, 128–29
during slow cooling, 133–38
solutions of heat flow equation, 139–43
Highlands, Moon, $^{40}Ar/^{39}Ar$ dating. *See* Lunar geochronology
Hornblende, 151–53. *See also* Amphiboles
    age spectra, showing Ar loss in, 95, 179
    $^{40}Ar/^{39}Ar$ and K-Ar dating of, 27–28, 117
    closure temperature for Ar, 27–28, 152
    complex age spectrum and exsolution in, 105–6
    diffusion of Ar in, 151–53
    effect of exsolution on, 106
    excess Ar in age spectra, 108–9

Illite, 34, 118–20
    Ar loss from, 34
    ^{39}Ar recoil loss from, 119
    $^{40}Ar/^{39}Ar$ dating of, 34, 118–20
    polymorphic forms of, 119–20
    provenance dating of, 34
Incremental heating, 6–8, 12, 106
Incremental total fusion age, 81
Inherited argon, definition of, 11
Interferences in $^{40}Ar/^{39}Ar$ dating
    from argon isotopes, 8, 51–57
    correction factors, 8, 51–57
    from reactions on Ca, 51–55
    from reactions on Cl, 52–57
    from reactions on K, 52–53, 55–56
    minimization of, 60–63
Interlaboratory standards, 42
Inverse isochron. *See* Correlation diagrams
Inyo Domes, California, 154–56
Iodine-xenon dating, 6
Irradiation, neutron
    facilities in nuclear reactor, 45–46
    lattice damage effects of, 66, 112
    optimization of, 57–63
Isobars, definition of, 15
Isochron. *See* Correlation diagrams
Isotope dilution, 10, 69

J, irradiation parameter, 19, 48, 61
    definition of, 19
    measurement of, 19
    variability of, 63, 65
Jilin, meteorite. *See* Kirin (Jilin) meteorite

^{39}K
    abundance of, 14–15

    neutron interaction with, 47–51
    transformation to ^{39}Ar, 5, 12, 47
    neutron cross section of, 50
^{40}K
    branching ratio of, 16
    decay constants of, 16–17
    decay scheme of, 4–5, 9, 15–17
    discovery of, 4
    electron capture by, 5, 16–17
    half life of, 9, 16
    isotopic abundance of, 14
Kaersutite. *See* Amphiboles
Kirin (Jilin), meteorite, age spectra, 100–101
Komatiites, $^{40}Ar/^{39}Ar$ dating of, 31

Langbeinite, 35
Laplace transform, 158
Laser heating, 71
Least-squares regression, 80, 83, 125–26
Lepidolite, 26
Leucite, 24
    Albion Hills, Italy, 24
    Ar diffusion in, 24
    $^{40}Ar/^{39}Ar$ dating of, 24, 116
    excess Ar in, 116
    Gaussberg, Antarctica, 24
Lunar geochronology, 163–78
    Apollo 11, Mare Tranquillitatis basalts, 163–71
    Apollo 12, Oceanus Procellarum basalts, 171–73
    Apollo 16, breccia 65015, 176–78
    highlands ages, histogram, 175
    highlands rocks, 174–78

Magnetization
    blocking temperature, 186–90
    dating acquisition of, 187–90
    diagrammatic representation of, 187
    theoretical blocking curves for magnetite, 189
Mass discrimination of mass spectrometer, 19, 90–81
Mass spectrograph, 4
Mass spectrometer, 73–85
    abundance sensitivity of, 76–77
    basic equations of, 74–75
    calibration of, 80–81
    collector, 74–76, 78–79
    data acquisition from, 82–83
    diagrams of, 74–75, 77
    first-order focusing, 75–76
    Fourier Transform-Ion Cyclotron Resonance, 78
    ion detectors in, 78–79
    ion sources in, 78
    magnetic analyzer of, 74–78

    magnetic focusing of, 74–75
    magnetic scanning of, 82
    mass discrimination of, 19–20, 78, 80–81
    memory effects in, 20, 80
    orifice correction, 81–82
    resolution, 76–77
    static operation of, 20, 79–80
    voltage scanning in, 82
Mass spectrometry. *See* Mass spectrometer
Memory effect. *See* Mass spectrometer
Metamorphic rocks, 21–22, 32–33
    cooling histories of, 184–86
    dating as whole, 32–33
    meaning of ages, 22
Micas, 25–27. *See also* Biotite; Glauconite; Lepidolite; Muscovite; Phengite; Phlogopite
Microcline, slow cooling of, 96, 115. *See also* Alkali feldspars
Minerals suitable for dating, 21–28
Mixed phases, effect of, 99–106
Moderator, nuclear reactor, 46
Monitor minerals, 41–43
Moon, $^{40}Ar/^{39}Ar$ dating. *See* Lunar geochronology
Muscovite, 26, 117–20
    age spectra from, 26, 117
    closure temperature for Ar, 26
    diffusion of Ar in, 153
    from Naxos, 101–3

Nambucca Slate Belt, N.S.W., Australia, 33
Naxos, Greece, age spectra, 101–3
Nepheline, 24, 116
Neutrons
    characteristics of, 44
    discovery of, 44
    distribution in nuclear reactor, 45–47
    energy spectrum from U fission, 45–46
    epithermal, 44
    fast, 12, 44–47
    flux gradients in nuclear reactor, 47, 63–66
    flux monitor, 6, 19
    lattice damage by, 66–67
    self-shielding, 66
    slow, 44
    thermal, 44
Noble gases, geochemistry, 4
Nonradiogenic argon
    blank, 35
    correction for, 11
    trapped, 35
Nuclear reactions, 43–44

interfering in $^{40}Ar/^{39}Ar$ dating, 51–57
Nuclear reactors, 44–47
  characteristics of various, 49
  core, 45
  irradiation facilities of, 45
  neutron flux gradients, 47, 63–66
  as neutron source, 12, 44–47
  principles of, 45–47
  schematic diagram of, 45
Numerical solutions
  for diffusion, 97–99
  for model age spectra, 97–99

Obsidian, dating of, 33
Omegatron, 8
Orifice correction, 81–82
Orthoclase. See Alkali feldspars

Paleomagnetism. See Magnetization
Pantar, meteorite, 6–7
Parent, radioactive, 3, 18
Phase changes, effects on age spectrum, 113–14
Phengite, 26, 117
  from Naxos, 101–3
Phlogopite, 25–26
  composition of, 25
  diffusion of Ar in, 148–51
Phyllite, K-Ar dating of, 28, 33
Plagioclase, 22–24, 116
  age spectra, Broken Hill, N.S.W., Australia, 103
  closure temperature for Ar, 23
  excess Ar in, 23–24, 102–5, 116
  exsolution in, 103–5, 114
  in lunar basalts, 24, 164–71
  lunar sample 65015, 176–78
Plateau, in age spectrum, 13, 88–89
  in Apollo 11 basalts, 164
Polyhalite, 35
Potassium
  atomic number of, 14
  atomic radius of, 14
  atomic weight of, 14, 17
  crustal abundance of, 10, 14
  decay constants for, 16–17
  decay products of, 4, 15–17
  decay scheme, 15–17
  discovery of, 4
  feldspar. See Alkali feldspars
  K-Ca dating method, 15
  ionic radius of, 14
  isotopes of, 4, 14–15
  isotopic abundances of, 14
  measurement of, 10
  radioactivity of, 3–4, 9, 15–17
  specific activity for, 16–17
Potassium-calcium dating method, 15

Pyroxenes, 28, 117
  Ar retention in, 28
  excess Ar in, 28
  K-Ar dating of, 28
  in lunar basalts, 164–67

Quartz, 21

Radiation damage, effect on ^{40}Ar retention, 66–67
Radioactivity
  decay law of, 17–18
  discovery of, 3
  of potassium, 9, 15–17
  safety aspects of, 68
  of uranium, 4
Radiogenic argon
  confirmation in minerals, 5
  definition of, 11
Reactor, nuclear, 12, 44–47
Recoil of ^{39}Ar, 110–12
Resolution
  in age spectrum, 106
  of mass spectrometer, 76–77
Riebeckite. See Amphiboles

Safety, neutron irradiated samples, 68
Sample preparation, 40–41
Sample size, 58
Sanidine. See Alkali feldspars
Sedimentary rocks
  dating of illite from, 34, 118–20
  dating of glauconite from, 26–27
  meaning of ages, 22
  provenance of, 23
  thermal history of, 23, 115, 179–81
Separation Point Batholith, New Zealand, age spectra, 183
Separation of variables, 156–58
Sericite. See Muscovite
Slate
  $^{40}Ar/^{39}Ar$ dating using laser, 71–72
  K-Ar dating, 28, 33
Slope
  Arrhenius relationship, 144–45
  of isochron, 120–26
  of isotope correlation line, 6
  least-squares regression, 125–26
Slow cooling, 96–97, 133–38
Smectite, 26. See also Clay minerals
Specific heat, 129
  definition of, 143
Standards for $^{40}Ar/^{39}Ar$ dating, 41–43
Static method, mass spectrometry, 20, 79–80
Stepwise heating, 7–8, 12–13, 106
Sylvite, 35

Tektites, dating of, 33–34
Temperature monitoring of sample, 72
Thermal conductivity
  definition of, 128, 143
Thermal diffusivity, 130
  components of, 143
Total Ar release date, $^{40}Ar/^{39}Ar$, 81
Total fusion $^{40}Ar/^{39}Ar$ ages, 8
Thermal histories
  of cooling pluton, 182–83
  of Haliburton Highlands, 188–89
  of metamorphic rocks, Appalachians, 184–86
  of sedimentary basins, 115, 179–81
  of Separation Point Batholith, 183
  from xenoliths in basalt, 181–82
Tracer, isotopic, 10, 69
Tranquillitatis, mare, Moon. See Lunar geochronology
Trapped argon, definition of, 11

Uplift
  of Appalachians, 184–86
  metamorphic cooling, 184–86
  thermal effect of, 139–40
Uranium
  fission energy spectrum of ^{235}U, 46
  radioactivity of, 4
  U-Pb dating, 4

Volcanic rocks
  ^{39}Ar recoil loss from, 32, 118
  $^{40}Ar/^{39}Ar$ dating of, 31–32
  dating as whole, 29–32, 117–18
  dating of submarine, 31–32, 118
  effects of alteration, 29–30, 118
  excess Ar in, 29, 118
  phenocrysts in, 29
  selection criteria for, 29–30, 118
  trapped Ar in, 29, 31, 36

Whole rock samples, 28–33, 117–18
  dating of lunar, 163–76
  selection criteria for dating, 118

Xenoliths, in volcanic rocks, 29, 181–82